Adaptive Computing in Design and Manufacture VI

Springer
London
Berlin
Heidelberg
New York
Hong Kong
Milan
Paris
Tokyo

I.C. Parmee (Ed.)

Adaptive Computing in Design and Manufacture VI

With 156 Figures

123

I.C. Parmee, PhD, CEng, MASCE, MEI
CEMS, University of the West of England, Coldharbour Lane, Bristol, BS16 1QY, UK

British Library Cataloguing in Publication Data
Adaptive computing in design and manufacture VI
1.Engineering design - Data processing – Congresses
2.Production management – Data processing – Congresses
3. Adaptive computing systems – Congresses
I.Parmee, I. C.
620'.0042'0285
ISBN 1852338296

Library of Congress Cataloging-in-Publication Data
A catalog record for this book is available from the Library of Congress.

ISBN 1-85233-829-6 Springer-Verlag London Berlin Heidelberg
Springer-Verlag is a part of Springer Science+Business Media
springeronline.com

Typesetting: Camera ready by editor
69/3830-543210 Printed on acid-free paper SPIN 10994962

Preface

This sixth evolutionary / adaptive computing conference is sponsored in 2004 by the University of the West of England, Bristol, the UK Engineering and Physical Science Research Council and the UK Network in Adaptive Computing in Design and Manufacture. The Conference again explores the utility of various adaptive search algorithms and complementary computational intelligence techniques within the engineering design and manufacturing domains.

Industrial interest in the integration of evolutionary computing (EC) with engineering design continues to increase as successful applications become more commonplace. Changes in design practice have also greatly facilitated the successful integration of EC with day-to-day activities. For example, powerful inexpensive desktop computing capability is increasingly available and more useable and accessible modelling and analysis tools within grid-based engineering problem solving environments increasingly support the designer in the coupling of techniques and solvers. Evolutionary algorithms now play a significant role within suites of optimisation software available to the engineer and their integration with routine design tasks can now be implemented relatively easily. Several commercial optimisation software packages include evolutionary algorithms and packages solely exploiting EC techniques are becoming available. The increasing utilisation of neural network-based meta-models generated from FEA and CFD analysis is now supporting evolutionary search and optimisation of detailed design problems previously inaccessible due to excessive computational expense.

Although significant potential has been realised extensive applied and fundamental research remains essential to ensure best utilisation of available computing resource by improving the overall efficiency of the various techniques and to promote the emergence of new paradigms. The potential of adaptive computing within the design field goes far beyond the optimisation of specific complex systems. The development of appropriate strategies involving the various techniques and technologies can provide extensive support in multi-disciplinary design domains characterised by uncertainty, poor problem definition, multiple objectives and variable constraint. A breadth first adaptive exploratory approach can enhance innovative problem solving and provide decision support. The integration of engineering heuristics with adaptive computation and a human-centred interactive role for adaptive search processes within design team activities is a growing area of research.

The book comprises selected papers that cover a diverse set of industrial application areas including engineering design and manufacturing process design, scheduling and control. The collection of papers will be of particular interest to both industrial researchers and practitioners in addition to the academic research communities across engineering, operational research and computer science.

I.C. Parmee
ian.parmee@uwe.ac.uk
http://www.ad-comtech.co.uk/ACDDM_Group.htm

Organisation:

Conference Chairman:

I C Parmee, University of the West of England, Bristol

Scientific Committee:

C. Anumba	Loughborough University, UK
T. Arciszewski	George Mason University, USA
T. Baeck	Informatik Centrum Dortmund, Germany
R. Balling	Brigham-Young University, USA
G. Bugmann	University of Plymouth, UK
E. Burke	Nottingham University, UK
B. Carse	University of West of England, UK
C. Coello Coello	CINVESTAV, Mexico
D. Corne	University of Exeter, UK
C. Dagli,	University of St. Louis, USA
K.. Deb	Indian Institute of Technology, Kanpur
K. De Jong	George Mason University, USA
M. Emmerich	ICD/CASA, Germany
J. Gero	University of Sydney, Australia
E. Goodman	Michigan State University, USA
D. Grierson	University of Waterloo, Canada
P. Hajela	Rensselaer Polytechnic Institute, USA
C. J. Harris	University of Southampton, UK
C. Hillermeier	Siemens Research, Germany
P. Husbands	University of Sussex, UK
A. J. Keane	University of Southampton, UK
F. Lohnert	Daimler Chrysler, Germany
M-L. Maher	University of Sydney, Australia
J. Miles	University of Cardiff, UK
C. Moore	University of Cardiff , UK
S Patel	Unilever Research Laboratory, UK
A. Pipe	University of the West of England, UK
C. Poloni	University of Trieste, Italy
W. Punch	Michigan State University, USA
D. Savic	University of Exeter, UK
M. Schoenauer	INRIA, France
H-P Schwefel	University of Dortmund, Germany
E. Semenkin	Siberian Aerospace Academy, Russia
P. Sen	University of Newcastle-upon-Tyne, UK
A. E. Smith	University of Pittsburgh, USA
I. Smith	EPFL, Switzerland
G. D. Smith	University of East Anglia, UK
J. Taylor	Kings College London, UK
G. Thierauf	University of Essen, Germany
A. Thompson	University of Sussex, UK
G. Walters	University of Essex, UK
D. Whitley	Colorado State University, USA
X. Yao	University of Birmingham, UK
A. Zalzala	Heriot-Watt University, UK

Sponsored by:

University of the West of England, Bristol
UK Engineering and Physical Science Research Council
UK Network in Adaptive Computing in Design and Manufacture

Supporting Bodies:

The Institution of Electrical Engineers, UK
The Institution of Civil Engineers, UK
The Institution of Mechanical Engineers, UK
The Institution of Engineering Designers, UK
The Institute of Energy, UK
The British Computer Society
The International Society of Genetic and Evolutionary Computation (ISGEC)
The European Network of Excellence in Evolutionary Computation (EVONET)

ACDM Lecture:

The Everyday Engineering of Organizational and Engineering Innovation
Professor D. E. Goldberg
Illinois Genetic Algorithms Laboratory (IlliGAL)
University of Illinois at Urbana-Champaign, USA.

Plenary Presentations:

Artificial Life: How Can it Impact on Engineering Practices of the Future?
Professor C. H. Dagli
Smart Engineering Systems Lab (SESL)
University of Missouri Rolla, USA.

Mixing Maths with Cases
Professor I. Smith
Head of the Applied Computing and Mechanics Laboratory
The Swiss Federal Institute of Technology, Lausanne, Switzerland

Agents, Evolution, and The Edge of Chaos: Possible Implications for Design and Manufacturing
Dr. R. E. Smith
University of the West of England, Bristol, UK

Contents

1. Plenary Papers

2. Engineering Design Applications

3. Manufacturing Processes

Chapter 1

Plenary Papers

The Everyday Engineering of Organizational and Engineering Innovation

David E. Goldberg
Department of General Engineering
University of Illinois at Urbana-Champaign
Urbana, Illinois 61801
deg@uiuc.edu

Abstract

This paper discusses how the routine study of the theory and design of genetic algorithms (GAs) has led to a number of unexpected results. Specifically, the paper considers how GA theory and design has led to (1) the design of genetic algorithms that solve hard problems quickly, reliably, and accurately, (2) a system for collaborative innovation, (3) methods for designing organizations more effectively, (4) GAs based on effective clustering in organizations, and (5) an extensible family of facetwise organizational models.

1 Introduction

When I was asked to give the ACDM Lecture, Ian Parmee wrote telling me that the "lecture can be philosophical, speculative, review, results oriented, theoretical, applied etc. or a completely mixed bag." Of course, as those of you who know me realize, much of my writing, for better or worse, is almost exactly all those things combined, almost all of the time; however, I was so shocked to have been given this much latitude explicitly, that I will try to outdo even my usual self.

In particular, I'd like to explore some loose threads that have run through my work going back to my dissertation [1] and bring you up to date on what those threads have sewn. Specifically, in my dissertation I argued that genetic algorithms have similarities to certain processes of human innovation. At the time, I used the argument to explain how such meager mechanisms as were contained in simple GAs might be doing something interesting, but this has led to a full-bore exploration of the connection, some important mathematics, and the hot algorithms that result from taking the connection seriously [2]. More recently, these musings on innovation have led my lab, the Illinois Genetic Algorithms Laboratory (IlliGAL), into an effort to create effective collaboration software for distributed innovation, integrating the benefits of interactive genetic algorithms [3] and human-based GAs [4]. Finally, these musings about improving organizations through GAs have found expression in the use of GAs to create better organizational structures and this has led to a different type of organizational modeling that permits useful quantification of matters that are usually only discussed qualitatively or empirically.

In the remainder of the paper, each of these things is discussed in somewhat more detail following a brief tour of GAs and the innovation connection.

2 The Nickel Tour of Genetic Algorithms

When I first started in this business, I had to preface every article by a fairly lengthy description of genetic algorithms, what they are, where they come from, and so forth. Today, audiences are more GA aware than ever, and less of this treatment is required. If you don't know what GAs are (and you're attending ACDM) what rock have you been hiding under? But more seriously, for those who need more background, I have written at length [5] about GA basics, and here we briefly review the elements of GA mechanics.

GAs search among *populations* of *chromosomes* that may be decoded to represent *solutions* to some *problem.* Basic to the operation of a GA is the ability to evaluate the *fitness to purpose* of one solution relative to another. With a population in place, *selection* and *genetic operators* can process the population iteratively to create a sequence of populations that hopefully will contain more and more good solutions to our problem as time goes on. There is much variety in the types of operators that are used in GAs, but quite often (1) *selection*, (2) *recombination*, and (3) *mutation* are used. Simply stated, selection allocates greater survival to better individuals, recombination combines bits and pieces of parental solutions to form, new, possibly better offspring, and mutation makes one or more changes to an individual's trait or traits.

2.1 The Fundamental Intuition

Our nickel tour doesn't do justice to GA mechanics, but even a longer explanation would leave us scratching our heads and asking how such simple operators might do anything useful, let alone promote an effective, robust search for good stuff. It is something of an intellectual mystery to explain why such individually uninteresting mechanisms acting in concert might together do something useful. Starting in 1983 (Goldberg, 1983), I have developed what I call the *fundamental intuition of genetic algorithms* or the *innovation intuition* to explain this apparent mystery. Specifically, I liken the processing of selection and mutation together and that of selection and recombination taken together to *different facets of human innovation*, what I call the *improvement* and *cross-fertilizing* types of innovation. Here we concentrate on the cross-fertilizing type of innovation exclusively.

2.2 Selection + Recombination = Innovation

To understand how selection and crossover might give us anything useful, we appeal to our own sense of human cross-fertilizing innovation. What is it that people do when they are being innovative in a cross-fertilizing sense? Usually they are grasping at a notion—a set of good solution features—in one context, and a notion in another context and juxtaposing them, thereby speculating that the combination might be better than either notion taken individually. My first thoughts on the subject were introspective ones, but others have written along similar veins, for example, the French poet-philosopher Valéry [6]:

> It takes two to invent anything. The one makes up combinations; the other chooses, recognizes what he wishes and what is important to him in the mass of the things which the former has imparted to him.

Verbal descriptions are far from the exacting rigor of computer code, but it is interesting that the innovation intuition has been articulated by philosophers of earlier times. Let's explore if we can push this notion further.

3 The Engineering of Engineering Innovation

Thus, if the GA-intuition connection is right, the design of effective genetic algorithms amounts to the design of effective engineering innovation processes. It would be nice to go beyond mere hand waving and understand some of the underlying critical issues of innovation. Elsewhere [2], these have been explored in a sevenfold decomposition paraphrased as follows:

1. Effective innovation decomposes solutions into subsolutions.
2. Good subsolutions must be supplied to the search or created during the search.
3. Innovation can solve problems that are nearly decomposable.
4. Decision making among different subsolutions in a population of solutions is statistical in nature.
5. Choosing among different solutions causes an increasing market share of better subsolutions.
6. The best subsolutions tend to take over a population quite rapidly.
7. Effective innovation demands that the best subsolutions mix from different solutions.

Each of these is explored more quantitatively in the original volume. Here we conncect with the ideas of the *race,* the *control map,* and the *sweet spot.*

3.1 The Race Between Innovation and Selection

Here we focus on the critical race that goes on in a competitive innovating system. In an evolving system acted upon by selection alone, we expect to see an S-shaped time history of the market share of the best individuals in the population (figure 1), and we may calculate the time it takes to go from a small market share to a large one as a characteristic takeover time or t^*. This seems reasonable enough, but real GAs have selection and recombination. What difference could it possibly make to understand the behavior of a competitive system under selection alone?

The answer to this question comes quickly and convincingly if we imagine another characteristic time, call it the innovation time t_i, which we shall define as the mean time for recombination or innovation operator to achieve a solution better than any achieved to this point. With such a characteristic time in mind there are two basic situations that we must be concerned with: the situation where the

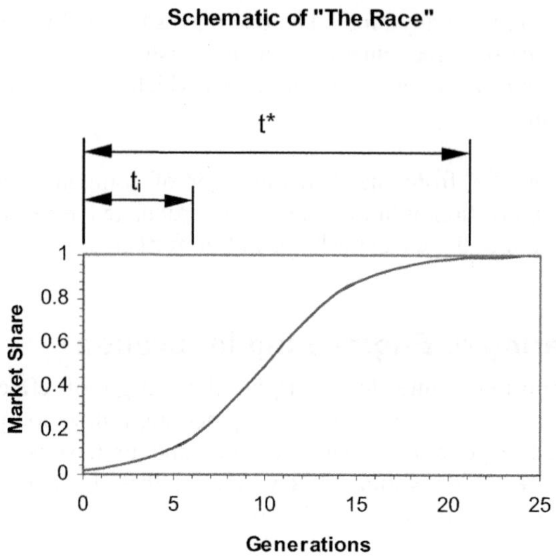

Figure 1. A schematic of the race illustrates the tension between between innovation and selection. When the innovation time t_i is less than the selection takeover time t^*, innovation proceeds apace (steady-state innovation). When the situation is reversed, innovation is stalled.

takeover time is greater than or equal to the innovation time, $t^* \geq t_i$, and that where the innovation time exceeds the takeover time $t^* < t_i$.

In thinking about these two situations, we immediately wonder which is the more advantageous for a selectorecombinative GA, and the answer is apparent with some straightforward reasoning as follows. The condition where innovation time leads (is less than or equal to) the takeover time is most advantageous for continuing innovation, because prior to the best individual dominating the population, recombination creates a better individual. Thereafter this better individual starts to dominate the population, and in essence, the innovation clock is reset. This cycle of partial takeover and continued innovation is repeated over and over again, resulting in the happy condition I have dubbed *steady-state innovation*.

Contrast this virtuous setting with the condition where innovation time lags (is greater than) takeover time. In such a situation, the current best guy continually increases in market share without serious competition and ultimately takes the population to substantial convergence, and now it is too late because diversity is a necessary condition of selectorecombinative success. This situation was called *premature convergence* [7] fairly early in the GA literature, but until the introduction of the above time scales argument [8], there was no means of analyzing the boundary between innovative success and failure. With the understanding of the crucial role of time scales and the race, rational analysis and design of competitive innovating GAs has advanced quite rapidly.

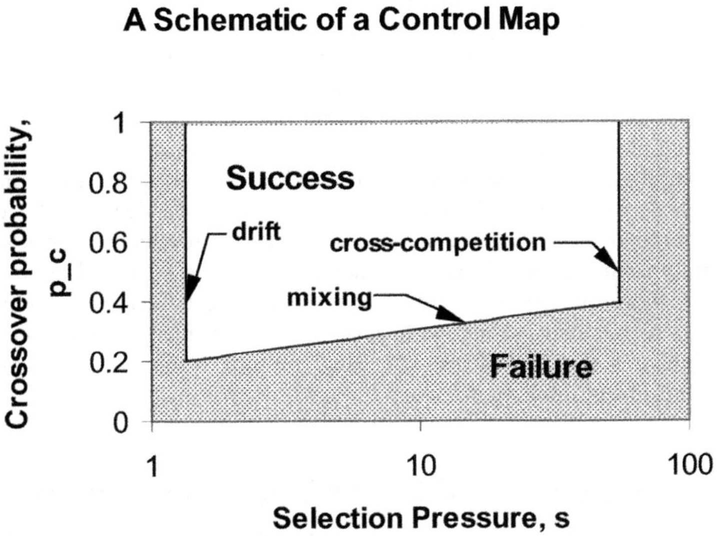

Figure 2. A control map shows the GA's sweet spot or zone of success for setting GA control parameters. The mixing boundary grows as the logarithm of the selection pressure *s*, as may be derived from the dimensional argument of the race presented in this paper.

3.2 A GA's Control Map and its Sweet Spot

One of the tools critical to these rapid advances is the so-called *control map*, which helps us delimit a genetic algorithm's *sweet spot.* Technical details of these developments are in the original papers [8-10], but here we strive for qualitative understanding of the key points. These can best be obtained by focusing on the schematic of the sweet spot of a genetic algorithm operating on an easy problem as shown in figure 2. In this map, we plot the feasible settings of the GA's control parameters, s, the selection pressure, and p_c, the probability of crossover. The selection pressure is simply the number of copies that are given to the best individual in the population under selection alone. The crossover probability is the frequency with which mated chromosomes actually undergo the exchange of crossover.

In the previous section, we discussed the race between innovation and selection. If we take this argument seriously and develop an equation from the condition when the takeover time is of the same order as the innovation time, we obtain the mixing or innovation boundary shown on the graph, where the crossover probability must increase as the logarithm of the selection pressure. Any value of crossover probability above this line is expected to succeed and any value below this boundary is expected to fail.

There are two other boundaries shown on the control map. To the left we see the region of success bounded by the so-called *drift boundary*, where convergence is controlled by the vagaries of random fluctuation when the selection pressure is small. To the right, we see the region of failure dominated by what we call *cross-competition*, when semantically independent traits end up competing with one another when the selection pressure is close to the population size. Understanding and applying the control map is central to the design of GAs that work and since 1993, much of my lab's work has been devoted to this endeavor. A complete discussion is beyond the scope of this treatment, but the results have been stunning and we are encouraged in adversarially designed test cases as well as in the real world that these procedures work. The hierarchical Bayesian optimization algorithm is a particular strong performer [11], and readers should consult that paper or Pelikan's excellent thesis [12]. Here we turn to some of the less obvious directions this line of thinking has led.

4 The Engineering of Organization Innovation

The creation of a scalable mechanism of computation innovation with supporting mathematical support has to be judged as a *technological* accomplishment, permitting the everyday solution of important problems that might otherwise languish. It also is a *scientific* contribution in that it gives us a tangible theory of innovation that can be used elsewhere. In this section, we examine three areas where this new science of innovation is leading. First, we consider information technology for collaborative innovation within organizations. Second, we consider how organizational thinking is leading to better GAs and how GAs are leading to better organizations. Finally, we consider how facetwise modeling used so successfully in GA design is leading to useful quantificative models of organizations.

4.1 From GAs to Collaborative Innovation Infrastructure

Innovation thinking has led immediately to the creation of a novel IT system called DISCUS (**D**istributed **I**nnovation and **S**calable **C**ollaboration in **U**ncertain **S**ettings) [13]. The goal of DISCUS is *to create a distributed and scalable environment for the integration of both human- and computer-generated knowledge in uncertain settings through effective human-human and human-machine collaboration.* Normal human collaboration and innovation involves human-human colloboration (HHC) through cross-fertilizing exchange between pairs of larger teams of individuals. The use of the tools of modern artificial and computational intelligence by individuals is a kind of human-machine collaboration (HMC). To achieve effective and scalable collaboration in difficult problem domains, this project envisions both types—HHC and HMC—as being important.

To achieve these goals, the DISCUS system proposes the integration of a number of key elements:

1. Interactive genetic algorithms
2. Human-based genetic algorithms (HBGAs)

3. Scalable genetic algorithms (machine based)
4. Flexible data and text mining (D2K/T2K).
5. Chance discovery using Key Graphs.

More detail about these efforts is available at the URL http://www-discus.ge.uiuc.edu/, but pilot results are promising and a fairly large project exploring the possibilities is underway.

4.2 GAs from and for Organizations

Another way innovation thinking has directed our work toward organizations is in the GA clustering of design structure matrices (DSMs) and in the use of such clustering at the heart of a new kind of competent GA, a design structure matrix GA (DSMGA). A design (or dependency) structure matrix (DSM) represents the strength of a relationship between items *i* and *j* in element d_{ij} of a matrix. Recent work [14] has demonstrated how fairly simple GAs can effectively cluster elements in such matrices using minimum description length (MDL) or other metrics. Follow-on work [15] has shown how to use such clusterings to build problem-effective crossover operators in a general manner. A detailed discussion of these matters is beyond the scope of this paper, and readers should consult the original work for more detail; however, the two-way nature of the application is intriguing, and both sides of the application perform usefully and well.

4.3 Organizational Theory from GA-Style Models

Even the style of modeling employed to create facetwise models of competent GAs has proven useful in learning to think more quantitatively about organizations. In writing the manuscript of a forthcoming book [16], I derived a simple model of *teamwork* as follows.

Imagine that in a team of size *n* that *deciding* and *doing* are processed differently. Consider that each of the *n* members takes T_1 units of time to decide what he or she would do and that this thinking is presented to the others serially. Thereafter, a simple vote is taken to determine which idea or combination of ideas will be done. Thereafter, the *n* members of the team share the task or doing time, T_2, required equally. This logic results in an equation for the time required for task completion by the team as follows:

$$T_{team} = nT_1 + T_2/n$$

Taking the derivative of this equation with respect to the team size *n* and setting to zero results in the following equation for optimal (fastest) team size, n^*:

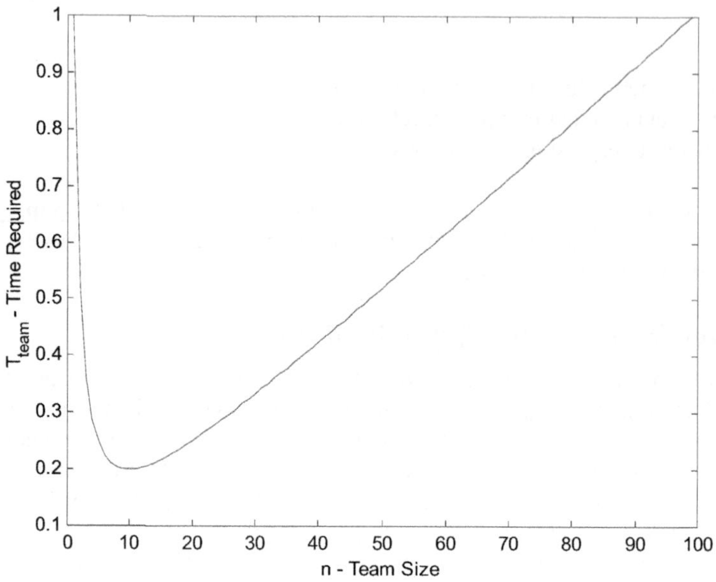

Figure 3. The time required for a team to complete a task depends on the decision time *T1* and the task time *T2*. In the figure, task time is approximately 100 times the decision time and the optimal team size is 10.

$$n^* = (T_2 / T_1)^{1/2}$$

For example, consider the case where $T_1 = 0.01$ and $T_2 = 0.99$. Then $n^* = (0.99 / 0.01)^{1/2} \approx 10$. Figure 3 shows the variation of total time required as a function of team size. The time required to perform the task with the optimally sized team according to this model is one-fifth of the time required by a person working alone. In other words, the speed up (T_{alone} / T_{team}) of working in a team is roughly five times that of working alone.

Extensions to this model have been worked out [17] and other topologies and types of optimization are being investigated. The point is that the substance and style of thinking that has led to computational models of innovation is leading to advances in organizational thinking at the same time.

5 Conclusions and an Invitation

This paper started by considering a connection between genetic algorithms and human innovation. Although GA operators were shown to have similarity to certain philosophical musings, the more interesting part of the discussion considered a dimensional argument regarding the relationship between characteristic time scales of takeover and innovation. This relationship led to the quantitative mapping of a

GA's parameter space as a control map, where a sweet spot of high quality convergence was illustrated.

From there, the paper shifted to a discussion of the ramifications of such *innovation thinking* for organizations, considering (a) the use of GAs in the design of collaborative IT infrastructure, (b) the use of GAs for organizational design and for the design of novel competent GAs themselves, and (c) the use of GA-style facetwise models for the creation of a novel type of quantitative organizational theory. In these topics, much challenging work remains to be done, but our initial forays have raised our spirits for the journey ahead.

That fairly straightforward inquiry into the nature of genetic algorithms should lead in such interesting directions should come as no surprise. Evolution and genetics are big ideas, having led to a revolution in human thought over the last century and a half across a spectrum of endeavor. It is rare that nuts-and-bolts engineers and hand-to-keyboard computer scientists find themselves so directly involved in pushing the boundaries on such big ideas, but the rareness of the occasion should not diminish our delight or detract from the opportunity to do good. Those who have not joined the party should consider this paper an engraved invitation to do so. For those who are engaged in these matters, I recommend the specific detours outlined above as well as the many others that await the mind accustomed to the delights and discipline of GA-innovation thought.

Acknowledgments

The work was sponsored by the Air Force Office of Scientific Research, Air Force Materiel Command, USAF, under grants F49620-94-1-0103, F49620-95-1-0338, F49620-97-1-0050, F49620-00-1-0163, and F49620-03-1-0129. Research funding for this work was also provided by the Technology Research Center (TRECC), a program of the University of Illinois at Urbana-Champaign, administered by the National Center for Supercomputing Applications (NCSA) and funded by the Office of Naval Research under grant N00014-01-10175. The US Government is authorized to reproduce and distribute reprints for Government purposes notwithstanding any copyright notation thereon.

The views and conclusions contained herein are my own and should not be interpreted as necessarily representing the official policies or endorsements, either expressed or implied, of the Air Force Office of Scientific Research, the University of Illinois at Urbana-Champaign, NCSA, the U. S. Navy, or the U. S. Government.

References

1 Goldberg, D. E. (1983). *Computer-aided pipeline operation using genetic algorithms and rule learning*. Doctoral dissertation, University of Michigan, Ann Arbor.

2 Goldberg, D. E. (2002). *The design of innovation: Lessons from and for competent genetic algorithms*. Boston, MA: Kluwer Academic.

3 Takagi, H. (1998). Interactive evolutionary computation: Fusion of the capabilities of EC optimization and human evaluation. *Proceedings of the IEEE, 89*(9), 1275-1296.

4 Kosorukoff, A., & Goldberg, D.E. (2002). Evolutionary computation as a form of organization, *Proceedings of the Genetic and Evolutionary Computation Conference (GECCO-2002)*, 965-972.

5 Goldberg, D. E. (1989). *Genetic algorithms in search, optimization, and machine learning.* Reading, MA: Addison-Wesley.

6 Hadamard, J. (1954). *The psychology of invention in the mathematical field.* New York: Dover.

7 De Jong, K. A. (1975). *An analysis of the behavior of a class of genetic adaptive systems.* Doctoral dissertation, University of Michigan, Ann Arbor.

8 Goldberg, D. E., Deb, K., & Thierens, D. (1993). Toward a better understanding of mixing in genetic algorithms. *Journal of the Society of Instrument and Control Engineers, 32*(1), 10–16.

9 Thierens, D. (1995) *Analysis and design of genetic algorithms.* Doctoral dissertation, Katholieke Universiteit Leuven, Leuven, Belgium.

10 Thierens, D., & Goldberg, D. E. (1993). Mixing in genetic algorithms. *Proceedings of the Fifth International Conference on Genetic Algorithms*, 38–45.

11 Pelikan, M. and D. E. Goldberg (2003). Hierarchical BOA solves Ising spin glasses and MAXSAT. *GECCO-2003: Proceedings of the Genetic and Evolutionary Computation Conference,* 1271-1282.

12 Pelikan, M. (2002). *Bayesian optimization algorithm: From single level to hierarchy.* Doctoral dissertation, University of Illinois, Urbana, IL.

13 Goldberg, D.E., Welge, M., & Llorà, X. (2003). *DISCUS: Distributed innovation and scalable collaboration in uncertain settings* (IlliGAL Technical Report No. 2003017). Illinois Genetic Algorithms Laboratory, University of Illinois at Urbana-Champaign, Urbana, IL.

14 Yu, T.-L. Yassine, A., & Goldberg, D. E. (2003). A genetic algorithm for developing modular product architectures. *Proceedings of the ASME 2003 International Design Engineering Technical Conferences.*

15 Yu, T.-L., Goldberg, D. E., Yassine, A. & Chen, Y.-P.(2003). A genetic algorithm design inspired by organizational theory. *Proceedings of the Genetic and Evolutionary Computation Conference (GECCO-2003)*, 1620 - 1621

16 Goldberg, D. E. (2004). *The entrepreneurial engineer.* Unpublished textbook manuscript.

17 Goldberg, D. E., Yu, T.-L., & Yassine, A. (2004). *Calculating efficient team size: Balancing deciding and doing as an elementary optimization problem.* Unpublished manuscript.

Mixing Maths with Cases

I.F.C. Smith

Applied Computing and Mechanics Laboratory
EPFL – Swiss Federal Institute of Technology in Lausanne
1015 Lausanne, Switzerland
Ian.Smith@epfl.ch

Abstract

There is much recent evidence to support the use of advanced search algorithms in engineering. Advantages have been demonstrated in fields of design, diagnosis, planning and control. However, when objective functions are incomplete and when environments are uncertain in the "open worlds" that are common in engineering, performance may fall drastically. Systems that include prior case knowledge provide opportunities for systems to adapt and learn in uncertain environments. This paper examines mixing stochastic search with case based reasoning through a recent study of a structure that learns.

1 Introduction

Adaptive computing methodologies are potentially useful for accommodating the dynamic characteristics of engineering tasks. Engineers are often required to design artefacts according to unique sets of requirements in particular contexts. Contextual conditions also influence planning, diagnostic and control tasks. Such conditions change with time and all possible situations may not be foreseen precisely and completely when software is written.

Adaptive computing methodologies are intended to meet these challenges. They inevitably include algorithms that perform machine learning using input at run time. What is learned may vary from simple parameter weights in most cases of neural networks and fuzzy systems to entire models when compositional approaches use input such as in-service measurements. Regardless of the approach, assumptions must be made during software development related to what will be learned. Since these assumptions often determine the scope of application of such software as well as their usefulness when information changes, they are best made as explicitly as possible and facilities should be available to change assumptions and have the software learn in a different manner.

Unfortunately, it is often impossible to change assumptions. Neural networks and fuzzy systems often contain fixed conceptual models of the engineering task. Even when adaptive methodologies act on higher level representations such as in composition modelling, there are a finite number of ways to learn with new information.

While such limitations are unavoidable, information representations should be made as independent as possible from the rest of the software. The greater the independence, the easier new information is accommodated. This paper begins with a fundamental discussion of why important engineering tasks are sensitive to contextual information. This is followed by a proposal for reducing the effects of such sensitivity in software use case based reasoning. Finally an example in structural engineering of a unique structural system that learns is used to illustrate the approach.

2 Fundamental Logic

Three types of logical inference are deduction, abduction and induction. The best way to understand inference types is through an example. Consider a bar of length, L, and area, A, that is fixed at one end and loaded at the other with a load Q, see Figure 1.

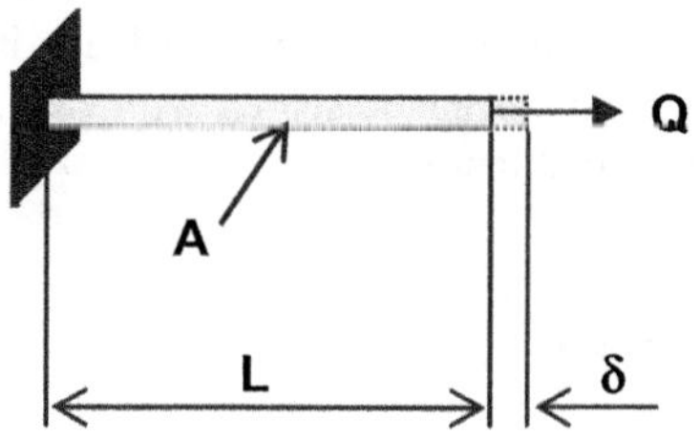

Figure 1: An axially loaded bar

From fundamental solid mechanics, we know that when a load, Q, is applied to a round bar having an area, A, this bar deforms axially by an amount $\delta = QL/AE$ where E is the Young's modulus of the material. Thus, for a given bar, we have the following pieces of information:

- two facts: Q and δ
- one causal rule: if Q then δ (a causal rule has the form, if *cause* then *effect*)

The difference between deduction, abduction and induction originates from what is given and what is inferred. For example, **deduction** begins with one fact, Q, and the causal rule. With these two items as given information, we infer that a deformation, δ, occurs. **Abduction** is the reverse of deduction; it begins with the effect, δ and the causal rule. Starting with these two items, we infer that a load, Q, is present. **Induction** begins with the two facts, Q and δ and then infers the causal rule. Table 1 summarizes the important elements of these three types of inference.

Type of Inference	Given Information	Inferred Information
Deduction	Cause and Rule	Effect
Abduction	Effect and Rule	Cause
Induction	Cause and Effect	Rule

Table 1: Three types of inference

Deduction is logically the only "legal" inference type in an open world. The use of abduction and induction requires a closed-world hypothesis. A closed-world hypothesis assumes that all facts and all rules are known for a given task. Such formulations are typically found at the beginning of mathematical proofs using symbols that stand for statements such as "for all x ..." and "y is an element of ...".

For illustration, refer to the example of the loaded bar. We can state with some confidence that if there is a load, Q, and when the causal rule, if Q then δ is known, we can infer that a deformation, δ, is expected (deduction). However, when we know the rule and only that a deformation, δ, is observed we cannot necessarily say in an open world that this deformation was caused by a load, Q (abduction). For example, the deformation may have been caused by an increase in temperature.

Similarly, just because we observe that there is a load, Q, and a deformation, δ we cannot necessarily formulate the rule, if Q then δ (induction). Other facts and contextual information may justify the formulation of a better rule. Abduction and induction become valid inferences when we make the closed-world hypothesis that the only relevant pieces of information are the two facts and one rule.

Scientists and mathematicians frequently restrict themselves to closed worlds in order to identify in a formal manner kernel ideas related to particular phenomena. Also, the closed-world idea is not limited to research. Job definitions in factories, offices and many other places often only require decision making within closed, well-defined worlds. *However, engineers (as well as other professionals such as doctors and lawyers) do not work in closed worlds.*

Society expects engineers to identify creative solutions in open worlds. For example, engineers may have to consider social, economic and political aspects of key tasks before reaching decisions. Such aspects are difficult to define formally and moreover, attempts to specify them explicitly, even partially, are hindered by environments where conditions change frequently.

Unfortunately, engineering tasks are not just deductive. While important activities such as simulation and structural analysis involve deduction, several key tasks are abductive. For example, many design activities can be thought of as reverse analysis since required behaviour is used to generate a geometrical description. In the same way, diagnosis can be thought of reverse simulation since an effect is used to infer a cause. In an open world, design and diagnostic tasks are dependent on context since they are abductive; they are thus not as reliable as simulation. Furthermore, there are one-to-many relationships between input and output.

Therefore, fundamental logic demonstrates that adaptive computing is a very important concept for engineers. Working in open worlds and performing abductive tasks mean that engineers must accommodate incomplete and changing information. Software that is able to adapt to particular contexts has the potential to provide valuable support. Such adaptation also helps reduce the size of search spaces that are created by one to many relationships.

3 Search in Open Worlds

When they are performed without simplification, full-size abductive-engineering tasks invariably generate large spaces of possible solutions. Therefore the challenge becomes one of searching efficiently for good solutions. Adaptive computing has the potential to modify how this search takes place using problem specific information.

Search strategies have been studied for decades. In mathematics they are studied in fields such operations research. In computing, artificial intelligence researchers have proposed hundreds of algorithms. Engineers use a combination of both mathematical and computer science based strategies. There are strategies for continuous, discrete and mixed variables; for finite and infinite domains; for bounded and unbounded variable values and for linear, non-linear, continuous and discontinuous objective functions [1]. There is no silver bullet; the best strategy depends on characteristics of the task [2].

In addition to the challenge of matching the search task with the best computational strategy, engineers must also determine how contextual knowledge is introduced into the computational support. When this knowledge is mixed into search algorithms, modification often requires much recoding and verification to ensure reliability. In such situations, adaptive computing is difficult to implement.

When case based reasoning is mixed with search algorithms, separation of contextual knowledge with search algorithms is clarified. While search algorithms take advantage of the calculation power of computers, case based reasoning benefits mainly from computer memory. Most importantly, adaptive computing is carried out transparently through addition (or removal) of cases.

Cases are also natural places for knowledge, data and relationships that originate from the open-world conditions that are common in engineering. Cases, being good solutions to previous tasks, contain characteristics that satisfy open-world conditions. Challenges of case-based reasoning involve finding the most appropriate case for the new task and adapting the old case to accommodate the new conditions. It is during the adaptation task that search strategies can provide support. Therefore a case is used to move close to a good solution and a search algorithm modifies the case to identify the most attractive solution. The following section gives an example of the advantages of mixing maths (search algorithms) with cases.

4 An Example in Structural Control

Controlling coupled features, such as adaptive geometry of complex structures, requires reasoning and advanced search techniques since local minima are usually present in the search space. Techniques can be combined to simulate the effects of simultaneously executed control actions and find a set of control actions that best improves system performance.

An example of a control action is a change in length of a structural element. Formulated as an optimization problem, variables in the optimization are the possible control actions, for instance ranges of feasible movements, and the objective function is determined from the performance goals and their priorities. An optimization method is then used to determine an optimally directed set of control actions, relative to the performance goals.

This set of control actions is used as input for controllers located on the structure such that the configuration (and thus, the behaviour of the structure) is modified. The behaviour of the new configuration is evaluated again and compared with performance goals to determine the current success of the control solution. This sequence is repeated until acceptable performance is obtained.

This framework only indicates the main information flow routes. As with any closed-loop control system, security triggers that are independent of the control loop are necessary in order to avoid exceeding limit values due to control malfunctioning. Although commands are checked computationally and application rates should be well dampened electronically, an errant signal may result in excessive stresses and deflections that lead to permanent structural damage.

Tensegrity (tensile-integrity) structures are lightweight, reusable structures. Since tensegrity structures are sensitive to asymmetric loads and small changes in external loading, active control helps make them suitable for practical use. The design consists of coupled active features that require continuous control in order to maintain and improve system performance. A system is comprised of tensegrity modules each of which consists of a self-stressing system of compression struts and tensile cables where the stress distribution and shape of the module is controlled by telescoping struts.

Modules are joined together so that struts of one module do not touch struts of adjacent modules. Therefore star-shaped strut configurations appear to float between modules. The control task involves determining the lengths of active struts that best distribute the stress in the structural system in a changing environment while maintaining a prescribed slope.

Static tests on a three-module adjustable structure at EPFL revealed that although the structure behaves linearly when subjected to a single point load, non linear behaviour is observed when there are multiple load positions. Most importantly, non linear behaviour also occurs when adjusting combinations of telescoping struts [3]. This means that prediction of behaviour under control commands cannot be carried out through superposition of contributions that are calculated from a sensitivity matrix. A complete structural analysis is required to verify each control command.

The dynamic relaxation method is used to analyse the structure since for these types of structures, avoiding matrix inversion and the direct form finding characteristics of the method are important advantages. For example, analysis of the three module structure can be carried out approximately one thousand times faster than finite element analysis when using the dynamic relaxation method.

Nevertheless, measurements under a range of loading conditions revealed that the analysis predictions were up to 25% different than measurement results. Differences were due to joint friction and other modelling inaccuracies. A correction strategy was devised to improve accuracy using measured response. This scheme improved response predictions most effectively when only two struts per module were used to adjust behaviour. When five struts were adjusted, the correction scheme was not as successful [3]. It is thought that five adjustable struts lead to over conditioning of the control system. This observation resulted in limiting further studies to two adjustable bars per module.

A subsequent study [4] investigated the combination of the dynamic relaxation method with artificial neural networks (ANNs). A two-hidden-layer (2 x 10 nodes) configuration with three input nodes and three output nodes was found to provide the best correction for the dynamic relaxation results. This combination resulted in increased accuracy of the analysis while maintaining the explicit structural knowledge within the analysis.

Completely replacing the analysis with an artificial neural net of a similar configuration (3-10-10-3) was not successful. Although a more complex network configuration may provide improved accuracy, the amount of training data required make this impracticable since requirements for training data grow exponentially with the complexity of the network. For example, increasing the number of input nodes to six would require increasing the number of training cases from 150 to over ten thousand.

Furthermore discussions with practicing engineers revealed that a complete reliance on a "black box" neural network was much less attractive to them than using the approach to obtain results that are applied as a correction to results that are obtained using a traditional structural analysis method. Finally, the experience gained during training and testing various network configurations can be reused for training and testing network configurations for other combinations of this type of tensegrity structure module. Such reuse would be less reliable in the absence of a structural analysis by dynamic relaxation.

Following these studies, a five-module structure was assembled and tested at EPFL, see Figure 2. This structure also contained improved joints and this resulted in increased stability and lower friction effects. Two electric motors per module were then installed and a closed-loop measurement-control system was built, see Figure 3. Increments of strut extensions of 0.1 mm are possible using these motors. Even though better joint design reduced the magnitudes of modelling inaccuracies, a similar configuration neural network configuration (3-10-10-3) provided the similar degrees of improvement in accuracy (percentage improvements) as was observed for the three module structure [5].

Figure 2: A five-module tensegrity structure

Figure 3: A five-module active tensegrity structure showing actuators on the structure and the closed-loop control configuration

Given a loading condition and corresponding structural deformations, the computational control system determines a set of movements of the active struts that will return the structure as close as possible to the initial slope. To achieve these structural control goals, the current computational control system is comprised of three modules: (1) state evaluation, (2) stochastic search, and (3) case-based reasoning. Stochastic search involves a global search of the combinatorial space of possible movements to produce control solutions.

The current state evaluation is comparison of slopes with and without loading and this feeds a single-criterion goal to minimize this difference. Other criterion exist and these will be added in subsequent research. Some control criteria can be expressed as limiting constraints and these are managed by the search algorithm, described next.

Stochastic search is required for two reasons. Firstly, there is no closed-form solution for strut movements given a target slope. The non-linear and closely coupled characteristics of this structure mean that that the approximations which would be necessary for a direct calculation would be unreliable. Therefore a generate-analyse-and-test strategy was adopted.

This strategy leads to the second reason. Determination of the total combinations of strut movements for a structure of this size results in a very large number. For example, assuming ten telescoping struts, a stroke of 50 mm, a minimum movement of 0.1 mm, there are 500^{10} possible solutions. If a structural analysis takes 0.1 seconds, this means that 3×10^{18} years are required to calculate all possibilities. Therefore brut-force search is not possible.

When brut-force search is impracticable and when the objective function has multiple local minima, stochastic search algorithms are useful. Three algorithms have been compared for this task [4, 5]; simulated annealing, genetic algorithms and PGSL – a probabilistic algorithm developed at EPFL [6]. PGSL was selected for this task because of its performance and because of the relatively few number of search parameters that required calibration for specific tasks. Using this algorithm, good control commands were identified after one to two hours of calculation.

Using feedback from measurements on the structure creates possibilities for improving performance during service. For each control solution carried out, a record of the initial structural state, corrective actions, and a success metric are stored for future retrieval. A "distance metric" is used to select the best case. This case is then adapted using a stochastic search algorithm. Further tests on the three search algorithms revealed that PGSL was much faster than the others for case adaptation. In this approach, cases are employed to come close to the solution and stochastic search is used to identify a good adaptation. This application of case based reasoning reduced calculation time to the order of tens of seconds. [5]. An example of computation times is given in Table 2.

Brute-force search	Stochastic search (PGSL)	Case based reasoning + PGSL
Up to 3×10^{18} years	1hour 56 minutes	97 seconds

Table 2: An example of computation times required to identify a good control commands using three strategies on a standard PC.

Addition of cases when good control solutions are identified allows the case base to grow during service. In this way, the structure learns with its experience. Also, successful yet unattractive sets of control commands are avoided through exclusion from the case base. For example, control commands that involve final positions close to maximum extension positions are not as desirable as those that include small movements close to the middle point of the actuators.

5 Conclusions

Adaptive computing provides important support for many engineering tasks, particularly those involving abduction since large search spaces need to be explored in situations of context dependence. A mixture of advanced search techniques with case based reasoning allows for explicit representation and use of contextual information in a manner that is independent of the details of the search algorithm. In this way, performance is improved easily with addition of good cases and changing contexts can be accommodated through different sets of cases. Such separation leads to a robust type of adaptive computing that engineering users can understand. Without understanding, there is no acceptance.

6 Acknowledgements

The author would like to thank the Swiss National Science Foundation, for financial support. Dr. K. Shea (Cambridge University) and Dr. B. Raphael are thanked for their scientific input into the structural control project. Dr. E. Fest and Dr. B. Domer have recently completed theses on this topic and this paper summarises some of their results.

References

1 Raphael, B. and Smith, I.F.C. (2003) Fundamentals of Computer-Aided Engineering, Wiley, UK, 306p.
2 Wolpert, D. H., and Macready, W. G. (1997). "No Free Lunch Theorems for Optimization." IEEE Transactions on Evolutionary Computation, 1(1), 67-82.
3 Fest, E., Shea, K., Domer, B. and Smith, I.F.C. (2003) "Adjustable tensegrity structures", J of Structural Engineering, Vol 129, No 4, pp 515-526.
4 Domer, B., Fest, E., Lalit, V. and Smith, I.F.C. (2003) "Combining the dynamic relaxation method with artificial neural networks to enhance the simulation of tensegrity structures", J Structural Engineering, Vol 129, No 5.
5 Domer, B. (2003) "Performance enhancement of active structures during service lives", EPFL Thesis No 2750, 151p.
6 Raphael, B. and Smith, I.F.C. (2003) "A direct stochastic algorithm for global search", J of Applied Mathematics and Computation, 146, No 2-3, pp 729-758.

Chapter 2

Engineering Design Applications

Morphogenic Evolutionary Design: Cellular Automata Representations in Topological Structural Design

Rafal Kicinger[1], Tomasz Arciszewski[2], Kenneth De Jong[3]

[1] Information Technology and Engineering School
George Mason University
Fairfax, Virginia 22030
rkicinge@gmu.edu

[2] Civil, Environmental and Infrastructure Engineering Department
George Mason University
Fairfax, Virginia 22030
tarcisze@gmu.edu

[3] Computer Science Department
George Mason University
Fairfax, Virginia 22030
kdejong@gmu.edu

Abstract

This paper provides the initial results of a study on the applications of cellular automata representations in evolutionary design of topologies of steel structural systems in tall buildings. In the paper, a brief overview of the state of the art in cellular automata and evolutionary design representations is presented. Next, morphogenic evolutionary design is introduced and illustrated by several types of cellular automata representations. Further, Emergent Designer, a unique evolutionary design tool developed at George Mason University, is briefly described. It is an integrated research and design support tool which applies models of complex adaptive systems to represent engineering systems and analyze design processes. The paper also reports the initial results of several structural design experiments conducted with Emergent Designer. The objective of the experiments was to determine feasibility of various types of cellular automata representations in topological structural optimization. Finally, initial research conclusions and recommendations for the further research are provided.

1 Introduction

Computationally simulated evolution has been successfully applied to many difficult problems in science and engineering. It is also becoming an increasingly attractive paradigm for civil and structural engineers. It reflects the ongoing transformation of computing in civil engineering from focus on mostly analytical to more holistic/conceptual aspects of design. This paradigm offers two major possibilities of enhancing traditional structural design. First, it provides engineers with a powerful optimization method that can be used to solve many difficult design problems that could not be solved with traditional methods. Second, it offers a true potential of developing novel/creative designs. Up to very recently, vast majority of applications of evolutionary methods in structural engineering was focused exclusively on structural optimization problems. Currently, we are observing an emergent trend of using the evolutionary paradigm to discover novel/creative designs [1]. One of the key issues in these types of applications is the choice of appropriate representations of engineering systems that support generation of novel/creative designs.

Nowadays, we are witnessing another emergent trend of replacing complicated models with distributed, or parallel, models based on simple rules and interactions among elements that can nevertheless generate very complex behavior [2]. Thus, even though the models studied in the conceptual design phase are becoming more and more complex, it is possible that this complexity can be simulated using only very simple rules and programs. Hence, the complex adaptive systems approach to conceptual design seems to be a plausible way of capturing the complex nature of the design process and may enable us to use simple mathematical and computational models to simulate this process. Also, presently available computing power opens new possibilities of designing and modeling complex engineering systems and their dynamic evolution.

2 Background

2.1 Cellular Automata

Cellular automata (CAs) are one of the simplest models of complex systems [3]. As such, they are useful idealizations of the dynamical behavior of various systems. CAs have been used to model complex systems and processes consisting of a large number of identical, simple, locally interacting components. They appear to capture many essential features of complex self-organizing cooperative behavior observed in real world systems. CAs can be used to study pattern formation and gain some insight into self-organization processes. The CAs research has generated great interest over the last forty years because of their ability to exhibit very complex patterns of behavior out of a set of relatively simple rules. Recently, Wolfram [2] suggested that cellular automata and other simple programs may better model nature's most essential mechanisms than traditional mathematical equations.

A cellular automaton is a deterministic system that is completely defined by its update rule (called in this paper a CA rule) and an initial configuration of cell state values. Figure 1 shows how simple cellular automata work. The process of iteration of an elementary CA is presented in Figure 1a). In this case, the individual cell states have only binary values and local neighborhoods affecting the iteration of the considered cell are formed by this cell and its immediate left and right neighbors. Therefore, three cells are considered in each local neighborhood and such situation is called a 'local neighborhood of size 3'. The bottom row of Figure 1a) consists of 6 squares (cells) denoting an initial configuration of cells (*t=0*). In this particular case, the initial configuration consists of cell state values 0 0 0 1 1 0. White squares in Figure 1 denote cell state values equal to 0 while black squares represents cell state values equal to 1. The particular CA rule used to iterate this initial configuration for 15 time steps is presented in Figure 1b). A CA rule is a complete set of decision rules whose conditions incorporate all possible combinations of cell state values in the given local neighborhoods (here of size 3) and outcomes determine the values of the considered cells (usually central cells in a local neighborhood) at the next time step. Graphical representation of the same rule is presented in Figure 1c). The CA rule shown in Figure 1b-c) is applied to the current configuration of cells and determines the new configuration at the next time step (*t=1, 2, 3,...*). Details of the process of determining the new configuration are presented graphically in Figure 1d).

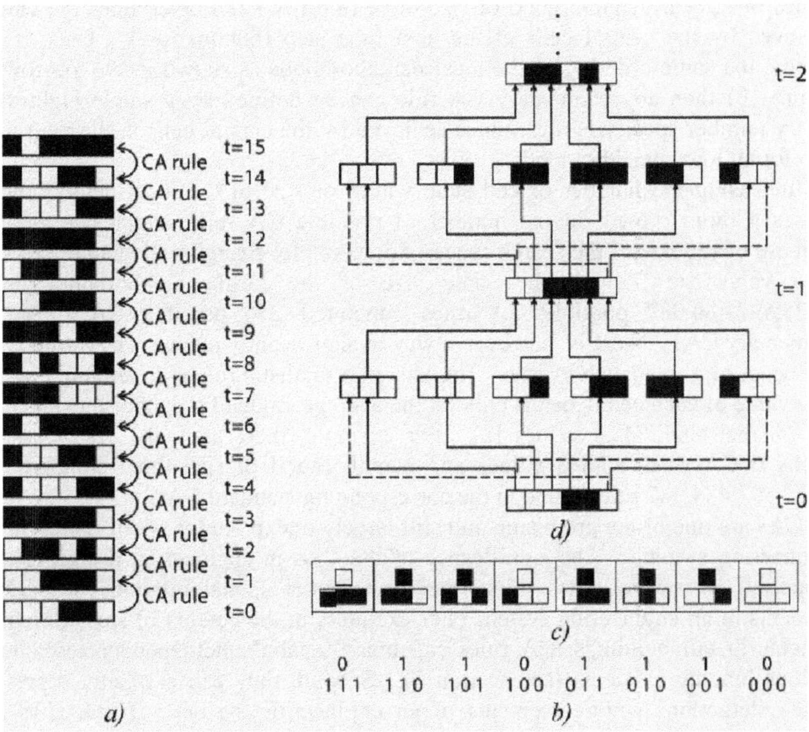

Figure 1: a) Process of iteration of an elementary CA staring with an initial configuration of cells, b) Numerical representation of the CA rule used in part a), c) Graphical representation of the same CA rule, d) Details of the mechanism of generation of subsequent cell configurations (*t=1,2,etc.,*)

Bottom part of Figure 1d) shows the same initial configuration (*t=0*) as in Figure 1a). First, a set of local neighborhoods of size 3 (it is an elementary CA) is constructed by taking each cell from the initial configuration together with its left and right neighbors and placing them respectively in the middle, left, and right of the lattice defining each local neighborhood (see the set of 6 local neighborhoods of size 3 placed above the initial configuration in Figure 1d)). In this particular example, so-called cyclic boundary conditions are used, meaning that the rightmost cell in the initial configuration becomes the left neighbor of the leftmost cell in the initial configuration, and vice versa (denoted by dashed lines in Figure 1d)). Next, the local neighborhoods created that way are compared to the local neighborhoods shown in the bottom row of Figure 1b-c). When the two match, the value shown in the top row of Figure 1b-c) defines the new value of the central cell at the next time step. This process is repeated for each local neighborhood and the values obtained are placed in appropriate slots in the new configuration of cells at time $t=1$ thus completely defining this configuration. The process is repeated for an arbitrary number of steps. Figure 1a) shows the results of the iteration process for the first 15 steps. Figure 1d) gives a detailed representation of the process of determining the new configuration at a subsequent time step of the first 2 iterations only.

The CA rule shown in Figure 1b) specifies all possible (8 in this case) cell state values of a local neighborhood of size three (top row) and determines the values achieved by the central cells at the next time step (bottom row). Thus, if we assume the same ordering of the local neighborhoods as shown in the top row of Figure 1b) then any elementary CA rule can be defined by a single eight-digit binary number specifying the values achieved by the central cells at the next time step for all local neighborhoods.

Increasing the number of cell state values or size of the local neighborhood causes a rapid growth in the number of possible CA rules and hence a rapid increase of the size of the search space. For example, changing the number of cell state values to 3 with the same size of the local neighborhood yields 7,625,597,484,987 possible CA rules compared 256 possible CA rules for elementary CAs. There is, however, a way to significantly reduce it by introducing a concept of a totalistic CA rule. The idea of a totalistic rule is to assume that the new value of each cell depends only on the average value of neighboring cells, and not on their individual values [2]. For example, there are only 2187 possible totalistic CA rules with 3 values and neighborhood of size three compared to 7,625,597,484,987 rules found in the corresponding standard CAs.

CAs are one of the promising and still largely unexplored ways of representing engineering systems. The significance of the CAs in engineering design can be supported by several facts. First, they can model spatial relations of various elements in an engineering system. For example, in the context of steel structural systems in tall buildings, CA rules can model spatial interdependencies among various bracing elements (see section 3). Second, they can explicitly represent local interactions among elements of an engineering system. Third, CAs are known to produce various kinds of emergent behavior and this property is highly relevant in many engineering design situations, e.g. when novel structural shaping patterns are sought. Finally, it is a fact that even designers of complex and sophisticated engineering systems (bridges, tall buildings, etc.) use only a small set of design/decision rules to develop their design concepts. This set of design/decision rules could be represented by the CA rules.

CAs have been a subject of significant research interests in various scientific disciplines, including structural design. Inou et al. [4] used cellular automata to investigate self-organization of topologies in mechanical structures. Kundu et al.

[5] applied CAs to shape optimization of structural plates. Kita and Toyoda used CAs to shape and topology optimization of two-dimensional elastic structures [6] and sizing optimization of truss structures [7]. Hajela and Kim [8] applied GAs to search the space of CA rules in structural analysis of 2D elastic structures.

2.2 Evolutionary Computation and Design Representations

A representation, or encoding, of an engineering design is as a computational/abstract description of an engineering system (that usually does not yet exist) expressed in terms of attributes [9]. Attributes can be symbolic (when they take values from an unordered or partially ordered set) or numerical (when they take numerical values representing quantities or measurements). Design concepts are typically described in terms of symbolic attributes. Numerical attributes are used for a detailed description of a design. In the most straightforward evolutionary computation (EC) applications, each gene corresponds to an attribute and represents a dimension of the design space. In such representations, called direct representations, each individual consists of a fixed-length binary string of genes, or a genotype, representing some subset of a given set of features. Often, in complex engineering applications, multi-valued attributes are more natural to use [10].

Representations of engineering designs are one of the key elements in any evolutionary design application and have been recently a subject of significant research investigations [11,12]. Vast majority of applications of EC in structural design, with few notable exceptions, used simple parameterizations of engineering systems encoded in terms of binary, real, or integer-valued attributes. Parameterizations are examples of direct representations which essentially encode actual design concepts. Recently, several researchers proposed and studied indirect and generative representations which do not encode complete design concepts but rather instructions on how to construct ("grow") these designs [12,13]. The inspiration for these types of representations comes again from nature which manipulates the *rules* for constructing complex objects, called genetic plans, rather than the complex objects themselves [10]. The objects are then built from the plans via a developmental process called *morphogenesis* [14]. Direct representations have been mostly used for evolutionary design optimization [15] and indirect/generative encodings for evolving creative/novel design concepts [12,16].

3 Morphogenic Evolutionary Design

This section proposes generative representations of steel structural systems in tall buildings based on cellular automata. Two types of CA representations of wind bracing systems in tall buildings are presented involving one-dimensional CAs (1D CA) and two-dimensional CAs (2D CA).

3.1 Generative Representations Based on One-Dimensional Cellular Automata

This representation uses a single one-dimensional initial configuration of cells and a single 1D CA rule to construct a design concept of a wind bracing system. It has been developed using the concept of division of the structural grid of a tall building (the system of vertical and horizontal lines of columns and beams, respectively) into cells. A cell is understood here as a part of the structural grid contained within the adjacent vertical and horizontal grid lines.

The structure of the genome is presented in Figure 2a). It consists of two parts: encoding of the "design embryo" and encoding of a "design rule" which is applied to the design embryo to build the structure of a wind bracing system from it.

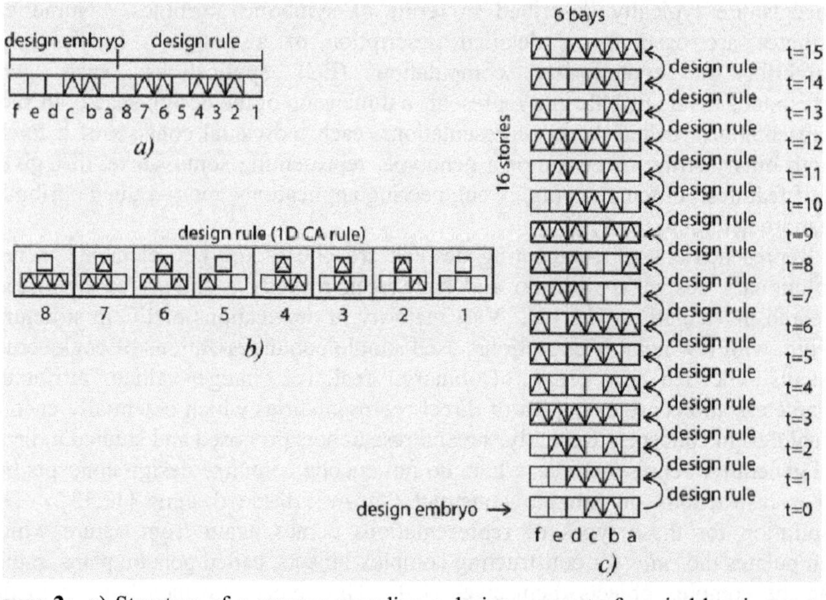

Figure 2: a) Structure of a genome encoding a design concept of a wind bracing system shown in part c), b) Design rule based on a 1D CA that "grows" a design concept from the design embryo, c) Process of constructing a design concept of a wind bracing system from the genome in part a)

Design embryo consists of a one-dimensional lattice representing the initial configuration of cell values and at the same time forms the configuration of the first story in a wind bracing system of a tall building (see the configuration at $t=0$ in Figure 2c)). It is encoded in the first part of the genome and represented by genes a-f in Figure 2a). Design rule, encoded in the second part of the genome (genes 1-8), is simulated by a 1D CA rule (see Figure 2b)). It consists of a complete set of decision rules whose conditions (bottom part of Figure 2b)) incorporate all possible combinations of cell state values (types of bracings) in the given local neighborhoods and outcomes specifying the values of the central cells of these neighborhoods at the next time step (top part of Figure 2b)). The length of the design embryo is equal to the number of bays in a tall building (see Figure 2c)). On the other hand, the length of the design rule depends on the number of possible cell state values and the size of the local neighborhood.

Figure 2b) shows a simple example of a design rule based on 1D CA with only two possible cell state values (empty cell denoting no bracing, and non-empty cell denoting K bracing), and neighborhood of size 3. The process of constructing ("growing") a design concept of a wind bracing system from the design embryo and using the design rule presented in Figure 2b) is shown in Figure 2c). The design rule is iterated for the number of steps that is one less than the number of stories in a tall building and thus forms a design concept which is subsequently evaluated. In this example, cyclic boundaries were used.

Bottom part of Figure 2b) shows all possible combinations of conditions for this design rule. In this particular example, they are ordered from 1 to 8. If this ordering is assumed fixed for the entire class of design rules with binary cell state values and local neighborhood of size 3, then the outcome values (shown in the top part of Figure 2b)) uniquely define every rule belonging to this class. This fact has been used in the definition of the structure of the genome shown in Figure 2a). Here, genes 1-8 encode the outcome values produced by the design rule presented in Figure 2b) and, given the assumed ordering, uniquely define it.

The advantages of this representation include compactness and excellent scalability. A genome encoding a wind bracing system shown in Figure 2c) is 14 genes long when 2 cell state values (types of wind bracing elements) are used and 349 genes long when 7 cell state values are used (as in the experiments reported in section 5 of this paper). In the case, when the design rule is encoded as a 1D *totalistic* CA rule, this representation is even more compact. When 2 cell state values are used the genome is 10 genes long, and when 7 values are used it has 25 genes.

The disadvantage of this approach is that a single design rule is applied at each story of a wind bracing system and hence it is impossible to diversify design rules for various parts of the wind bracing system, e.g. in traditional design different design rules may be used in the bottom part of the structure, where internal forces are the largest, compared to the upper part of the structure where internal forces are the smallest but local stiffness requirements are the same.

3.2 Generative Representations Based on Two-Dimensional Cellular Automata

The second type of the generative representations of steel structural systems proposed in this paper is based on two-dimensional cellular automata (2D CA). Design rule simulated by a 2D CA acts upon a design embryo that is now a two-dimensional array representing an initial configuration of a wind bracing system. Figure 3 shows a schematic view of the structure of the genome for this representation, two of the most popular local neighborhood shapes (called Moore neighborhood and von Neumann neighborhood) used in 2D CAs, and the process of constructing a design concept of a wind bracing system using this representation.

Figure 3a) shows a schematic view of the structure of a genome that encodes a design embryo (gray squares) and design rule based on 2D CA (white squares). This time the design embryo is a two-dimensional array encoding the entire initial configuration of a wind bracing system in a tall building. Initial part of the genome encodes a linearized version of this array and its length is equal to the total number of cells in a configuration of a wind bracing system in a tall building, i.e., number of bays * number of stories. In the particular example of a tall building with 6 bays and 30 stories shown in Figure 3c), the initial configuration consists of 180 cells, and hence 180 leftmost genes in the genome encode the design embryo.

The number of genes encoding the design rule depends on a few parameters. As it was the case with design rules based on 1D CAs, the number of possible cell state values and neighborhood size significantly affect the length of the design rule encoding. In the 2D CA case, however, one additional parameter, namely shape of the local neighborhood, has to be specified to completely define the design rule. The two most popular and most frequently used two-dimensional shapes of the local neighborhood include von Neumann neighborhood, and Moore neighborhood. The two shapes with radii equal to 1 and 2 are shown in Figure 3b). In the design experiments reported in this paper only Moore neighborhood was used.

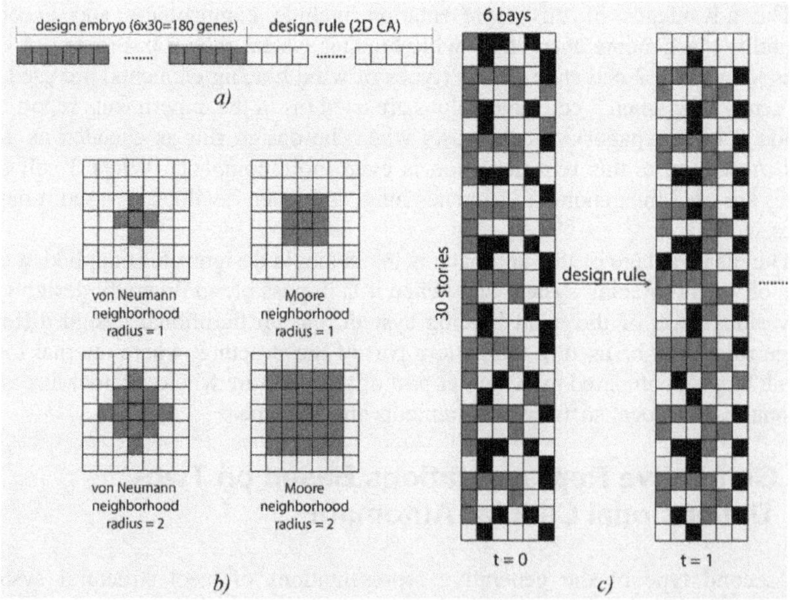

Figure 3: a) Schematic view of the structure of a genome based on 2D CAs encoding a design concept of a wind bracing system shown in part c), b) Two most popular local neighborhood shapes used in 2D CA (Moore, and von Neumann), c) Process of constructing a design concept of a wind bracing system from the genome in part a)

The process of constructing a design concept of a wind bracing system using this representation is presented in Figure 3c). The design embryo encoded in the genome is converted into a 2D array and iterated an arbitrary number of times using a 2D CA rule. Thus, the number of iteration steps becomes an additional parameter for this representation. The final configuration obtained during this process of iteration forms a design concept which is subsequently evaluated. In the particular example shown in Figure 3c), cells have 3 possible cell state values denoting 3 types of bracing elements (e.g., white cells denote no bracing, gray cells K bracings, and black cells X bracings).

One of the major advantages of this representation is that it allows for an explicit representation of two-dimensional interactions among design elements. It is possible to investigate various ranges of interaction by using different shapes of neighborhood as well as by changing the radius of the local neighborhood. This property might be particularly important in modeling complex engineering systems where local and highly nonlinear interactions among design elements are impossible to describe using traditional mathematical formulas.

Disadvantages of this approach include the problems of scalability and that of customized genetic operators. Increasing a number of values in a 2D CA or a size of the local neighborhood causes a rapid growth in complexity. Hence, 2D totalistic CA rules are used for most practical purposes. They were also employed in the design experiments reported in this paper.

4 Emergent Designer

In order to investigate the ideas presented above, an experimental research and design tool, called *Emergent Designer*, has been developed at George Mason University. It is a Java-based system intended for both the design experiments in the area structural design as well as the analysis of the complex design processes from the perspective of complex adaptive systems and dynamical systems. The system has seven major components:

1. Domain Selection and Problem Specification Component
   - Selection of a domain, e.g. steel skeleton structures in tall buildings.
   - Selection of a problem, e.g. weight minimization of a wind bracing system.
2. Representation Component
   - Selection of a representation for a given problem (implemented representation types include binary, real, integer-valued and cellular automata).
   - Specification of additional parameters for some types of representations (e.g., resolution for binary representations, or neighborhood shape for CA representations).
3. Concept Generation and Optimization Component
   - It implements the mechanisms of design concept generation including various types of cellular automata using *Mathematica*, developed by Wolfram Research.
   - It implements mechanisms of design concept optimization including a unified approach to EA proposed in [17].
4. Evaluation Component
   - It uses *Wind Load V2.2.S*, developed by Novel CyberSpace Tools, to determine wind forces specific to a given design case.
   - It uses structural analysis, design and optimization package called *SODA*, developed by Waterloo Systems in Waterloo, Ontario, Canada, to produce

complete and detailed structural designs, test the feasibility of the produced designs with respect to predetermined design codes, and provide fitness measure that guides the evolutionary optimization process.

5. Statistical and Dynamical Systems Analysis Component
   - It conducts statistical analysis of experimental results including e.g. means, medians, and confidence intervals, analysis of sample distributions, using *JMSL Numerical Library* developed by Visual Numerics.
   - It conducts dynamical systems analysis of complex design processes, including e.g. identification of fixed-points, attractor reconstruction using delay coordinates, and nonlinear time series analysis of experimental data.
6. Visualization Component
   - It implements mechanisms to visualize generated design concepts and simple landscapes.
7. Report Generation Component
   - It uses *OpenOffice.org* to generate reports of design experiments, including full description of experimental settings, visualization, graphical and numerical results of individual runs, and final statistical analysis of the experimental results.

5 Experimental Design and Initial Results

The objective of the initial experiments reported in the paper was to determine the feasibility of morphogenic evolutionary design in structural engineering. It has been accomplished through the analysis of the results of a number of experiments using the two types of CA representations described in sections 3.1 and 3.2 and a comparison with the results produced using the direct integer-valued representations, similar to those used in [18].

In the performed experiments, the following parameter settings have been used:

1. Domain parameters (tall buildings):
   - number of bays: 6
   - number of stories: 30
   - bay width: 20 feet (6.01 m)
   - story height: 14 feet (4.27 m)
   - distance between transverse systems: 20 feet (6.01 m)
   - structural analysis method: first order
   - beams: fixed
   - columns: fixed
   - supports: fixed
   - wind bracings: no bracing, top-down diagonal bracing (/), bottom-up diagonal bracing (\ ), K bracing, V bracing, simple X bracing, and X bracing (see Figure 4)
2. CA representation parameters:
   - CA dimension: 1D totalistic and 2D totalistic
   - CA neighborhood radius: 1
   - number of cell state values: 7 (7 types of bracing elements, see Figure 4)
   - CA neighborhood shape (2D CAs): Moore
   - CA iteration steps (2D CAs): 30

3. Evolutionary computation parameters:
   - evolutionary algorithm: ES
   - population sizes (parent, offspring): 5,25
   - generational model: overlapping
   - selection (parent, survival): uniform stochastic, truncation
   - mutation rate: 1/L (L - the length of genome)
   - crossover (type, rate): uniform, 0.2
   - fitness: weight of the steel skeleton structure (minimization problem)
   - initialization method: random
   - constraint-handling: death penalty (infeasible designs assigned 0 fitness)
4. Simulation parameters:
   - termination criterion: 1000 fitness evaluations
   - number of runs: 5 (in each experiment)

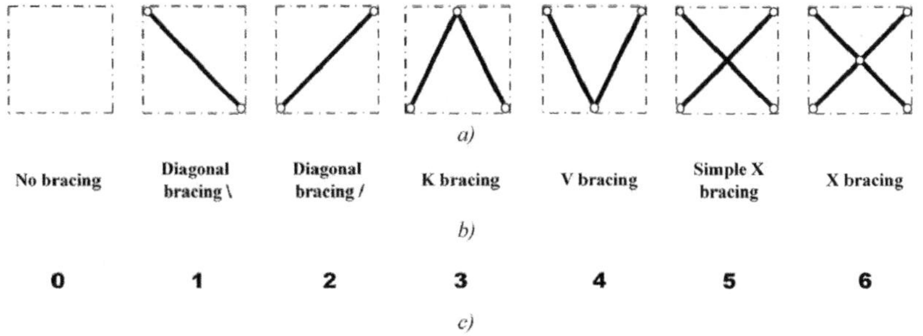

Figure 4: a) Phenotypic representation of the evolved bracing elements, b) Their symbolic values, c) Genotypic values corresponding to appropriate bracing types

The optimization of steel skeleton structures was conducted in two stages. In the first stage, evolutionary algorithm optimized the topology of steel structural systems in tall buildings. The evolutionary algorithm manipulated genomes encoding design embryos and design rules as it is shown in Figures 2a) and 3a). In the reported experiments only topologies of wind bracing configurations were evolved. All other structural elements, including beams, columns, and supports were assumed fixed for all designs. The second stage of optimization (sizing optimization) was conducted by *SODA*. Here, cross-sections of all structural members (including beams, columns, and wind bracings) were optimized with respect to the total weight of the steel structure.

Figure 5 shows the results of 3 experiments conducted using Emergent Designer. In these experiments 1D totalistic and 2D totalistic CA representations were used and compared to direct integer-based representations (also shown in Figure 5), similar to the ones used in [18]. All other parameter values were held constant. It is clearly visible, and at the same time statistically significant, that 1D totalistic CA representations outperformed both 2D totalistic CA representations and integer-valued representations. Figure 6a) shows the best design produced in the reported experiments. Its fitness (weight) was equal to 533,201 lbs. (241,856 kg) compared to the best design produced using integer-valued representations which had fitness of 567,894 lbs. (257,592 kg).

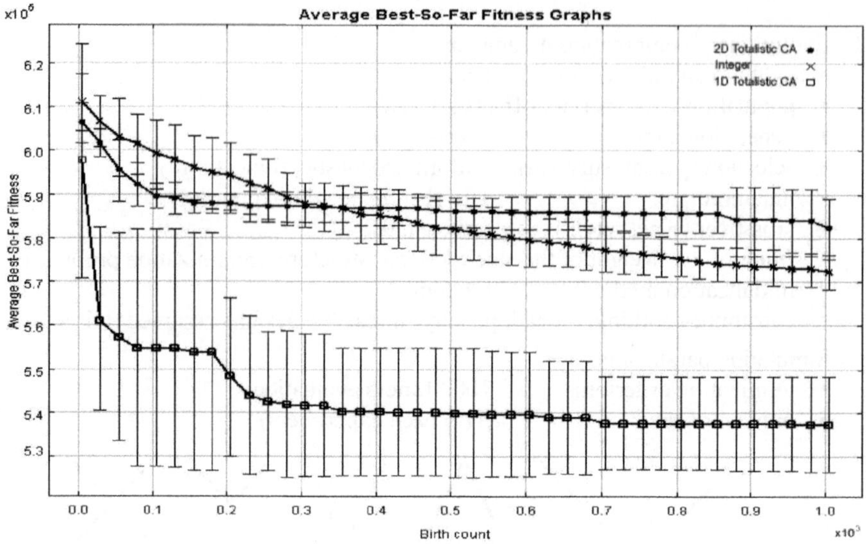

Figure 5: Average best-so-far fitness for CA representations (1D totalistic and 2D totalistic) and integer-based representations (vertical bars show 95% confidence intervals calculated using Johnson's modified t test)

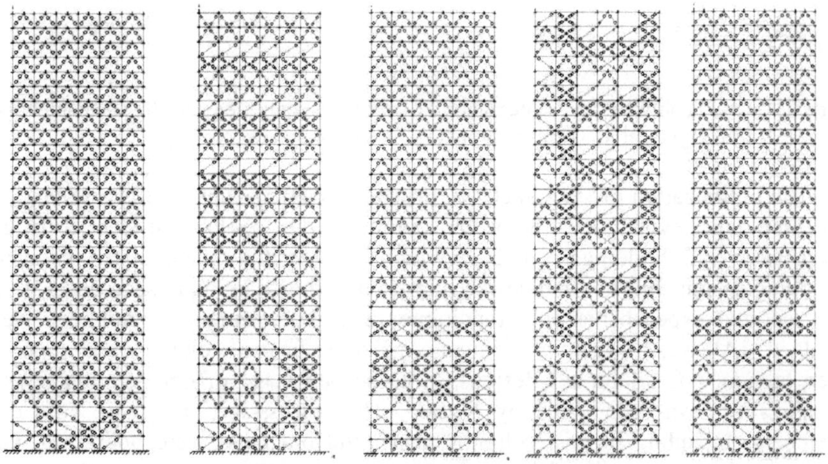

a) *b)* *c)* *d)* *e)*

Figure 6: a) Best design produced in the reported experiments, b-e) Examples of interesting structural shaping patterns generated by Emergent Designer using 1D totalistic cellular automata representations

In the conducted experiments, the authors also observed the formation of interesting structural shaping patterns. Emergent behavior has been observed in several instances. Examples of the patterns generated by one-dimensional totalistic CA representations are presented in Figure 6b-e). It is worth mentioning that the structural shaping patterns generated by the CA representations are qualitatively different than the ones observed in the experiments with standard integer-valued representations [19].

6 Conclusions

In this paper, only the preliminary results have been reported. However, the initial results are encouraging and clearly demonstrate the feasibility of morphogenic evolutionary design in structural engineering. Cellular automata representations proved to perform well both in generating optimal design concepts as well as in producing emergent structural shaping patterns.

The conducted research will be continued. First, various sensitivity analyses will be conducted involving both various CA representation parameters (neighborhood shape (e.g. von Neumann, etc.), neighborhood radius, CA iteration termination criteria (2D CAs)) as well as evolutionary computation parameters (population sizes, selection mechanisms, genetic operators and their rates). Next, the experiments reported in this paper will be repeated for longer evolutionary processes with several thousands of fitness evaluations. Further, the authors are planning to investigate other types of CA encodings representing not only wind bracing elements but complete steel structural systems in tall buildings. Finally, the authors intend to improve the design fitness evaluation by incorporating several other criteria (deflection, cost, etc.) into the fitness function and by using evolutionary multi-objective approaches.

Acknowledgements

Initial part of the project presented in this paper was conducted during the first New Kind of Science Summer School 2003 at Brown University, Providence, RI, the first author attended. He would like to thank Dr. Stephen Wolfram and Dr. Todd Rowland for the possibility of taking part in the Summer School, and for many valuable discussions and helpful suggestions. He would like to thank all the faculty members and other students for inspiring exchange of ideas and wonderful time spent there! He also acknowledges continuous support from George Mason University, in a form of the Interdisciplinary Scholarship from the Office of the Provost and Graduate Research Assistantship from the IT&E School.

References

1 Bentley, P. J., & Corne, D. W. (Eds.). (2002). *Creative evolutionary systems.* San Francisco, CA: Morgan Kaufmann Publishers.

2 Wolfram, S. (2002). *A new kind of science.* Champaign, IL: Wolfram Media.

3 Wolfram, S. (1994). *Cellular automata and complexity: collected papers.* Reading, MA: Addison-Wesley.

4 Inou, N., Uesugi, T., Iwasaki, A., & Ujihashi, S. (1998). Self-organization of mechanical structure by cellular automata. In P. Tong, T. Y. Zhang & J. Kim (Eds.), *Fracture and strength of solids. Part 2: Behavior of materials and structure*, 1115-1120.

5 Kundu, S., Oda, J., & Koishi, T. (1997). *A self-organizing approach to optimization of structural plates using cellular automata.* In W. Gutkowski &

Z. Mroz (Eds.), Proceedings of the Second World Congress on Structural and Multidisciplinary Optimization (WCSMO-2), Zakopane, Poland, 173-180.

6 Kita, E., & Toyoda, T. (2000). Structural design using cellular automata. *Structural and Multidisciplinary Optimization, 19*, 64-73.

7 Kita, E., & Toyoda, T. (2001). *Truss structural design based on concept of cellular automata.* In Proceedings of the International Association for Shell and Spatial Structures Symposium (IASS2001), Nagoya, Japan.

8 Hajela, P., & Kim, B. (2001). On the use of energy minimization for CA based analysis in elasticity. *Structural and Multidisciplinary Optimization, 23*(1), 24-33.

9 Arciszewski, T., Michalski, R. S., & Wnek, J. (1995). Constructive induction: the key to design creativity. In *Preprints of the Third International Round-Table Conference on Computational Models of Creative Design.* Heron Island, Queensland, Australia, 397-426.

10 Arciszewski, T., & De Jong, K. A. (2001). *Evolutionary computation in civil engineering: research frontiers.* In B. H. V. Topping (Ed.), Proceedings of the Eight International Conference on Civil and Structural Engineering Computing, Eisenstadt, Vienna, Austria.

11 Bentley, P. J. (2000). *Exploring component-based representations.* In I. C. Parmee (Ed.), Proceedings of the Fourth International Conference on Adaptive Computing in Design and Manufacture (ACDM'2000), University of Plymouth, UK, 161-172.

12 Hornby, G. S. (2003). *Generative representations for evolutionary design automation.* Ph.D. Dissertation, Department of Computer Science, Brandeis University, Waltham, MA.

13 Bentley, P. J., & Kumar, S. (1999). *Three ways to grow designs: a comparison of embryogenies for an evolutionary design problem.* In W. Banzhaf, J. Daida, A. E. Eiben, M. H. Garzon, V. Honavar, M. J. Jakiela & R. E. Smith (Eds.), Proceedings of the Genetic and Evolutionary Computation Conference (GECCO'99), Orlando, Florida, USA, 35-43.

14 Thompson, D. A. W. (1942). *On growth and form.* Cambridge: Cambridge University Press.

15 Jakiela, M. J., Chapman, C. D., Duda, J., Adewuya, A., & Saitou, K. (2000). Continuum structural topology design with genetic algorithms. *Computer Methods in Applied Mechanics and Engineering, 186*, 339-356.

16 Hamda, H., Jouve, F., Lutton, E., Schoenauer, M., & Sebag, M. (2002). Compact unstructured representations for evolutionary topological optimum design. *Applied Intelligence, 16*, 139-155.

17 De Jong, K. A. (to appear). *Evolutionary computation: a unified approach.* Cambridge, MA: MIT Press.

18 Murawski, K., Arciszewski, T., & De Jong, K. A. (2001). Evolutionary computation in structural design. *Journal of Engineering with Computers, 16*, 275-286.

19 Kicinger, R., Arciszewski, T., & De Jong, K. A. (2004). Evolutionary designing of steel structures in tall buildings. *Journal of Computing in Civil Engineering*(tentatively approved).

Design search and optimisation using radial basis functions with regression capabilities

A. J. Keane[1]

[1]School of Engineering Sciences, University of Southampton, Southampton, SO17 1BJ, U.K.
ajk@soton.ac.uk

Abstract

Modern design search and optimisation (DSO) processes that involve the use of expensive computer simulations commonly use surrogate modelling techniques, where data is collected from planned experiments on the expensive codes and then used to build meta-models. Such models (often termed response surface models or RSMs) can be built using many methods that have a variety of capabilities. For example, simple polynomial (often linear or quadratic) regression curves have been used in this way for many years. These lack the ability to model complex shapes and so are not very useful in constructing global RSM's for non-linear codes such as the Navier Stokes solvers used in CFD - they are, however, easy to build. By contrast Kriging and Gaussian Process models can be much more sophisticated but are often difficult and time consuming to set up and tune. At an intermediate level radial basis function (RBF) models using simple spline functions offer rapid modelling capabilities with some ability to fit complex data. However, as normally used such RBF RSM's strictly interpolate the available computational data and while acceptable in some cases, when used with codes that are iteratively converged, they find it difficult to deal with the numerical noise inevitably present. This paper describes a modification to the basic RBF scheme that allows a systematic variation of the degree of regression from a pure linear regression line to a fully interpolating cubic radial basis function model. The ideas presented are illustrated with data from the field of aerospace design.

1 Introduction

Design search an optimisation is concerned with improving designs by modifying parameters under the designer's control while at the same time meeting any constraints imposed by requirements such as structural integrity, safety, cost margins, etc. Commonly such improvements are based on the use of computer simulations. If used to justify important design changes these simulations must be accurate and they are thus often computationally expensive: runs times ranging from tens of minutes to days are not uncommon. When multiple design parameters are to be investigated, searching over design spaces directly with the full simulation code becomes unworkable. In such cases surrogate models are often used. These surrogate, response surface or meta-models seek to reproduce the behaviour of the full simulation code at much reduced computational cost and usually take the form of some kind of multi-dimensional curve fit through pre-computed results obtained from the expensive code. Figure 1 illustrates just such a design process (see Keane [1], for example).

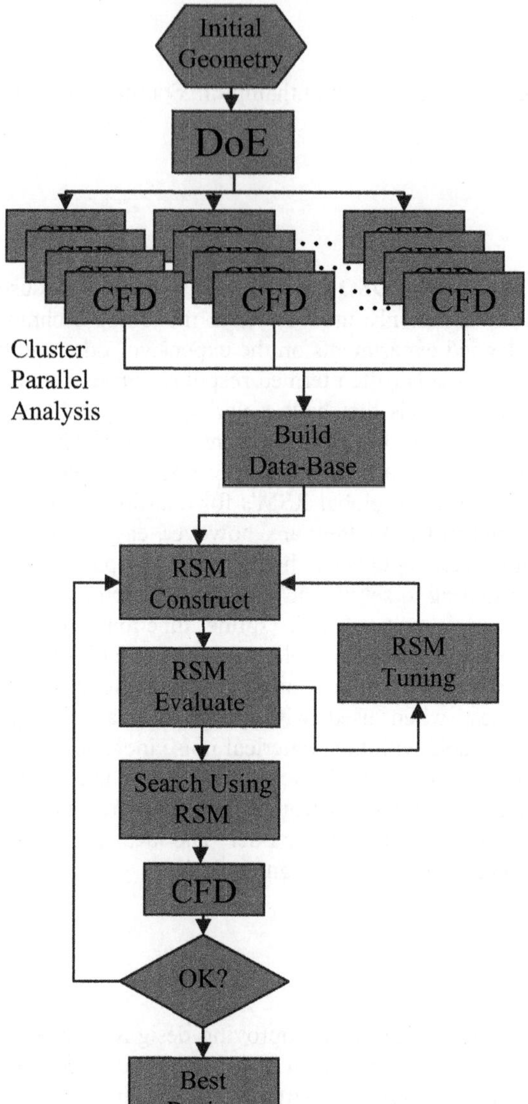

Figure 1 – a surrogate based search process for CFD.

In Figure 1 an initial geometric model is varied parametrically (e.g., by use of a parametric CAD system) according to some Design of Experiments (DoE) plan and then the resulting geometries are analysed (in parallel) to construct a data-base of results. A multi-dimensional RSM is then used to curve fit this data and this can then, in turn, be searched using optimisers to suggest improved designs for analysis. If the improved designs are not adequate the RSM can be updated and the search repeated until some form of convergence is achieved. There are a range of RSM types that can be used in this way (see Myers and Montgomery [2] or Jones [3]) and a range of update strategies (see Jones et al [4] for example).

One issue that arises in this process concerns any noise that may be present in the computed results. Such noise commonly arises when iterative solvers are used to carry out performance analysis, since the final convergence must usually be truncated before absolute precision is achieved if realistic run times are to be maintained. Also, most expensive PDE solvers use discretization schemes that must usually be changed with any geometric changes so that discontinuities can arise in the resulting data. In such circumstances the RSM curve fit must allow for this noise if over-fitting is to be avoided. It is also the case that response functions of many variables can have highly complex behaviour and any attempt to curve fit by interpolation with limited data can be misleading without some form of smoothing to establish overall trends, as opposed to a focus on local features.

In this paper a simple and robust technique is introduced to allow designers to control the degree of regression involved in radial basis function RSM construction in an easy to use fashion. While there are numerous more sophisticated approaches available in the statistics and neural networks literature, the suggested approach neatly fits between two existing methods widely used in the DSO RSM community, i.e., linear regression and interpolating cubic spline radial basis functions.

1.1 Theory

When carrying out optimisation a designer is seeking a solution vector in parameter space, $\boldsymbol{x}^*$ such that $y(\boldsymbol{x}^*)$ is a minimum of the objective function $y(\boldsymbol{x})$, subject to k constraints $g_k(\boldsymbol{x})>0$. Here $\boldsymbol{x}$ is a vector of m design variables, $\boldsymbol{x}=(x_1, \ldots, x_m)^T$. If a surrogate or response surface model $y'(\boldsymbol{x})$ is to be used, this model is set up using n training points $y_j(\boldsymbol{x}_j)$, $j=1, \ldots, n$, $\boldsymbol{x}_j=(x_{1j}, \ldots, x_{mj})^T$, taken from a pre-computed data-base of results (as per Figure 1). To begin with consider two such models for $y'(\boldsymbol{x})$: linear regression and cubic radial basis functions. In linear regression the RSM is given by

$$y'(\mathbf{x}) = a_0 + \sum_{i=1}^{m} a_i x_i \qquad (1)$$

where the coefficients a are chosen to minimize the sum of the squares of the errors between the surrogate and the training data, typically using an SVD least squares approach, see Figure 2.

By contrast an interpolating radial basis function model is constructed from a series of functions φ based at the sample points:

$$y'(\mathbf{x}) = \sum_{j=1}^{n} b_j \varphi(\mathbf{x} - \mathbf{x}_j). \qquad (2)$$

Here the coefficients b are chosen so that the model passes through the training points, i.e., n equations are set up using the n training points to solve for the n unknown coefficients b. The functions, $\boldsymbol{\varphi}$, can take many forms (see for example Jones[3]), here we use cubic splines, i.e.,

$$\varphi(\mathbf{x}-\mathbf{x}_j)=\|\mathbf{x}-\mathbf{x}_j\|^3 . \qquad (3)$$

As Jones notes, however, this method can be numerically poorly conditioned and so it is better to use a modified approach which combines a regression line with the basis functions:

$$y'(\mathbf{x})=a_0+\sum_{i=1}^{m}a_i x_i+\sum_{j=1}^{n}b_j\varphi(\mathbf{x}-\mathbf{x}_j). \qquad (4)$$

In this form the model has m+n+1 unknowns and so a set of m+1 additional constraints are added:

$$\sum_{j=1}^{n}b_j=0 \qquad (5)$$

and

$$\sum_{j=1}^{n}b_j x_{ij}=0 \qquad (6)$$

for x_{ij} = 1,..., m. In this form the basis functions model the differences between the regression line and the actual data points. The resulting model still strictly interpolates the data, but is now numerically better behaved (n.b., this approach can easily be extended to higher order regression models if desired). Figure 3 illustrates this approach applied to the same data as used in Figure 2.

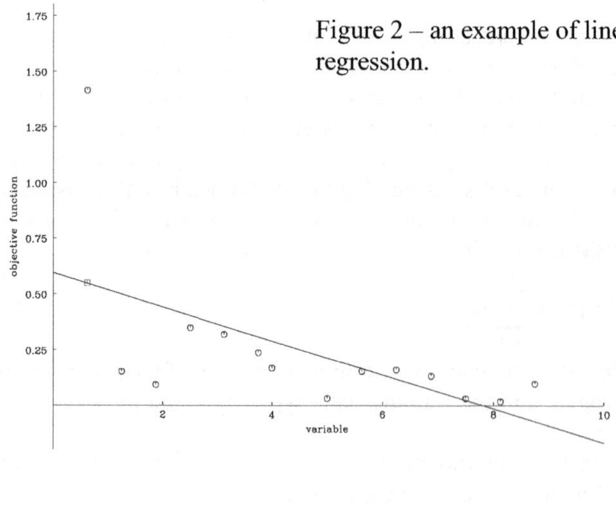

Figure 2 – an example of linear regression.

It is clear that these two approaches take a fundamentally different view of the training data: the linear regression approach assumes that there are errors in the data which cause it to deviate from a basically very simple trend; the RBF model assumes the data to be error free and the underlying trends to be much more complex. As has already been noted, however, engineering calculations usually lie somewhere between these two extremes. Certainly the results of engineering calculations are usually repeatable and often give complex trends, but they also contain noise. A number of sophisticated curve fitting schemes with regression abilities have been proposed to cater for this kind of issue, such as Kriging with regularisation, see for example Jones et al [4]. Such schemes often contain hyper-parameters which must be tuned and so it is natural to ask whether a simpler approach may be possible. That proposed here is based on the previous two models.

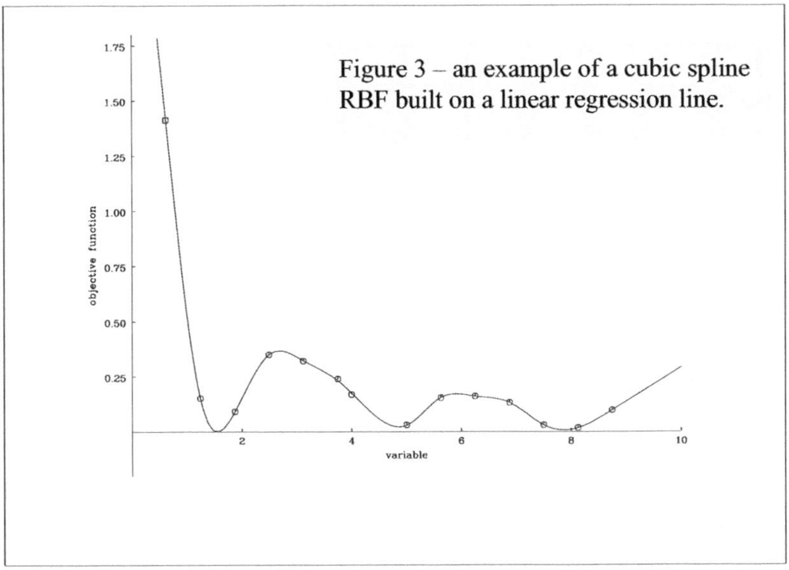

Figure 3 – an example of a cubic spline RBF built on a linear regression line.

To see how to proceed we note that in equation (4) we may regard the summation over the n basis functions as a corrector to the underlying linear regression. If we include only a subset of these functions but still use all the training data to solve for the m coefficients a and reduced number of coefficients b we have an over-determined set of equations that can still be solved in a least squares sense using SVD. Similarly the summations in equations (5) and (6) are taken over the reduced numbers of basis functions. Therefore, as the number of basis functions increases from zero to n the resulting model moves from being one of simple linear regression up to full RBF interpolation. Figures 4 and 5 illustrate this process for one and seven basis functions, respectively. Figure 6 shows members of the family spanning the space between Figures 2 and 3.

It is clear from figures 4 to 6 that a range of models can be built in this way. When setting them up the user must first decide how many basis functions to include and at which training data points to centre them. These decisions open up a degree of model tuning that can be exploited by the user according to the needs of the problem in hand. Choosing the number of functions to use, essentially addresses the problem referred to as "over-fitting" in the regression literature (see Ralston [5] for example): if too few functions are used the resulting model may fail to represent important trends in the data; conversely, if too many are used the resulting model may just be driven by the noise in the data.

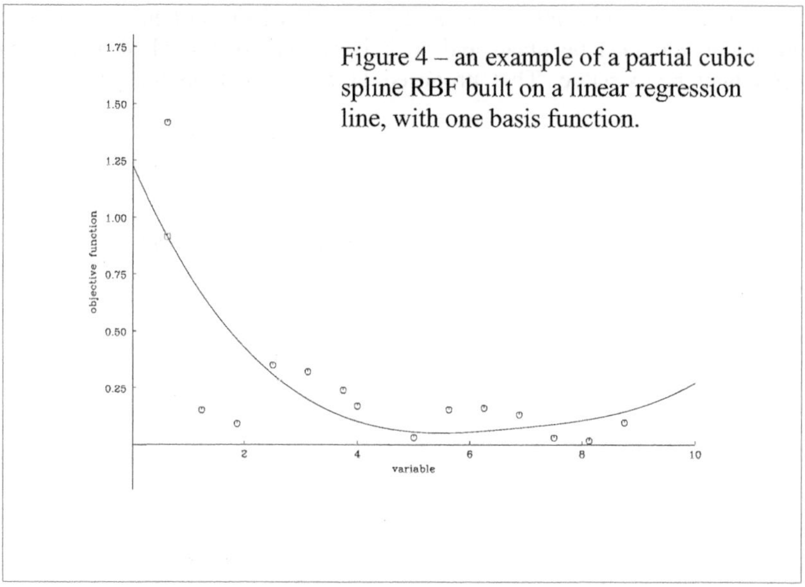

Figure 4 – an example of a partial cubic spline RBF built on a linear regression line, with one basis function.

One scheme for addressing this issue may be readily built from the Null hypothesis used in polynomial curve fitting (see Ralston [5] pp 254). When dealing with polynomials this hypothesis allows the order of the model to be tensioned against errors in the model, so that increasing the order must significantly reduce errors between the model and the data. Here we monitor the following expression:

$$\sigma^2 = \delta^2 / (n - n') \qquad (7)$$

where n' is the reduced number of basis functions being used and δ^2 is the sum of the squares of the errors between the model and the function values at the training points (note that both denominator and numerator become zero when all n basis functions are used, i.e., the model then interpolates the data). When this expression is nearly constant for several vales of n' we stop adding more basis function terms. Figure 7 shows a plot of this function for the data used so far and it can be seen that values around seven give a good fit., i.e., as per Figure 5. This might surprise the reader until it is realised that the data used for this test is, in fact, taken from the so called "bump" function (ref Keane [6]) which is oscillatory in just this sort of way.

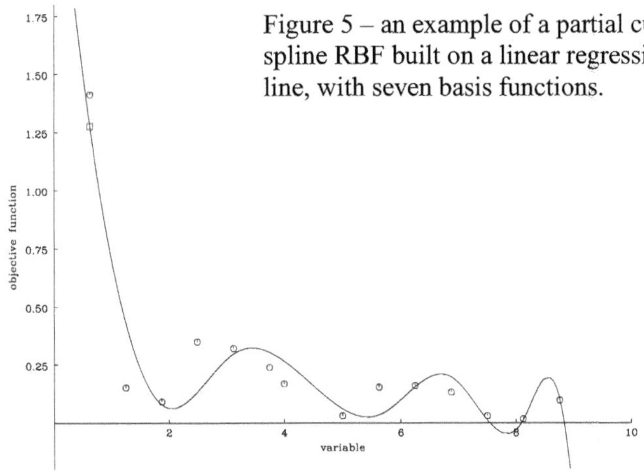

Figure 5 – an example of a partial cubic spline RBF built on a linear regression line, with seven basis functions.

The second issue that must be addressed in setting up the model concerns selection of the basis functions to include. Note that whichever functions are used, all the training data are still used in the SVD least squared process used to construct the model. It is just that some training points additionally become the centres for RBF construction, and thus have more impact on the resulting curve fit. In fact once fewer than n basis functions are used there is no need to place their centres at the training points at all. There are various schemes available in the literature for building models with arbitrary centres, such as those using clustering algorithms to select good locations, radial basis function neural networks (where similar models are used but complex training strategies are used to solve for the desired weights) or support vector machines. Such approaches lie outside the scope of this paper; suffice to say that they all involve a good deal of statistical analysis and tuning that rather defeats the desire for building simple models that can be rapidly set up – here the centres are selected randomly from the locations in the training set. If more complex models are required then more statistically soundly based approaches such as Kriging and stochastic process models should be considered. The approach adopted here simply relies on the fact that the model is bounded on one side by linear regression and on the other by an interpolating cubic spline RBF. Both of these schemes are well understood and simply to apply – by just tracking equation (7) for various degrees of model fit the user can decide if what is being proposed makes sense. A characteristic of the proposed approach is that the degree of fit, σ^2, tends to be oscillatory showing significant increases if the regression process is poor, because gradual addition of further basis functions then fails to consistently improve the model with diminishing returns at each stage.

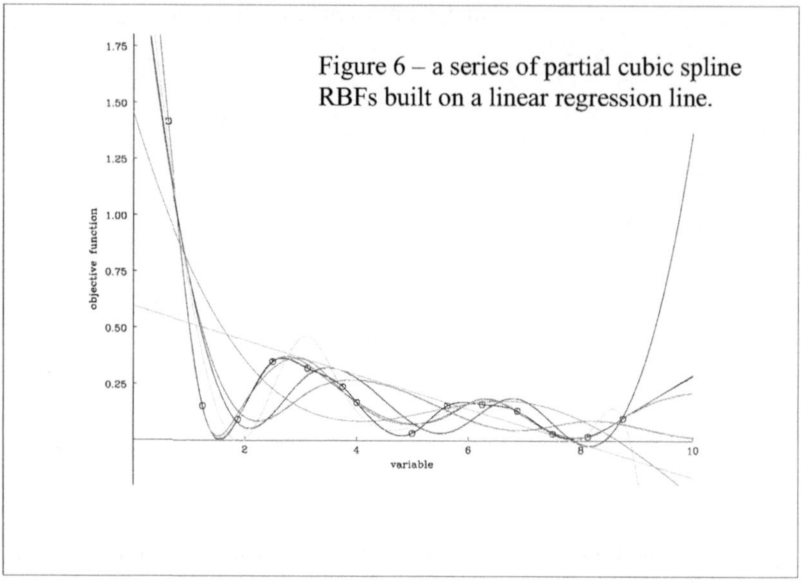

Figure 6 – a series of partial cubic spline RBFs built on a linear regression line.

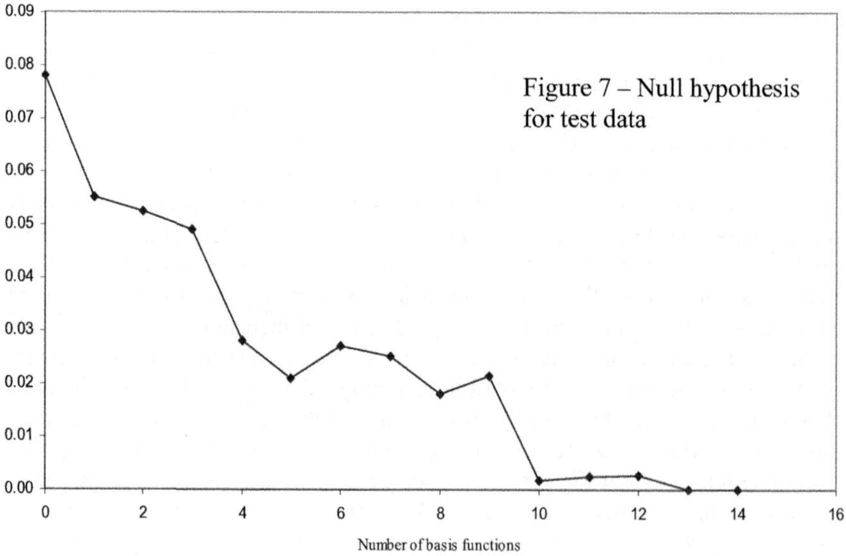

Figure 7 – Null hypothesis for test data

1.2 Example

To demonstrate this approach on a more practical example the wing design case presented by Keane[1] is re-analysed here. The problem concerns the design of a transonic civil transport aircraft wing defined by 11 variables, see Table 1. In the reference both empirical and CFD results were used; here only the empirical "Tadpole" model is tested for simplicity [7]. In the work by Keane a Krig was used to model the results coming from Tadpole and this was able, after careful tuning, to predict 250 unseen test data points with a correlation coefficient of 0.991. This test data is re-analysed here using various numbers of basis functions to regress the data. First the Null Hypothesis test of equation (7) is applied, see Figure 8. The figure shows that several levels of regression may be tried: there being plateaus in the curve at around 35, 100 and 150 basis functions. Figure 9 shows a correlation plot for a model with 150 basis functions and as can be seen this fits the unseen trial data quite well, yielding a correlation coefficient of 0.938. Adding more basis functions will, of course make the RSM fit the training data better but not necessarily improve its ability to predict unseen test data. Table 2 shows how the correlation coefficient for the test data varies with different numbers of basis functions. As can be seen the plateau in the Null Hypothesis at around 35 basis functions is matched by a good correlation, while that at 215 is slightly worse. Using a full set of basis function, which forces the model to interpolate the training data, massively over-fits the inputs and yields a very poor ability to predict unseen test data – a simple linear regression model is much better.

Conclusions

This paper has introduced a simple modification to the well known cubic radial basis function response surface model to enable it to have controllable regression capabilities. The proposed approach can be used to model noisy data without recourse to more sophisticated methods with little loss in accuracy. The only parameter tuning required is the selection of the degree of regression and this can set up either with a simple Null hypothesis approach or by using unseen test data to check for over-fitting. The advantage of the Null hypothesis is that it does not require additional unseen test data when setting up the model and is thus much cheaper to use. The approach taken lacks some statistical rigour but trades this for simplicity of definition and ease of use. It has been demonstrated on a simple one-dimensional test case and a realistic 11-dimensional aircraft wing design problem.

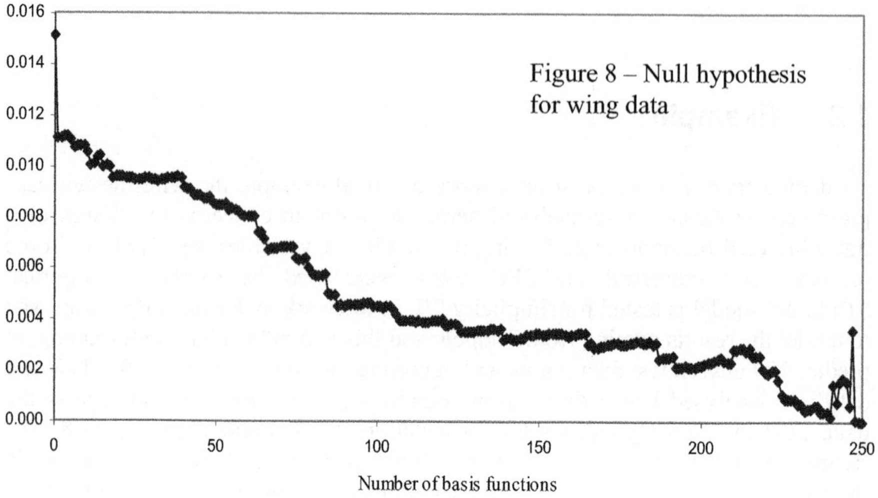

Figure 8 – Null hypothesis for wing data

Lower limit	Value	Upper limit	Quantity (units)
100	168	250	Wing area (m^2)
6	9.07	12	Aspect ratio
0.2	0.313	0.45	Kink position
25	27.1	45	Sweep angle (degrees)
0.4	0.598	0.7	Inboard taper ratio
0.2	0.506	0.6	Outboard taper ratio
0.1	0.150	0.18	Root t/c
0.06	0.122	0.14	Kink t/c
0.06	0.122	0.14	Tip t/c
4.0	4.5	5.0	Tip washout (degrees)
0.65	0.75	0.84	Kink washout fraction
	3.145		D/Q (m^2) – Tadpole

Table 1 – Design variables and objective function value

Number of basis functions	Correlation coefficient
0 (simple linear regression)	0.869
35	0.930
100	0.928
150	0.938
215	0.924
250	0.257

Table 2 – Correlation data

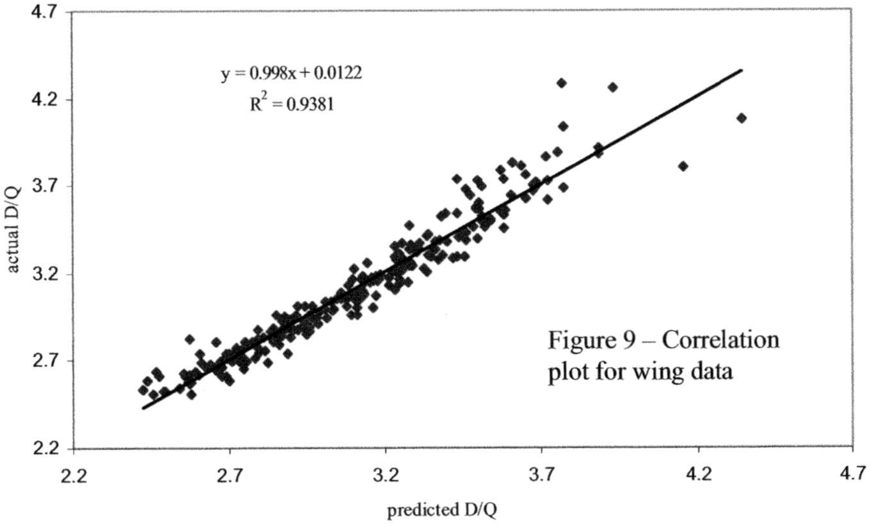

Figure 9 – Correlation plot for wing data

References

1 Keane, A.J., (2003) "The Use of Design of Experiment, Response Surface and Data Fusion Methods in Transonic Aircraft Wing Design Optimisation", *J. of Aircraft*, **40**(4) pp. 741-750.

2 Myers, R.H. and Montgomery, D.C., (1995) "*Response surface methodology: process and product optimization using design of experiments*", John Wiley and Sons inc.

3 Jones, D.R., (2001) "A taxonomy of global optimization methods based on response surfaces", *Journal of Global Optimization*, **21**, pp. 345-383.

4 Jones, D.R., Schonlau, M. and Welch, W.J., (1998) "Efficient global optimization of expensive black-box functions", *Journal of Global Optimization*, **13**, pp. 455-492.

5 Ralston, A. and Rabinowitz, P., (1978) "*A first course in numerical analysis*", McGraw-Hill, New York.

6 Keane, A. J., (1995) "Genetic algorithm optimization of multi-peak problems: studies in convergence and robustness", *Artificial Intelligence in Engineering* **9**(2) pp. 75-83.

7 Cousin, J., and Metcalfe, M., (1990) "The BAE Ltd Transport Aircraft Synthesis and Optimization Program", AIAA-90-3295.

Genetic Programming within Civil Engineering

David Shaw[1], John Miles[1] and Alex Gray[2]
[1]Cardiff School of Engineering, Cardiff University, UK
ShawD2@cardiff.ac.uk
[2]Department of Computer Science, Cardiff University, UK

Abstract

Genetic programming is a relatively new method of evolutionary computing with few published applications in civil engineering. This paper both describes and demonstrates how GP can be applied to structural optimisation and design problems to produce results that offer significant improvements over traditional GA based methods. The paper concludes by presenting the direction of the work currently being undertaken at Cardiff University, which includes the conceptual design of frame structures and the visualisation of results using VRML/X3D to produce virtual reality simulations that can be used as a collaborative design environment over the Internet.

Keywords: Civil Engineering, Genetic Programming, Structural Optimisation, X3D.

1 An introduction to genetic programming.

This section provides a general background to the topic of genetic programming (GP) that is expanded in subsequent sections to demonstrate how GP can be used in structural engineering. GP is a domain-independent, problem-solving approach in which computer programs are evolved to find solutions to problems. The solution technique is based on the Darwinian principle of 'survival of the fittest' [1] and is closely related to the field of genetic algorithms (GA) [2]. However three important differences exist between GAs and GP:

- *Structure*: GP usually evolves tree structures while GAs evolve binary or real number strings.
- *Active vs Passive*: Because GP usually evolves computer programs, the solutions can be executed without post-processing i.e. active structures, while GAs typically operate on coded binary strings i.e. passive structures, which require post-processing.
- *Variable vs fixed length*: In traditional GAs, the length of the binary string is fixed before the solution procedure begins while in GP, a parse tree can vary in size throughout the run. However it is recognised that in more advanced GA work, variable length strings are used.

The ability to search the solution space and locate regions that potentially contain optimal solutions for a given problem is one of the fundamental components of most artificial intelligence (AI) systems. There are three primary types of search; the blind search, hill climbing and beam search [3]. GP is classified as a beam search because it maintains a population of solutions that are smaller than all of the available solutions. GP is also usually implemented as a weak search algorithm containing no problem specific knowledge, although some research has been directed towards 'strongly typed genetic programming' [4]. However while GP can find regions containing 'good' solutions, no stochastic search method can guarantee to find the optimal solution. Memetic algorithms can fulfil this role by combining an evolutionary algorithm with a problem specific search algorithm to locate the optima [5].

1.1 The primitives of genetic programming.

Every solution evolved by GP is assembled from two sets of primitive node; terminals and functions. The terminal set contains nodes that provide an input to the GP system while the function set contains nodes that process values already in the system [3]. Constants can be used in GP by including them in the terminal set.

Once the evolutionary process is started, the GP system randomly selects nodes from either set and thus may not utilise all of the available nodes. GP is also very good at evolving new functions. Therefore it is usual to initially only provide relatively simple node sets and add nodes only if required because increasing the size of each set also enlarges the search space.

1.2 Tree based genetic programming.

The primitives of GP, the function and terminal nodes, must be assembled into an executable structure before applying them to the problem. Three main types of structure exist: tree, linear and graph. Within this work, the input (the structure to be optimised or designed) actually forms a graph network. However by the duplication of joint data i.e. the same 'joint node' can exist in the same tree on more than one occasion, this graph network is converted into a tree structure (see Section 3).

1.3 Genetic operators.

There are three major evolutionary operators within a GP system:

- *Reproduction:* selects an individual from within the current population to be copied exactly into the next generation. There are several ways of deciding which individual is to be copied including 'fitness proportionate', 'rank' and 'tournament' selection.
- *Crossover:* mimics sexual recombination in nature, where two parent solutions are chosen and parts of their sub-trees are swapped. Every crossover operation should result in the formation of a legal structure because each function exhibits the property 'closure' (each function is able to process all possible argument values).
- *Mutation:* causes random changes in an individual before it is introduced into the subsequent population. Unlike crossover, mutation is asexual and thus only operates on one individual. During mutation either all nodes are removed beneath an arbitrarily defined point and a new sub-branch is randomly created, or a single node is swapped for another.

1.4 Generational genetic programming.

GP has developed two main approaches to dealing with the issue of its generations; generational and steady-state. In generational GP well-defined and distinct generations exist, with each generation being represented by a complete population of individuals. Therefore each new population is created from the older population, which it then replaces. Steady-state GP does not maintain these discrete generations but continuously evolves the current generation using any available genetic operators [3]. This work uses the steady-state approach to GP.

2 Genetic programming in civil engineering.

Procedural programming techniques have been successfully used for the detailed design of structures where the path to the final solution is known in advance. However other design stages lack this predefined approach and thus it becomes necessary to adopt self- learning/ organising computing techniques. This work applies GP at the conceptual design stage, of the design process, with the aim of providing the designer with a series of possible solutions that can be carried forward to the later stages. Table1 lists other published applications of genetic programming in civil engineering.

3 Genetic programming in structural engineering.

All but one of the published applications in civil engineering to date, utilise a GP system to evolve relationships between variables as suggested by Koza [1] e.g. symbolic regression functions. This section describes how GP can be applied to structural optimisation problems by using the tree structure of GP, to represent a building or structure.

In 2000 Soh and Yang [18] published an approach that solved the key issue of how to represent the 'node-element diagram' of structural analysis as a 'point-labelled tree' (as used in GP). However because the structure (the phenotype) is now different from the GP tree (the genotype), an additional decoding step must be included in the solution procedure before any fitness evaluation can occur (Fig.3). This step was not previously required when evolving regression functions, as these solutions could be applied directly to the problem. Although this is a departure from traditional GP, by utilising this representation it can be demonstrated that GP can produce better results when attempting to simultaneously optimise the geometry, sizing and topology of a truss than work using other search techniques. It is also important to note that this tree will not degenerate into a linear search, as the corresponding structure would be a mechanism. A mechanism is not a viable structure in engineering and thus will be heavily penalised by the fitness function and therefore should not be sustained within the population.

3.1 Structural encoding.

Soh and Yang [18] propose that the GP tree should compose two types of node: inner nodes which are GP functions representing the cross-sectional areas of the members A_p (p= i,j,k,l,m,n) and outer nodes which are GP terminals representing various node points N_i (i= 1,2,3,4) (Fig.1). To create a GP parse tree, one member must be selected from the structure to be the root node. This member then has its' corresponding start and end joints represented by children nodes to its left and right. Then from left to right the members associated with each joint node are removed

Table 1: Applications of GP in Civil Engineering

Application	Author	Year	Details
Shear strength prediction of deep RC beams	Ashour et al [6]	2003	Estimation of the shear strength of deep RC beams, subjected to two point loads, from 141 published experimental results.
Modelling of wastewater treatment plants	Hong and Bhamidimarri [7]	2003	Use of genetic programming to model the dynamic performance of municipal activated sludge wastewater treatment plants.
Detection of traffic accidents.	Roberts and Howard [8]	2002	Detection of accidents on motorways in low flow, high-speed conditions i.e. late at night based on three years of traffic data whilst producing a near zero false alarm rate.
Flow through a urban basin	Dorado et al [9]	2002	Construction of sewage network model in order to calculate the risk posed by rain to the basin and thus provide prior warning of flooding or subsidence.
Prediction of journey times	Howard and Roberts [10]	2002	Investigation of GP to forecast the motorway journey times.
Estimation of design intent	Ishino and Jin [11]	2002	Using GP to automatically estimate design intent based on operational and product-specific information monitored throughout the design process.
Modelling of water supply assets	Babovic et al [12]	2002	In order to determine the risk of a pipe burst, a GP is evolved to 'data mine' a database containing information about historic pipe bursts.
Identification of crack profiles	Kojima et al. [13]	2001	Detection of cracks inside hundreds of heat exchanger tubes in a nuclear power plant's steam generator via analysis of data measured via quantitative non-destructive testing.
Modelling rainfall runoff	Whigham & Crapper [14]	2001	Discovery of rainfall-runoff relationships in two vastly different catchments.
Improving engineering design models	Watson and Parmee [25]	1998	Symbolic regression and Boolean induction to model engineering fluid dynamics systems.
Prediction of long-term electric power demand	Lee et al [15]	1997	Symbolic regression via genetic programming to predict the long-term electric demand of Korea (based on training data from 1961 to 1980).
Systems identification	Watson and Parmee [24]	1996	Symbolic regression to calibrate Rolls Royce preliminary design gas turbine cooling systems software.
Evolution of traffic light control laws	Montana and Czerwinski [16]	1996	Evolution of a new type of adaptive control system for a network of traffic signals depending on variations in traffic flow.
Identification of crack profiles	Köppen and Nickolay [17]	1996	Agent generation to detect and track dark regions that could be cracks in greyscale images of textured surfaces.

from the structure and used to replace the relevant joint nodes in the GP tree (Fig.1). This procedure continues until every structural member is represented in the GP tree. However it is important that the left-right relationship of child and parent node is maintained as the GP tree is constructed. Therefore each member's start and end joints are represented by the far left and far right children. For example function A_j connects nodes N_1 and N_2 (Fig.1).

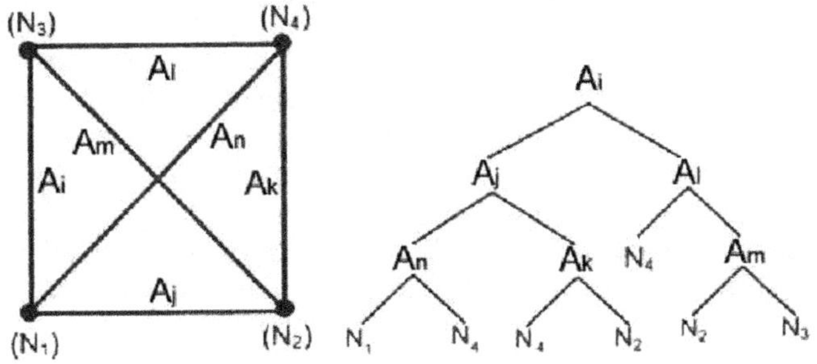

Figure 1: GP tree encoding for structural analysis

This approach to structural encoding appears very simple when compared to the complex binary string used by a GA to represent member properties. For example Soh and Yang [18] have published their results for the evolution of a truss that was capable of carrying six specified loads. Soh and Yang required a GP tree containing 29 nodes (16 joint nodes and 15 member nodes) where as the GA representation required a chromosome of 25,200 bits [18].

3.2 Architecture of the current work.

The work currently being undertaken at Cardiff University uses the structural encoding approach listed above to produce a tree based representation of an engineering structure, however many important differences do exist. This work does not use a finite element package (FEM) to analyse the each possible structure as FEM is computationally very intensive which slows down the generation of results. It was considered more important to deliver a range of possible solutions in a shorter period of time to the designer at the conceptual design stage that could be analysed at a later date by FEM. This work also uses an eXtensible Markup Language (XML) interface to allow the designer to input structures to the system and the same interface also enables the visualisation of results in 3D, using models created by the virtual reality modelling language (VRML) and X3D. The work is delivered using Java and at all levels the principle of objects is utilised so that each tree is an object composed of separate node objects.

4 An Example of Structural Optimisation.

This section will highlight the differences and improvements between the current work and any existing GA and GP based work. The following example aims to minimise the overall weight of the structure by selecting progressively lighter members that can sustain the neseccary forces and moments. Whilst it is acknowledged that other factors contribute to the overall cost of a structure e.g material cost and fabrication, strucutral weight does make a substantial contribution to a structure's cost and therefore it is used here as a first level approximation to a minimum cost structure.

4.1 10 Member Planar Truss.

An illustrative example of a 10-member planar truss is now provided (Fig.2) that compares the current GP based work to the results generated by a previous GA and GP based systems. In this example, given the initial structure, the goal was to minimise the structure's weight by varying its' shape and topology. The solution is reported by splitting it into three sections the Model, View and Controller (MVC). These three sections relate to the 'MVC' architecture that this work is based around. MVC separates the underlying GP system from the input GUI and the visualisation output.

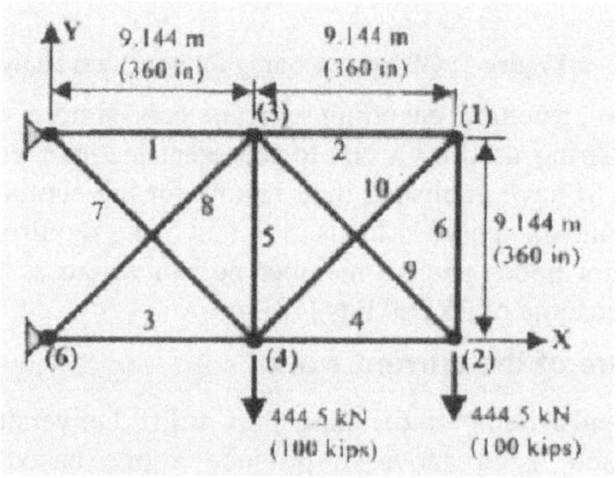

Figure 2: 10 member planar truss

4.2 Controller- GP Tableau.

The Controller allows the user to input both data and constraints into the system. Within this work, the user can alter the variables controlling the evolutionary process, the population size and number of generations. As optimisation problems do not usually have an explicitly defined end point, the GP system runs for a stipulated number of generations before returning the 'best-so-far' individual as the solution. The other essential input for structural optimisation problems is an initial structure to improve.

The structure is encoded as a tree during the evolutionary phase but can also be represented by an XML Document Type Definition (DTD) created by MIT [19] called 'StructureXML'. StructureXML not only provides a semantic, storage format but also provides an interface that helps this system to separate its' structural analysis, evolutionary and visualisation modules. However the StructureXML DTD does not provide any information about a structure's response due to loading. Therefore a new XML DTD was created called 'GPStructureXML' that includes the forces and moments experienced by a structural member and the displacement, rotation, reaction forces and moments experienced by each individual joint. The following GP Tableau summaries the main choices made when applying the major preparatory steps of GP to the problem.

Table 2: GP Tableau for the 10 member planar truss problem

Objective	For a given 10 Member Planar Truss, subjected to two loads acting simultaneously the objective is to minimise its weight by varying geometry, shape and topology.
Terminal Set	Structural joints.
Function Set	Structural members.
Fitness Case	-
Raw Fitness	Multiplying the structure's objective function (overall weight) by the structure's corresponding penalty factor.
Standardised Fitness	Equals the raw fitness for this problem.
Hits	-
Wrapper	-
Parameters	M=500, G=51
Success Predicate	None.

4.3 Model.

The Model represents the underlying data of the object, in this instance the structural analysis and evolutionary parts of the overall system. As stated previously each GP tree should be composed of two types of node: the inner nodes which are GP functions representing the members and the outer nodes which are GP terminals representing the structural nodes. This system uses Java based objects to represent each node therefore encapsulation can be used to simplify the genetic operators as the actual data is 'hidden'. This is only possible when a GP parse tree is created using a fully object orientated language.

Two approaches can be taken for designs that fail to meet functional requirements: outright rejection or a penalty-based approach. The penalty-based approach adds a penalty value to the fitness score and although this is a conservative strategy, the increase in convergence time is considered to be potentially less dangerous than the potential for premature loss of genetic material [26]. A penalty-based approach was employed in this work, because the optimum solution will typically be located on the boundary between the feasible and infeasible regions. Therefore the penalty-based approach allows the GP search to search for the optimum solution from the direction of both the feasible and infeasible regions.

The structural analysis module is composed of the 'Trussworks' package designed using Java by MIT [19] that allows users to create and analyise 3D structures using the Direct Stiffness Method. Trussworks was selected because its source code is freely available and it is written in the same language as the all other parts of the system (Java). A finite element package was also considered but this would result in a significant increase in computational power and thus slow down the solution procedure. Any other package would also require a StructureXML interface to be created. Therefore it was decided that at the conceptual stage a faster but more approximate answer would be advantageous as FEM could be used during the later stages of design.

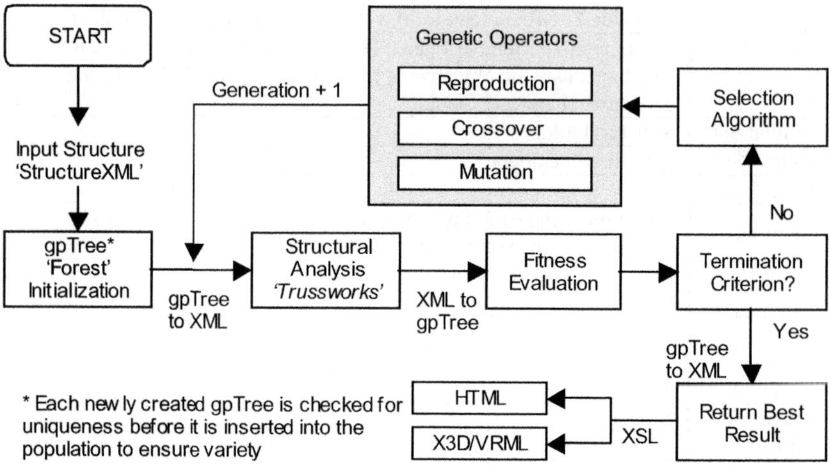

Figure 4: Schematic representation of the solution procedure

The results from this work indicate that for the 10 bar truss, the GP system produced a weight saving of 0.5%, when compared to a GA based method [20] however it only produced a saving of 0.01% over the results published by Soh and Yang [18]. Moreover the more complicated the structure, the greater the potential weight saving of a GP based method over a GA. For example this work found a weight saving of 0.1% when optimising a 25 bar space truss, over the existing GP system and 3.5% over the best GA work [20]. Whilst these improvements may not seem significant it must be remembered that the solution space for these problems is relatively small and that these case studies have been repeatedly used as benchmarks.

4.4 View- Visualisation.

The View accesses the results data from the Model and specifies how it is displayed. Visualisation is very important at the conceptual design stage as it allows all interested parties e.g. client, architect and engineer to collaborate more closely. Four types of visualisation are to be employed using this system. The first two relate to the visualisation of the GP parse tree: the traditional and radial while the third and fourth display the actual structure using virtual reality augmented with more traditional, HTML based documentation.

The traditional approach to viewing GP trees starts with the root node and continues downward, with the terminal nodes at the base of the tree (Fig.4). However if the total number of nodes or the tree depth does not remain small then these this method of depiction does not readily allow for the viewer to understand the tree's shape or balance as the representation becomes very large. Radial visualisation as proposed by Daida et al [21] solves these problems by mapping the GP tree nodes onto a set of concentric rings according to depth with each node being equally spaced onto the perimeter (Fig.5).

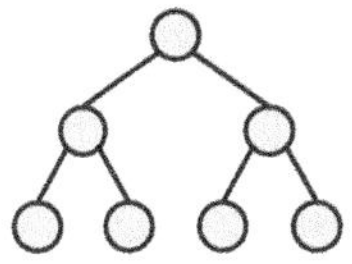

Figure 4: Traditional

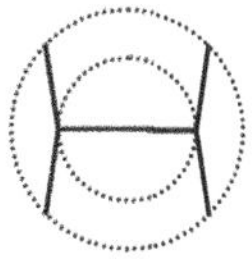

Figure 5: Radial

The third visualisation transforms any GPStructureXML document produced by the system into a X3D based, virtual reality model. The X3D specification has been developed by the Web3D consortium [22] and is currently under ISO Standard review. Touted as 'next-generation open standard for 3D on the WWW' X3D essentially extends and upgrades the capabilities of VRML97 using XML and allows the user to not only visualise the structure created but also to 'fly' around it to view it from any angle. X3D's extrusion node has been utilised in this visualisation process because given a cross-section and length, X3D can produce an object of any size. The main advantage of X3D as a file format is that the visualisations produced are very small in size typically under 10kB and thus can easily be transmitted over the internet with ease and secondly they do not require much processing power to view as they can use any Internet browser with the correct 'plug-in' installed.

The fourth uses XSL to convert a structures' XML tag tree into a HTML table detailing information about each joint and member. This more conventional approach can be used to directly compare individual designs and allow them to be browsed or printed in the normal way. These design tables are essential as they allow the experienced designer to rapidly assess the solution and should complement the VR visualization rather than be replaced by it.

5 Conclusion.

This GP based system has been shown to provide better results to an optimisation problem when compared to an equivalent GA system while using a method of representation that is significantly more simplistic than that used by a GA. The use of XML not only provides an interface between the components of this system and other programs but also allows for the visualisation of the structure using a virtual reality model and more traditional design documentation.

6 Future Aims.

Current literature contains only one existing application of GP in structural analysis and that is related to the optimisation and design of trusses. This project aims to extend GP to the conceptual stage of multi-storey office building design. Here the GP will be used to generate several novel solutions some of which the designers may take forward into subsequent design stages. This will provide many challenges as initially any GP parse tree must contain two types of member node i.e. column and beam but both of these members allow the GP tree to remain as a binary tree. However if a slab is also to be included because a slab must span between more than two nodes, the GP tree is no longer a binary tree (Fig.6).

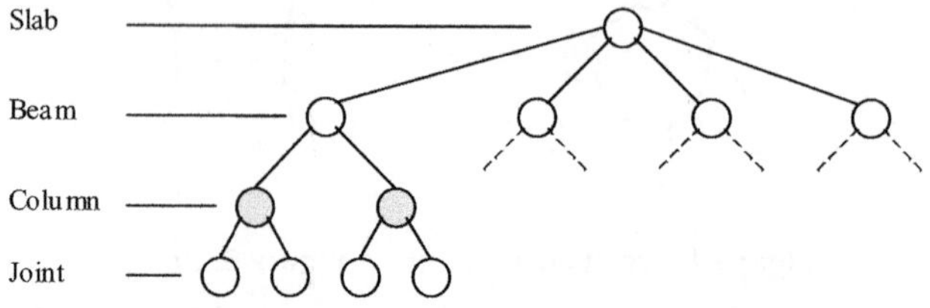

Figure 6: Slab/ column/ beam binary tree

The other main area of research is to allow the X3D visualisations to show the deformation of the structure under load and by using server software designed by Blaxxun create a structure that can be viewed both independently and in collaboration over the Internet. This would enable a distributed design teams to view the structure in a collaborative virtual reality environment although issues related to tethering and reattachment [23] will be beyond the scope of this project.

7 References.

[1] Koza J.R., Genetic Programming: On the programming of computers by means of natural selection, Cambridge MA: MIT Press, ISBN 0-262-11170-5, 1992.

[2] Holland J.H, *Adaptation in natural and artificial systems*, Ann Arbor: The University of Michigan Press, 1975.

[3] Banzhaf W et al, *Genetic Programming- An introduction (On the automatic evolution of computer programs and its applications)*, Morgan Kaufmann Publishers, ISBN 1-55860-510-X, 1998.

[4] Montana D.J, "Strongly typed genetic programming", *Evolutionary computation*, 3(2), 1995, pp199-230.

[5] Radcliffe N.J and Surry P.D, "Formal memetic algorithms", *Lecutre Notes in Computer Science 865*, 1994.

[6] Ashour A.F et al, "Empirical modelling of shear strength of RC deep beams by genetic programming", *Computers and Structures*, Pergamon, 81 (2003), pp331-338.

[7] Hong YS and Bhamidimarri R, "Evolutionary self-organising modelling of a municipal wastewater treatment plant", *Water Research*, 37(2003), pp1199-1212.

[8] Roberts S.C. and Howard D, "Detection of incidents on motorways in low flow high speed conditions by genetic programming", *Cagnoni S et al (eds): EvoWorkshops 2002, LNCS 2279*, Springer-Verlag, 2002, pp245-254.

[9] Dorado J et al, "Prediction and modelling of the flow of a typical urban basin through genetic programming", *Cagnoni S et al (eds): EvoWorkshops 2002, LNCS 2279*, Springer-Verlag, 2002, pp190-201.

[10] Howard D and Roberts SC, "The prediction of journey times on motorways using genetic programming", *Cagnoni S et al (eds): EvoWorkshops 2002, LNCS 2279*, Springer-Verlag, 2002, pp210-221.
[11] Ishino Y and Jin Y, "Estimate design intent: a multiple genetic programming and multivariate analysis based approach", *Advanced Engineering Infomatics*, 16(2002), pp107-125.
[12] Babovic V et al, "A data mining approach to modelling of water supply assets", *Urban Water*, 4(2002), pp401-414.
[13] Kojima F. et al, "Identification of crack profiles using genetic programming and fuzzy inference", *Journal of Materials Processing Technology*, Elsevier, 108 (2001), pp263-267.
[14] Whigham P.A. and Crapper P.F, "Modelling rainfall-runoff using genetic programming", *Mathematical and Computer Modelling*, 33(2001), pp707-721.
[15] Lee D.G et al, "Genetic programming model for long-term forecasting of electric power demand", *Electric power systems research*, Elsevier, 40, 1997, pp17-22.
[16] Montana D.J. and Czerwinski S, "Evolving control laws for a network of traffic signals", *Proceedings of the Firs Annual Conference: Genetic Programming, July 28-3, 1996. Stanford University*, pp333-338.
[17] Köppen M and Nickolay B, "Design of image exploring agent using genetic programming", *Proceedings of IIZUKA'96 Japan*, 1996, pp549-552.
[18] Yang Y. and Soh C.K, "Automated optimum design of structures using genetic programming", *Computers and Structures,* Pergamon, 80 (2002), pp1537-1546.
[19] MIT Trussworks.: 2002, Trussworks home page, [ONLINE], Available: http://web.mit.edu/emech/dontindex-build/java/trussworks/index.html [20/10/03].
[20] Yang J and Soh C.K, "Structural optimization by genetic algorithms with tournament selection", *Journal of Computing in Civil Engineering*, July 1997, pp195-200.
[21] Diada J.M et al, "Visualizing tree structures in genetic programming", *Lecture Notes in Computer Science 2724*, 2003, pp1652-1664.
[22] Web3D Consortium.: 2003, X3D Working Group, [ONLINE], Available: http://www.web3d.org/x3d.html [13/10/03]
[23] Wernert E.A and Hanson A.J, "Tethering and reattachment in collaborative virtual environments", *Proceedings of IEEE Virtual Reality 2000,* IEEE Computer Society Press, 2000, pp292.
[24] Watson A.H and Parmee I.C, "Systems identification using genetic programming", *Proceedings of ACEDC'96*, 1996.
[25] Watson A.H. and Parmee I.C., "Improving engineering design models using an alternative genetic programming approach", *Proceedings of International Conference on adaptive computing in design and manufacture*, 1998, pp193-206.
[26] Hudson M.G. and Parmee I.C.: 1995, "The application of genetic algorithms to conceptual design", *in* Sharpe, J. (eds), *AI System Support for Conceptual Design,* Springer-Verlag, pp. 17-36.

Enhancing Automated Process Design with Cognitive Agents, Distributed Software Components and Web Repositories[1]

Iain D Stalker and Eric S Fraga
Centre for Process Systems Engineering, Department of Chemical Engineering, University College London (UCL), London WC1E 7JE
{i.stalker,e.fraga}@ucl.ac.uk

Abstract

We present a novel problem solving environment for automated process design, integrating cognitive agents, web repositories and distributed software components with an existing automated process design tool. The approach is portable and addresses two key aspects of design: problem formulation and innovation. We model problem formulation and implement the resulting strategy through a pair of cognitive agents. The agents are situated within a larger framework which provides access to a wealth of resources. The approach supports innovation in design by freeing the designer to focus on creative aspects.

1 Introduction

The aim of preliminary design for chemical processes (process design) is to develop the best flowsheet — a selection of processing units and interconnections — for the manufacture of desired products from identified raw materials, according to some set of ranking criteria [1]. Process design benefits from an array of automated tools and techniques (see, for example, [2]), however, effective use of these requires a concise and comprehensive problem definition. Process design problems are typically underdefined: the development of a suitable problem definition (problem formulation) requires preliminary exploration, which demands a certain expertise of a would-be user. The difficulty of problem formulation potentially limits the use of automated tools in process design.

We present a novel problem solving environment (*c.f.* [3]) which integrates cognitive agents, web repositories and distributed software components with an existing automated process design tool, Jacaranda [4]. The Synthesis COGents[2] System assists

[1]Work supported by Project COGENTS, Agent-Based Architecture For Numerical Simulation, funded by the European Community under the Information Society Technologies Programme (IST), under contract IST-2001-34431.

[2]The term COGent derives from CAPE-OPEN Agent. In the sequel, we use this to denote the agents developed in the COGents Project. Details of the CAPE-OPEN effort can be found at www.colan.org.

the user in the difficult task of problem formulation: clarifying problem structure, recommending particular solution strategies, and identifying and obtaining appropriate resources; it then generates a ranked list of design alternatives based on the resulting problem definition.

The Synthesis COGents System comprises a pair of communicating agents, a Design COGent and a Synthesis COGent, developed within COGents, a European Project to use cognitive agents to support the dynamic and opportunistic interoperability of CAPE-OPEN compliant software [5] (see `www.cogents.org`). CAPE-OPEN [6], developed a set of interfaces with standard specifications to promote interoperability of heterogeneous process modelling software components. However, CAPE-OPEN address only syntactic issues, leaving it to the (human) user of CAPE-OPEN systems to determine which software component best suits a specific application. In COGents, we have developed a FIPA compliant (see `www.fipa.org`) agent based framework to assist a user in this task. Specifically, the cognitive agents in the framework communicate using concepts formalised in the project ontology, *OntoCAPE* [5], to identify relevant CAPE-OPEN compliant resources software resources. The agents take a number of forms, including catalogue agents and wrappers for CAPE-OPEN software, and are typically (geographically) distributed. In the sequel, we illustrate those aspects of the COGents framework which form a problem solving environment for automated process design.

2 The Synthesis COGents System

We summarise the functional aspects of the Synthesis COGents System in Figure 1. The user initiates the design process which is then driven by the Synthesis COGent. The Synthesis COGent acts to obtain a well-formed problem definition, however, its knowledge is limited and so it collaborates with the Design COGent. The Design COGent communicates with the user and other external agents to assemble appropriate information. Both agents exhibit intelligent behaviour (as characterised, for example, by [7]): namely, *proactivity* in the pursuit of some end, *social ability* through interaction with other and *reactivity* in responding appropriately and in a timely fashion to changes perceived. Significantly, the communication between the user, the Synthesis COGents System and the larger COGents framework is in the language of the problem, an important aspect of problem solving environments [3]. The (conceptual) structure of the Synthesis COGents System in shown in Figure 2.

The Synthesis COGent is composed of a Synthesis Tool COGent and a CAPE-OPEN (CO) compliant Synthesis Tool (CO Synthesis Tool). The Synthesis Tool COGent is wrapper agent, that is, it wraps a CO Synthesis Tool to bring it into an agent

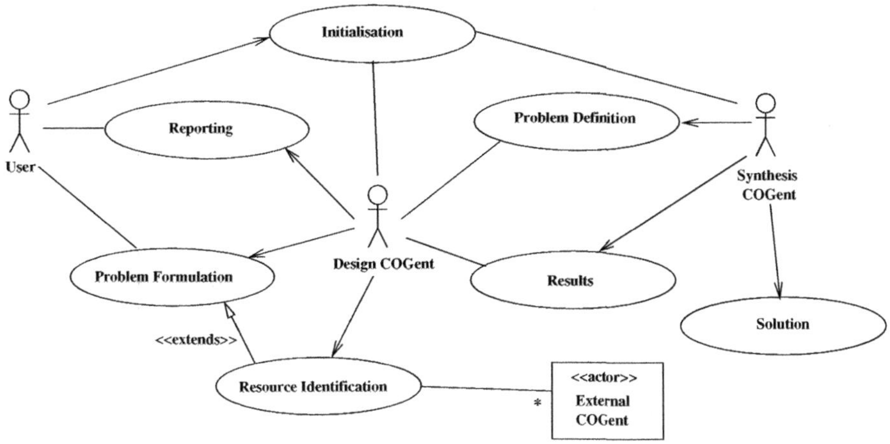

Figure 1: Functional Aspects of the Synthesis COGents System.

based environment: endowing it with communication capabilities and knowledge of what services it provides and what information it needs to function. The CO Synthesis Tool is derived from the Jacaranda automated design system [4]. Jacaranda uses implicit enumeration to simultaneously generate and search an appropriate superstructure. The approach can generate process flowsheets with heat integration and recycle structures, handle under-specified problems and supports multicriterial optimisation.

The Design COGents System provides a user interface and assists the user in problem formulation. We have modelled this as the culmination of three main activities, illustrated in Figure 3. In *Problem Statement* the user identifies the problem to be solved by the Synthesis COGents System. The user is free to provide as much information as he chooses subject to a minimum of desired products, raw materials, appropriate reactions, reaction conditions and a set of ranking criteria; he may, in fact, give a complete problem definition. The level of detail reflects the assistance needed and the extent of the following activities. For example, an expert user has a greater knowledge than a novice and will require less assistance in problem formulation: indeed, he may even bring with him all needed resources. The precise format of the user input depends upon the interface used, however, the form is restricted by the requirements of the underlying automated design tool; to a subset of the project ontology. *Problem Clarification* denotes an interactive treatment of the input information with the user, including the identification of feedstream composition, appropriate discretisations of continuous parameters and a consideration of recycle structures. Problem clarification makes use of resource identification. *Resource Identification* occurs at two stages: during problem clarification when physical properties information is sought and after

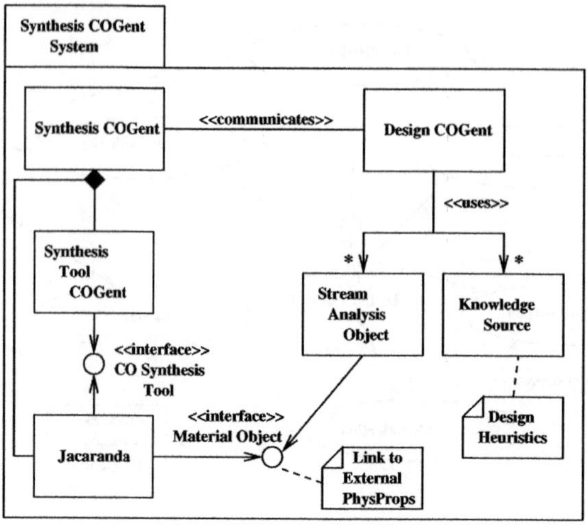

Figure 2: Structure of the Synthesis COGents System

problem clarification when appropriate processing technologies are sought. In either case, resource identification involves the preparation of a *Modelling Task Specification* (MTS): a concise description of the resource needed, including details of the intended function of the resource, the behaviours expected of it, and in some cases, the actual physical structure. This description is passed onto a matchmaking agent, which consults a number of library agents, which in turn, return the addresses of appropriate resources (*c.f.* Figure 3). The Design COGent uses this communication and collaboration to to obtain copies of the needed resources.

Case Study: an Example Problem Formulation

Our Case Study concerns the Hydrodealkylation of Toluene to produce Benzene (see for example, [8]) (HDA Process): Benzene is derived from Toluene using Hydrogen as a dealkylating agent; a secondary reversible reaction leads to a conversion of some of the Benzene into Diphenyl. We use this to provide a partial example walk-through of a problem formulation. Recall Figure 3.

Problem Statement For the HDA Process our problem statement (ideally) includes the following information [8]:

- Desired product: benzene, with purity and rate of production requirements.
- Raw materials: pure toluene at ambient temperature and pressure; 95% hydrogen (H_2) with 5% methane (CH_4) as an impurity at a temperature of 100^oF.

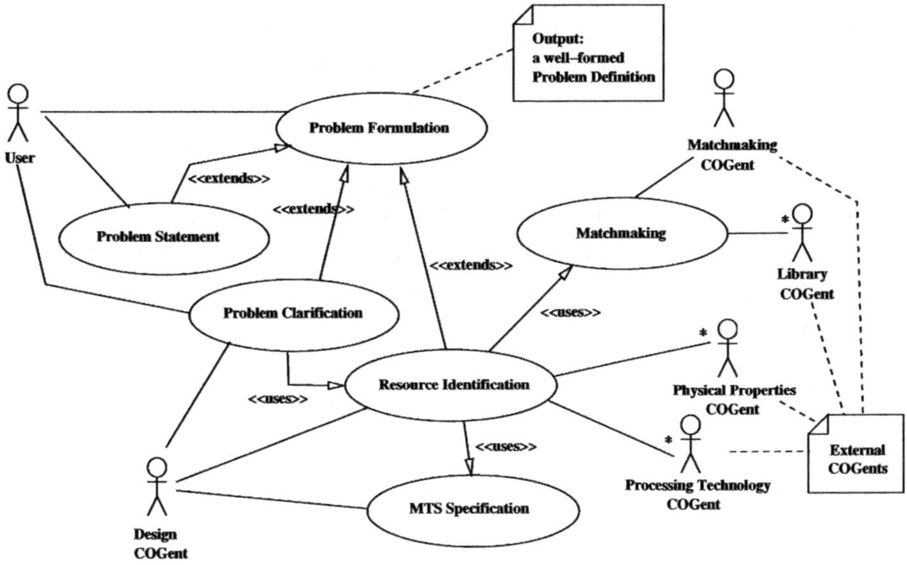

Figure 3: Problem Formulation.

- Appropriate reactions:
  - primary reaction: $Toluene + H_2 \rightarrow Benzene + CH_4$.
  - secondary (reversible) reaction: $2Benzene \leftrightarrow Diphenyl + H_2$
  - reaction conditions: inlet temperature 1150^oF; reactor pressure 500 psia.
- Constraint information for feed ratios, for temperature at the reactor outlet and the permissible temperature of the reactor effluent. For example, the ratio of hydrogen to toluene at the reactor inlet is at least 5:1.
- Information about selectivity and conversion of toluene.
- Ranking criterion: (minimise) annual costs = annual operating costs + appropriate depreciation charges of fixed assets.

The user may also provide his own processing unit models or physical properties information and methods, or both. These must implement the appropriate CAPE-OPEN compliant interfaces. From the problem statement, the system can identify all the chemical components in the process. and seeks information about these.

Resource Identification 1 If the user has provided his own physical properties packages, then this phase of resource identification is not required. The Design COGent prepares an MTS for a physical properties package which requires that suitable physical properties packages will have physical constants information, such as normal boiling points and molecular weight, for the chemical components identified, that is, benzene, diphenyl, hydrogen, methane and toluene. Additional requirements include methods for the calculation of appropriate physical property information, such as relative volatility; see [8] for details. The MTS is included in a request, using FIPA compliant communication, to the matchmaking COGent. The matchmaking COGent communicates with a number of library COGents to determine whether these agents are acquainted with physical properties packages which match the MTS. Assuming that a library COGent has such an acquaintance, it returns, via the matchmaking COGent, the address of the physical properties package. The Design COGent negotiates directly with this physical properties COGent to download to a local repository a copy of the the physical properties software. The local address becomes part of the problem definition and any further communication with the physical properties software is through appropriate CAPE-OPEN compliant interfaces (*c.f.* Figure 2). The physical properties information now available for use in problem clarification.

Problem Clarification Identifying a potential for recycle structures forms part of the Problem Clarification. The recycle structure knowledge source embodies design heuristics in the form of Prolog statements. The Design COGent uses a stream analysis object to obtain physical properties for the reactor effluent and uses conclusions, in the form of Prolog predicates, to drive a recycle structure knowledge source. For example, following [8], to assess the relevance of a gas recycle and purge, the normal boiling points of components in the reactor effluent are compared with that of propylene: *Whenever a light reactant and either a light feed impurity or a light by-product boil lower than propylene (-55F,-48C), use a gas recycle and purge stream.* (p.120). This is expressed in the recycle knowledge source as follows:

```
% destination/3
% destination(Component1,Component2,Destination).
%
destination(Component1,Component2,gasRecyclePurge):-
      refNBP(lowerThanPropylene,Component1),
      reactant(Component1),
      refNBP(lowerThanPropylene,Component2),
      feedImpurity(Component2).
destination(Component1,Component2,gasRecyclePurge):-
      refNBP(lowerThanPropylene,Component1),
      reactant(Component1),
      refNBP(lowerThanPropylene,Component2),
      product(secondary,Component2).
```

For the HDA Process, we infer from the above that Hydrogen and Methane should be recycled with purge to prevent a build of the impurity Methane in the system. Additional heuristics suggest that Toluene be recycled; and Diphenyl is a candidate for recycling, but recommend that it be taken as a byproduct.

Additionally, since the automated design tool, Jacaranda [4], discretises continuous variables and parameters, during the problem clarification, the user may provide the desired discretisation increments or elect to use the system defaults.

Resource Identification 2 In the second phase of resource identification, the Design COGent, uses the problem statement and physical properties information, obtained through the stream analysis object, to preselect appropriate technologies for the process at hand. For example, in the HDA Process the two chemical reactions occur within the same temperature and pressure ranges. Thus, we need only one reactor. We immediately constrain and reduce our space of potential solutions, by including in the unit definition the key word *limit*, as illustrated in the following excerpt from the specification of the reactor as part of the problem definition.

```
Reactor reactor
  reaction primaryReaction
  reaction secondaryReaction
  RealSet conversion 0.75 # conversion
  Stream makeup makeup # available makeup stream
  Real pressure "500 *psia"
  Real temperature "(1150+Fzero)*R" # 1150 Fahrenheit
  ...

  limit
end
```

For units for separation, the Design COGent considers the phase of each component, relative volatilities and so forth. The following are extracts from the appropriate knowledge source:

```
% separation/3 (Simple)
% separation(GenericTechnology,Component1,Component2):-
%
separation(flash,Comp1,Comp2):-
     volatility(distant,Comp1,Comp2),
     phase(vapour,Comp1), phase(liquid,Comp2).
separation(distillation,Comp1,Comp2) :-
     volatility(distant,Comp1,Comp2).
```

On the basis of the above, the Design COGent obtains software components for flash and distillation, as for physical properties packages. We include additional heuristics for alternative simple and complex separation technologies.

3 Discussion

Potentially, access to a larger framework offers a wealth of information unlikely to be available within a single design system or even design team: best-of-breed processing units, sophisticated models, recent technological innovations, up-to-date and consistent physical properties, and robust (numerical) solution mechanisms. Furthermore, the Synthesis COGents System provides a systematic approach to automatically integrate these additional resources with existing design tools. Appropriate management of available resources frees the designer to focus on more efficient and creative design choices, dramatically increasing the likelihood of truly innovative designs [1].

Preselection of generic processing technologies prunes the space of potential design solutions: we consider only processing units which we know to be applicable to our design problem; and, on occasion, we can limit the number of a given unit. This has the important incidental benefit of reducing that computational effort. Nevertheless, an automated approach not only removes the onus from the user, it also removes a certain amount of control. Consider, when an experienced designer interacts directly with an automated process design tool he controls the information employed by the design tool, through level of discretisation, values for variables and the constants used, and so forth. He can make use of the design tool for preliminary explorations of a given solution space, a key to successful design [9, 10]; for example, through the use of partial solutions [11]. Too high a level of automation prevents this creative use of the design tool: the designer either accepts the results of the system without question or seeks an alternative; should a design problem remain unsolved, often no indication of nearness to a solution nor of those constraints which may have restricted particular design alternatives. Thus, there is little information available to guide a reuse of the system, or to take on board when preferring an alternative design tool. We seek to realise the full potential of an agent based approach by using the technology to reduce the burden and including mechanisms through which to re-introduce the designer into the loop. We promote an interactivity at each stage, for example, feeding back design choices for confirmation by the user, which allows the designer to supervise the system, returning control without the burden. To ensure this we appeal to the distinctions of Engineer Computer Interaction (ECI), a methodology for coordinating aspects of HCI with domain specific knowledge to facilitate the development of software systems to support engineers [12].

To communicate effectively our agents must use a common vocabulary and agree on the semantics of the terms in it: this is *ontological commitment* (see, for example, `www.ontology.org`). To ensure this, the Synthesis COGents System motivates concepts for process design which are included in the project ontology, *OntoCAPE*

[5]. Thus, the system interface ensures that requests are expressed in a subset of the concepts of the project ontology: for example, through the use of selection from predefined options and including widely accepted terminology in the project ontology. *OntoCAPE* specifies a conceptualisation of process modelling, simulation, and design, in the form of detailed class hierarchies, class definitions, attributes, relations and constraints.

4 Related Work

Approaches which combine computational software with intelligent support arise in a number of domains. Example include: the combination of finite element methods with artificial intelligence (AI) techniques, such as [13] which uses an expert system to create appropriate meshes, an important task in problem formulation using a finite element approach; and the combination of solid modelling techniques with aspects of AI which led to the inception of the field of Intelligent Computer Aided Engineering [14] and yielded forth a number of hybrid systems commonly referred to as ICAD systems, see, for example, [15, 16]. Intelligent Product Design Environments take ICAD a step further by suggesting that computational techniques are incorporated. [17] discusses integration strategies as a step towards a possible platform for such a system. Indeed, such platforms are an active research area which retain the goal of systematic support for product design [18].

The domain of Computer-Aided Process Engineering has witnessed a number of research pursuits in relevant areas. There have been a number of applications of mathematical programming approaches which are surveyed in [2]. Several knowledge based approaches have been made, including design support for process engineering which seeks to capture design rationale [19]; and the layering of strategic and decision-making expertise above calculational aspects [20, 21]. Access to common information for function, structure and behaviour of chemical process systems has been suggested by a researchers outside of the CAPE-OPEN efforts, for example [22]. The combination of declarative and computational paradigms has become of interest: for example, [23] and [24] use logic programming and constraint logic programming, respectively, to combine heuristic knowledge with optimisation approaches, Agent based approaches have been tried in supply chain modelling [25] and the design of separation systems [26].

However, there are two related works which have particular relevance: these use the heuristics of [8] and employ the HDA process as the primary case study. The *Process Invention Procedure (PIP)* [27] is an expert system which progressively develops a solution: if the design choice is found to be profitable at the given level of detail, then more is added; otherwise, it is discarded. *ConceptDesigner* reported in [28] implements a generic agent based approach for computer aided design: it is a self-contained system with technologies hardwired into the agents.

The Synthesis COGents System distinguishes itself from previous work in three fundamental ways: the resources used by the Synthesis COGents System are *distributed*; the resources are *assembled dynamically* and *opportunistically*; the agents are used solely for *problem formulation*, with the responsibility for solution remaining ultimately with the Synthesis Tool.

5 Summary

We have introduced the Synthesis COGents System, a novel problem solving environment which integrates cognitive agents, web repositories and distributed software components with an existing automated process design tool. The framework is portable, extensible and addresses two important themes. It offers a user substantial help in the difficult and important stage of problem formulation. It also enlarges the resources available and provides a systematic way for integrating these with existing tools. This resource management frees the designer to focus on more efficient and creative design choices and increases the potential for innovative designs.

References

[1] L. Biegler, I.E. Grossman, and A.W. Westerberg. *Systematic methods of Chemical Process Design*. Prentice Hall International Series in the Physical and Chemical Engineering Sciences. Prentice-Hall, Inc., 1997.

[2] I E Grossmann, J A Caballero, and H Yeomans. Mathematical programming approaches to the synthesis of chemical process systems. *Korean J Chem Eng*, 16(4):407–426, 1999.

[3] E. Gallopoulos, E.Houstis, and J.R.Rice. Problem solving environments for computational science. *IEEE Comput. Sci. Eng*, 1:11–23, 1994.

[4] E. S. Fraga, M. A. Steffens, I. D. L. Bogle, and A. K. Hind. An object oriented framework for process synthesis and simulation. In M. F. Malone, J. A. Trainham, and B. Carnahan, editors, *Foundations of Computer-Aided Process Design*, volume 323 of *AIChE Symposium Series*, pages 446–449, 2000.

[5] B.L. Braunschweig, E.S. Fraga, Z. Guessoum, W. Marquardt, O. Nadjemi, D. Paen, D. Pinol, P. Roux, S. Sama, M. Serra, I.D. Stalker, and A. Yang. CAPE

web services: The COGents way. In *European Symposium on Computer Aided Process Engineering 14 (ESCAPE 14)*, 2004. Accepted.

[6] J-P Belaud, B Braunschweig, and M Pons. Open software architecture for process simulation : The current status of CAPE-OPEN standard. In *European Symposium on Computer Aided Process Engineering 12 (ESCAPE 12)*, 2002.

[7] M.J.Wooldridge. *An Introduction to MultiAgent Systems*. John Wiley & Sons, 2002.

[8] J. M. Douglas. *Conceptual Design of Chemical Processes*. McGraw-Hill International Editions, 1988.

[9] D Navinchandra. *Exploration and Innovation in Design: Towards a Computational Model*. Springer-Verlag, 1991.

[10] T Smithers. Towards a knowledge level theory of design process. In J S Gero and F Sudweeks, editors, *Artificial Intelligence in Design '98*, pages 3–21. Kluwer, 1998.

[11] E S Fraga. The generation and use of partial solutions in process synthesis. *Chemical Engineering Research and Design*, 76(A1):45–54, 1998.

[12] R. Stalker and I. Smith. Structural monitoring using engineer-computer interaction. *Artificial Intelligence for Engineering Design, Analysis and Manufacturing*, 16(5), November 2002. Special Edition, Human-Computer Interation in Engineering Contexts.

[13] B Dolsak and S Muggleton. The application of inductive logic programming to finite element mesh design. In S Muggleton, editor, *Inductive Logic Programming*, pages 453–472. Academic Press, 1992.

[14] K D Forbus. Intelligent computer aided engineering. *AI Magazine*, 9, 1988.

[15] S H Chung and J E Sneckenberger. An innovative design scheme for product development using AI-type engineering. In *International conference on CAD/CAM, robotics and factories of the future*, St Petersberg, Russia, 1993.

[16] H Grabowski and S Rude. Methods of finding solutions in CAD systems. In *Proceedings of ICED'88*, pages 148–155, Zurich, 1988.

[17] U Roy, B Bharadwaj, S Sarathy, and P Graham. Development of an intelligent product design system: Integration strategies. *Applied Artificial Intelligence*, 9:563–585, 1995.

[18] T W Simpson, J R A Maier, and F Mistree. Product platform design: Method and application. *Res Eng Design*, 13:2–22, 2001.

[19] R Bañares-Alcántara. Design support for process engineering III. Design rationale as a requirement for effective support. *Computers and Chemical Engineering*, 21:263–276, 1997.

[20] D R Myers and J F Davis. A task-oriented approach to knowledge-based systems for process engineering design. *Comput. chem Engng*, 12(9), 1988.

[21] L Beltramini and R L Motard. KNOD - a knowledge based approach for process design. *Comput. chem. Engng*, 12(9):939–958, 1988.

[22] D C Miller, J R Josephson, M J Elsass, J F Davis, and B Chandreskaran. Sharable engineering knowledge databases for intelligent system applications. *Computers chem. Engng*, 21(Supplement):S77–S82, 1997.

[23] J A Caballero and I E Grossmann. Logic-based methods for generating and optimizing thermally coupled distillation columns. In J Grievink and J van Schijndel, editors, *European Symposium on Computer Aided Engineering (ESCAPE 12)*, pages 169–174, Amsterdam, 2002. Elsevier.

[24] H A Abbass, G A Wiggins, R Lakshmanan, and W Morton. Heat exchanger network retrofit by constraint logic programming. *Journal of Computers in Chemical Engineering*, 23(Supplement):S129–S132, June 1999.

[25] R. Garcia-Flores, X. Z. Wang, and G. E. Goltz. Agent-based information flow for process industries supply chain modelling. *Computers chem. Engng*, 24:1135–1142, 2000.

[26] K H Tyner and A W Westerberg. Multiperiod design of azeotropic separation systems. I. An agent based solution. *Computers chem. Engng*, 25:1267–1284, 2001.

[27] R L Kirkwood, M H Locke, and J M Douglas. A prototype expert system for synthesizing chemical process flowsheets. *Computers chem. Engng*, 12(4):329–343, 1988.

[28] C Han, J M Douglas, and G Stephanopoulos. Agent-based approach to a desgin support system for the synthesis of continuous chemical processes. *Computers chem. Engng*, 19S:S63–S69, 1995.

Robust Solutions in Engineering Design: stochastic simulation versus DACE

R A Bates and H P Wynn,
Decision Support and Risk Group,
London School of Economics and Political Science,
Houghton Street, London WC2A 2AE,
United Kingdom.
r.a.bates@lse.ac.uk
h.wynn@lse.ac.uk

Abstract

This paper compares two different methods for robust design improvement. The first method, called stochastic simulation, combines traditional Computer-Aided Engineering (CAE) simulation tools with variation in the simulation model parameters in order to estimate the resulting uncertainty in system behaviour for design improvement. The second method, called DACE, employs traditional Design of Experiments (DOE) methodologies to build statistical models of CAE simulation tools, called emulators because they emulate the behaviour of the simulator. The emulators are much faster to compute than the corresponding simulation model and can therefore be used to search the design space for robust solutions in an efficient way.

The two methods can therefore be characterized by their computational cost, flexibility and accuracy. Two example problems are used to highlight the methods and their advantages. The use of measures of variation in responses is carried forward to be included in multi-objective optimization, so that robustness is naturally considered as a design objective.

1 Introduction

The design of engineering products and processes can be roughly divided into the activities of conceptual design, redesign (design improvement or design optimization) and routine design (minor modification) [1]. Because conceptual design requires creative input, any mathematical framework imposed on the design process must result in a restricted set of possible design solutions. However, the imposition of a framework does allow a rigorous search for the best designs, but it must be recognized that the scope of the search is restricted once the framework has been defined.

We consider the redesign problem in the following mathematical framework. The design is defined by a set of d parameters known as *design factors*, $\mathbf{x} = x_1, \ldots, x_d$ with ranges defined by their lower and upper bounds $\mathbf{x}^{lb} = x_1^{lb}, \ldots, x_d^{lb}$ and $\mathbf{x}^{ub} = x_1^{ub}, \ldots, x_d^{ub}$ respectively. The range of values for these factors defines the *design space*, $\mathcal{D}$, the set of all possible design solutions. The engineering system under consideration is represented by a function f which has outputs or *responses* $\mathbf{y} = y_1, \dots, y_m$ so that we write

$$\mathbf{y} = f(\mathbf{x}) . \tag{1}$$

The key restriction here lies in the choice of the design factors and their ranges as this defines $\mathcal{D}$. Another important point to consider is whether the design factor values are allowed to vary in a continuous or a discrete manner. The framework that we propose here assumes continuous factors, but that is not to say that discrete factors cannot be considered, just that their inclusion will require further work.

1.1 Robustness

The introduction of noise into the redesign problem to achieve robust solutions is the subject of ongoing research. The field of Robust Design focuses on this area and usually employs traditional Response Surface Methodology (RSM) to find robust solutions, see [2] for a review of RSM and [3] for a recent example in Robust Design. However these methods usually involve specifying a special class of (usually) polynomial model, such as linear, or linear with specific interactions, to estimate the behaviour of the underlying physical system *before* any experimentation takes place. Often in engineering this is not practical as there is either not sufficient a priori information to justify such assumptions, or the underlying system is highly nonlinear and cannot be accurately approximated by linear or second-order polynomial models. In such cases a different approach to modelling may be required, see for example [4] for a general polynomial model fitting strategy, or it may be the case that no adequate model of the system can be identified. The approaches adopted by the two methods presented here do not require any model assumptions but instead focus on the practical implementation of statistical methodologies in the CAE environment.

Robustness in engineering design can mean many things. A set of working definitions can be found at: `http://discuss.santafe.edu/robustness/`. A handful of these definitions are listed here.

1. Robustness is the ability of a system to maintain function even with changes in internal structure or external environment.

2. Robustness is the degree to which a system is insensitive to effects that are not considered in the design.

3. Robustness is a design principle of natural, engineering, or social systems that have been designed or selected for stability.

4. A robust solution in an optimization problem is one that has the best performance under its worst case (max-min rule).

5. Robustness of an analytical procedure is a measure of its ability to remain unaffected by small, but deliberate variations in method parameters, and provides an indication of its reliability during normal use.

Of these definitions, the first two cannot be considered in the mathematical framework defined for this paper. However the final three definitions can be expressed mathematically and it is these notions of robustness that will be considered here.

Separate *noise factors* can be introduced to the mathematical framework to help define these variations, but often in engineering design problems, factors can be both design factors and noise factors, which can cause some confusion. Factors can also be categorized according to whether they are controllable (design) or uncontrollable (noise) but this does not account for design factors that can also experience some variation in value. Mathematically there are three possible categories.

1. Design factor without noise: the factor has a point value somewhere within the defined range.

2. Design factor with noise: the factor value varies according to a statistical distribution defined to be within the overall range of the factor.

3. Noise factor: factor value varies according to a given statistical distribution of fixed range.

These categories are illustrated in Figure 1 where a Normal distribution is assumed.

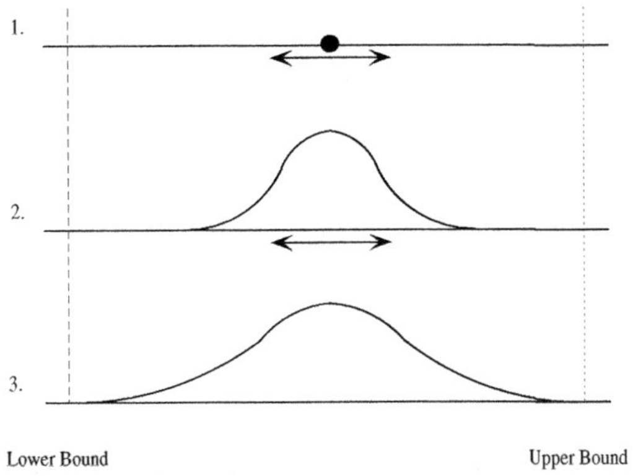

Figure 1: Different categories of factor: 1. design factor, 2. design factor with noise, 3. noise factor.

In general, noise factors are not set at particular values but are sampled from statistical distributions dictated for example by use, or by nature. In order to estimate the effect of the noise on the response(s), the system must either be evaluated many times while varying noise factor values (Monte Carlo-based simulation), or some other approach must be taken such as analytic propagation of the statistical distribution through the system, see for example [5].

The decision on which factors to include in a redesign activity are guided by engineering knowledge and, possibly to a greater degree, available resources.

The inclusion of a large number of design factors can lead to very large search or optimization problems and is often impractical. Add to this several noise factors and the problem quickly becomes too large to handle.

This paper is concerned with how to implement strategies for searching a design space using computer simulation models to find solutions that are robust to noise factors and to variations in design factor values brought about by noise.

2 Computer simulation

One way of considering robustness is to use Monte-Carlo style simulation methods to estimate the variation in responses due to imposed variation in factor values.

For simplicity, consider a design improvement problem where there are two design factors x_1, x_2 and two responses, y_1, y_2. Assigning values to the factors produces a specific design that needs to be evaluated to determine the associated responses. If the two design factors are subject to noise, such as a manufacturing tolerance, then their values will change and the responses must be recomputed. Figure 2 illustrates this situation with the circle representing a single design point and the solid dots representing the same design point with four different instances of noise added.

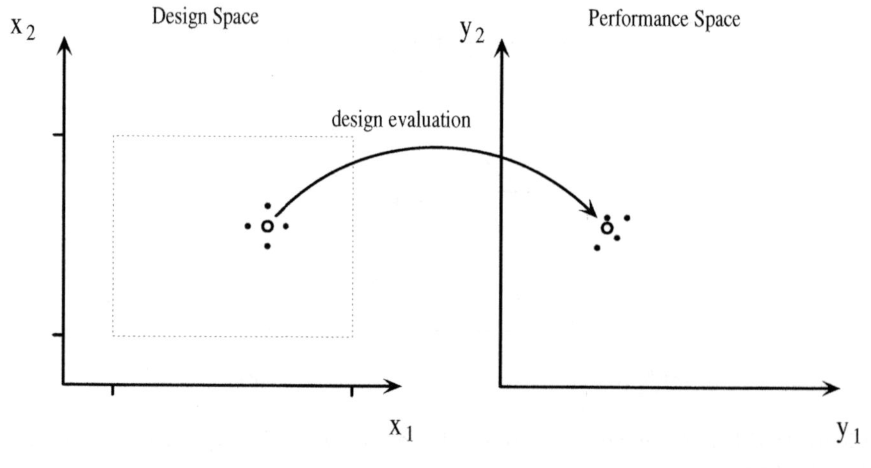

Figure 2: Design and performance spaces: a simple example.

Minimizing the variation in response due to noise will improve the robustness of the design, therefore a method of estimating this variation would be very useful. Taken together, the five evaluations shown in Figure 2 give a rough estimate of how variation in factor values translates to variation in performance. Increasing the accuracy of this estimate of variation will require more evaluations and this significantly increases computational cost. If the variation of y_1 and y_2 are measured, e.g. by estimating their standard deviations, this can be included as an additional response (or pair of responses) to be minimized. An optimization strategy can then be applied that uses the simulator to compute both the responses and the variation in responses due to noise at given design factor values.

This idea is explored using a simple example of a combustion engine piston simulator taken from [6]. The simulator has seven input factors (all design factors with noise) and one response: cycle time. The robust objective is to minimize both cycle time and the standard deviation of cycle time in the presence of noise associated with the design factors. Starting with the nominal value of all design factors, $\mathbf{x}^n$, an initial set of 50 designs are generated at different factor values. The points are chosen uniformly in a restricted range defining a sub-region of the entire design space, chosen here as $\mathbf{x}^n \pm \frac{(\mathbf{x}^{ub}-\mathbf{x}^{lb})}{10}$, where, as stated in Section 1, $\mathbf{x}^{lb}$ and $\mathbf{x}^{ub}$ are the lower and upper bounds, respectively, of the design factor values. For each design, 50 simulations are made, varying the values of noise at each simulation. The noise is assumed to be Gaussian with standard deviations as quoted in [6]. For each of these 50 simulations, the mean and standard deviation of the response are calculated. The best design from the original 50 is then chosen as the new starting point and the process is repeated. Each stage therefore requires $50 \times 50 = 2500$ simulations. Figure 3 illustrates the process over seven stages. The total number of simulations performed is $7 \times 2500 = 17500$. Over successive stages the 'cloud' of 50 points moves closer to the origin, representing an overall design improvement.

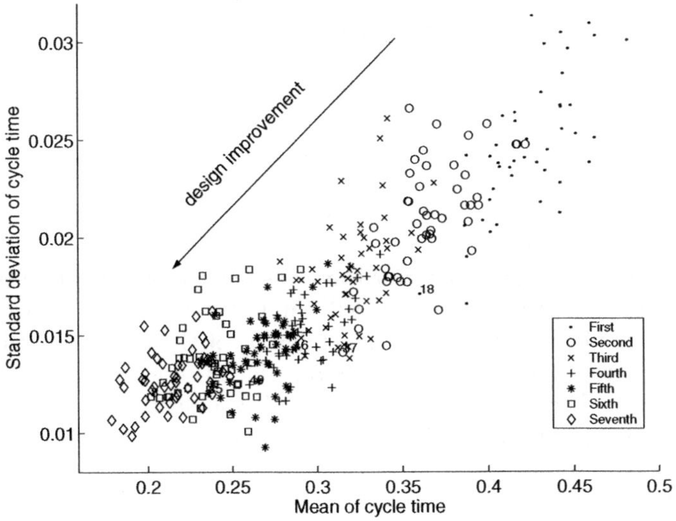

Figure 3: Design improvement of piston simulator.

The main feature of this approach is that no statistical analysis of the data is required other than computation of the two responses (mean and standard deviation). Improvement may not be optimal and the computational cost may be very high, but it is likely that some improvement will be made. In cases where the number of noise factors is very high (for example, problems with over a hundred noise factors are possible) this may be the only practical solution. This methodology is embodied in a software package called ST-ORM (for details see: `www.easi.de`).

3 Statistical modelling: emulation

Instead of estimating variation using Monte Carlo, a more systematic approach is to build a spatial model, an emulator, that mimics the behaviour of the simulator, but that runs much faster. If an accurate enough emulator is built then it can be used to estimate variation due to noisy factors. The extension proposed here is to make the emulator the subject of further experimentation where the inputs are the same factors but the response is now the desired measure of variability. A second *stochastic* emulator can then be fitted and used to estimate this variability.

This technique is demonstrated by using the piston example again, where the seven input factors are labelled A to G and, as in the previous section, each factor is assigned a lower bound and an upper bound. A 64-point Latin Hypercube Sampling [7] design $S = \mathbf{s}_1, \ldots, \mathbf{s}_{64}$ is then used to conduct an experiment on the simulator f where the response $Y = y_i, i = 1, \ldots, 64$ (cycle time) is calculated at each design point $\mathbf{s}_i$. Using the DACE kriging model [8], an emulator of the piston simulator is fitted to the experimental data to estimate the response y at untried inputs $\mathbf{x}$.

$$\hat{y(\mathbf{x})} = \hat{f}(\mathbf{x}, S, Y) \tag{2}$$

Note here that the fitted emulator is an exact interpolator of the data as there is no associated measurement error. The process of generating experimental data and fitting the emulator model takes only 1-2 minutes on an Intel Pentium 4 2.0Ghz processor machine.

Figure 4 shows the generalized or leave-one-out Cross Validation (CV) and main effects plots for the model. The left-hand plot in Figure 4 shows that the fitted DACE model is a very accurate emulator of the simulator, with a CV root mean squared error (RMSE) of 0.0073, which is less that 1% of the range of the response. The plot also shows the 95% confidence intervals associated with each cross-validation prediction. The main effects plot shows the effect of each factor on the response, with the effects of the other factors removed. Looking at the right-hand plot, one can see that the greatest effect is from factor B. The effects of factors F and G are very small and are not plotted here. In fact, if it was desirable to reduce the number of factors, this would indicate that it may be possible to remove factors F and G from the analysis, but this is ignored here. For most simulation models, the emulator will be much faster to run than the simulator and can therefore be used to evaluate

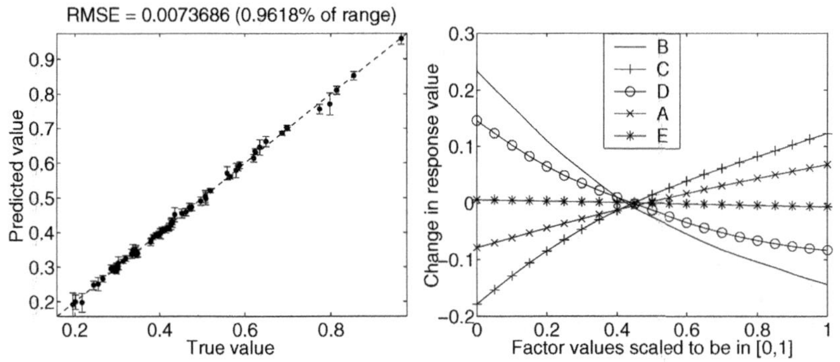

Figure 4: Piston emulator Cross Validation (left) and main effects (right) plots.

the response at many more factor values, making it possible to generate good estimates of variability due to noise.

The next step is to use the emulator to generate a second set of experimental data to which a second tier of DACE models is fitted. A second 128-point LHS design, $T = \mathbf{t}_1, \ldots, \mathbf{t}_{128}$, is generated and, at each point, a 200-run Monte Carlo experiment, $N = \mathbf{n}_1, \ldots, \mathbf{n}_{128}$, is conducted to estimate the effect of input factor noise on the response. The crossed experiment $T \times N$ therefore consists of 128×200 points. From this experiment, at each of the 128 points, there will be estimates of both the mean and the variance of the response. Recalling Equation (2) the experiment is described by

$$\begin{aligned}
\hat{y}_{ij} &= \hat{f}(\mathbf{t}_i + \mathbf{n}_j, S, Y) \\
\hat{y}_i^m &= \frac{1}{200} \sum_{j=1}^{200} \hat{y}_{ij} \\
\hat{y}_i^s &= \sqrt{\frac{1}{200-1} \sum_{j=1}^{200} (\hat{y}_{ij} - \bar{y}_i)^2}
\end{aligned}$$

where $Y^m = \hat{y}_i^m, i = 1, \ldots, 128$ and $Y^s = \hat{y}_i^s, i = 1, \ldots, 128$ represent estimates of the mean and standard deviation of y, respectively. Using this data, two *stochastic* emulators can be built:

$$\hat{y}^m(\mathbf{x}) = g(\mathbf{x}, T, Y^m) + \varepsilon \tag{3}$$

$$\hat{y}^s(\mathbf{x}) = h(\mathbf{x}, T, Y^s) + \varepsilon \tag{4}$$

Note the error term ε, indicating that the stochastic emulators are not exact interpolators of the data. This is because this data contains measurement error due to the fact that the responses are estimates of mean and standard deviation. Again, this process takes only a couple of minutes. Figures 5 and 6 show plots of CV and main effects for both stochastic emulators. The CV plots show that the accuracy of both stochastic emulators is good and they are carried forward for multi-objective optimization. By choosing an objective composed

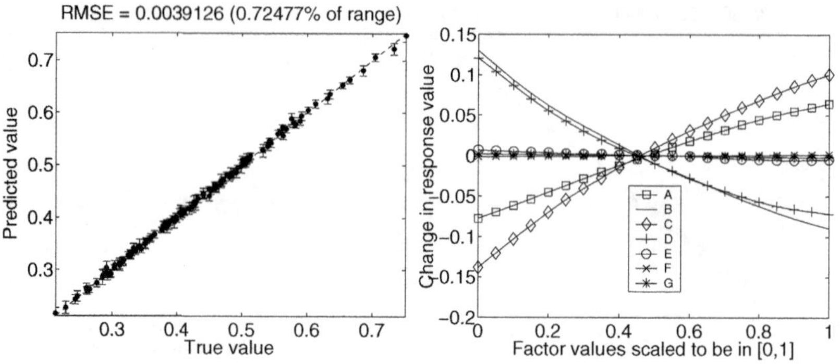

Figure 5: Piston stochastic emulator of mean.

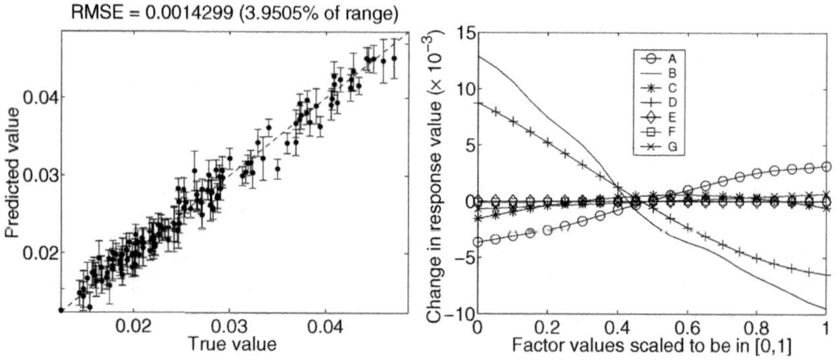

Figure 6: Piston stochastic emulator of standard deviation.

of a weighted sum of the two responses, points on a Pareto-optimal boundary (see for example, [9]) can be identified, as shown in Figure 7, using mono-objective methods. The optimization method used here involves a two-stage process. First a global optimizer [10] is used to find a solution from a randomly-chosen starting point. This solution is then used as the stating point for a second local optimization using the MathWorks Matlab optimization toolbox. The optimization is repeated at a set of different weights to produce the Pareto boundary which can be used to study the tradeoff between minimizing cycle time and minimizing variation due to noise.

4 Satellite simulation data

A second example consisting of a simulation of a satellite is now studied. The problem is to design the satellite to be robust against small vibrations during use. This is important as vibration can disrupt the communication signal between the satellite and the ground station. In the example there are two responses, LOS (Line of Sight) and T, and seven design factors. Both responses represent micro-scale vibrations experienced by the satellite. As the satellite

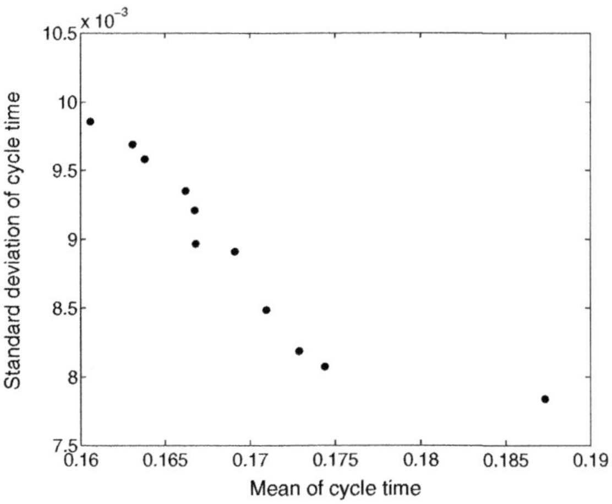

Figure 7: Pareto-optimal design points.

communicates via laser beam, any small vibration may deflect the beam and disrupt communication. In this case the noise is injected directly into the simulation and the responses are time series signals, representing displacement over time in each x-, y- or z- direction. For a fixed time segment, the standard deviation of each x- y- and z- signal is computed and the sum of the squares of these values is taken to be an estimate of the magnitude of vibration experienced.

An experiment consisting of a 999-point Monte Carlo simulation was conducted, and the responses calculated at each point. The data was then divided into a 96-point set for model building and an 903-point set for prediction. Two DACE emulators were fitted to the 96-point data set, one emulator for each response. Again, this process took only 1-2 minutes of computation time for each emulator. Figures 8 and 9 show the CV and main effects for each emulator. The CV plots show that reasonable emulators have been fitted to the data,

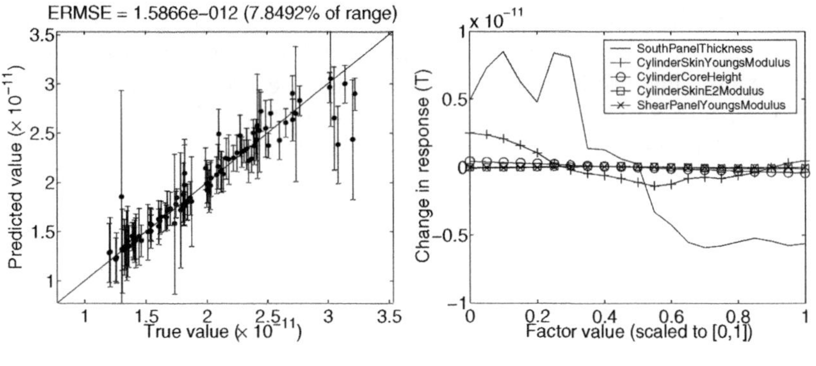

Figure 8: DACE emulator of satellite data, response T.

while the main effects plots show that the factor 'SouthPanelThickness' has a

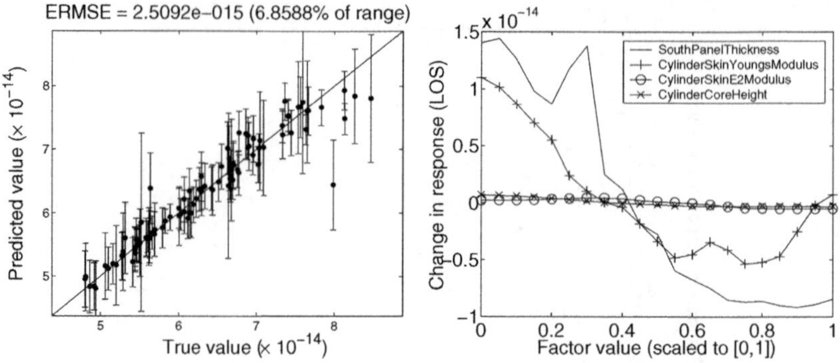

Figure 9: DACE emulator of satellite data, response LOS.

highly nonlinear effect on both responses. As an extra model validation step, the emulators were used to predict the response at the remaining 903-point prediction data set. The results of this are shown in Figure 10. The prediction

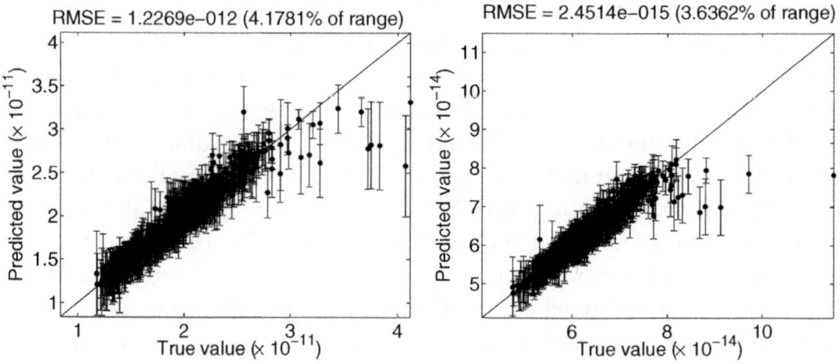

Figure 10: Emulator prediction (left-hand plot: T, right-hand plot: LOS).

results show that both emulators are able to provide reasonable estimates of the responses at untried input factor values and at least as well as the CV results suggest. The two fitted emulators are then used, as in the piston example, with a multi-objective optimization algorithm to find a Pareto-optimal set of solutions. The results of this are shown in Figure 11.

5 Conclusions

This paper has presented two frameworks for design improvement for robustness, direct improvement using simulators and stochastic emulation using spatial models (emulators).

Measures of robustness need to be considered when performing design improvement if robust and practical solutions are the be found. However this can be costly and may increase the size of the problem dramatically. For exam-

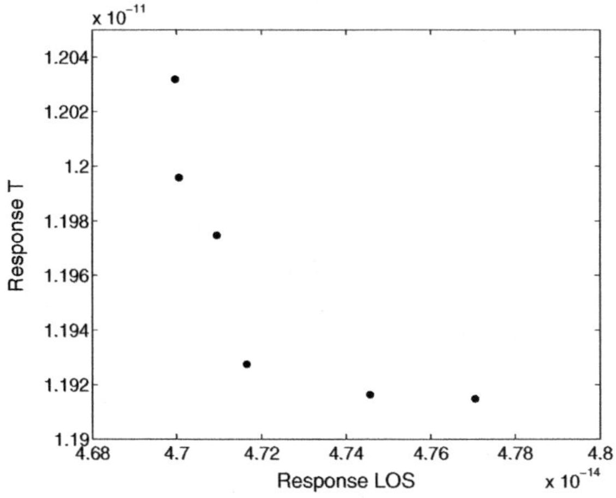

Figure 11: Satellite case study Pareto-optimal solutions.

ple, including small variation in material thickness when performing computer simulations can lead to the creation of many noise factors, increasing the dimensionality of the search space. The transformation from small deterministic problems to larger stochastic problems may force a re-think in terms of which design improvement strategy to follow.

Comparing the two methods presented using the piston example, direct optimization using the computer simulator required 17500 simulation runs, whereas the stochastic emulator approach required only 64 runs. However, the latter approach requires accurate statistical models to be built which entails additional computational cost and the risk of not achieving the required level of accuracy. In both methods it may be possible to reduce the total number of simulations required, particularly for stochastic simulation, as the number of simulations for each iteration was chosen purely from experience. However, the simplicity of directly using the simulator clearly comes at a computational cost and appropriate use of statistical modelling methods can help reduce this burden.

The stochastic simulation approach produced a reasonable improvement in both the response and its standard deviation, but, compared with the stochastic emulator approach, the improvement was not optimal and was achieved at much greater computational cost. The additional computation time required to build emulator models and generate a Pareto-optimal boundary is in the order of minutes which is often less than the cost of a single CAE simulation run.

The inclusion of measures of robustness such as standard deviations forces the use of multi-objective optimization methods and in this case, the use of Pareto-optimality is useful in deciding tradeoffs between objectives.

Lastly, direct optimization using the simulator does not assume any restriction on possible correlation between noise factors, whereas the stochastic emulator approach assumes independent noise. At present this is not generally a problem in engineering as information on correlated noise is usually not available.

6 Acknowledgements

The authors would like to acknowledge the support of the European Union (TITOSIM: Project No. GRD1-2000-25724) and would also like to thank EASi engineering GmbH for providing the satellite data set and the reviewers for their very helpful comments and suggestions.

References

[1] N F O Evbuomwan, S Sivaloganathan, and A Jebb. A survey of design philosophies, models, methods and systems. *Proc. Inst. Mech. Engrs. Part B*, 210, 1996.

[2] Raymond H Myers, Douglas C Montgomery, G Geoffrey Vining, Connie M Borror, and Scott M Kowalski. Response surface methodology: A retrospective and literature survey. *Journal of Quality Technology*, 36(1):53–77, Jan 2004.

[3] D Romano, M Varetto, and G Vicario. Multiresponse robust design: A general framework based on combined array. *Journal of Quality Technology*, 36(1):27–37, Jan 2004.

[4] R A Bates, B Giglio, and H P Wynn. A global selection procedure for polynomial interpolators. *Technometrics*, 45(3):246–255, August 2003.

[5] R A Bates and H P Wynn. *Advanced polynomial emulation for robust engineering design*, pages 29–34. MCB University Press, 2002.

[6] R S Kenett and S Zacks. *Modern Industrial Statistics.* Duxbury Press, 1998.

[7] M D McKay, W J Conover, and R J Beckman. A comparison of three methods for selecting values of input variables in the analysis of output from a computer code. *Technometrics*, 21:239–245, 1979.

[8] J Sacks, W J Welch, T J Mitchell, and H P Wynn. Design and analysis of computer experiments. *Statistical Science*, 4:409–435, November 1989.

[9] P L Yu. *Multiple Criteria Decision Making: Concepts, techniques and Extensions.* PlenumPress, NewYork, 1985.

[10] L Pronzato, E Walter, A Venot, and J F Lebruchec. A general-purpose global optimizer - implementation and applications. *Mathematics and Computers in Simulation*, 26(5):412–422, 1984.

Adaptive Search for Optimum in a Problem of Oil Stabilization Process Design

A. Žilinskas[1,2], E. S. Fraga[3], A. Mackutė[1], A. Varoneckas[1]

[1] Vytautas Magnus University
8 Vileikos str.
Kaunas
LT 3035
Lithuania
antanasz@ktl.mii.lt

[2] School of Mathematics
Cardiff University
Cardiff CF24 4YH
United Kingdom

[3] Centre for Process Systems Engineering
Department of Chemical Engineering
UCL (University College London)
London WC1E 7JE
United Kingdom
e.fraga@ucl.ac.uk

Abstract

The formulation of a model for an industrial problem in process design leads to an optimization problem with a small, implicitly defined, feasible region, a region which is difficult to identify *a priori*. The difficulties of obtaining a good solution with conventional optimization methods are discussed. A novel method is proposed, based on the paradigm of evolutionary computing and a two stage search: the first stage aims to find a set of points covering the feasible region and the second stage is a search for the optimum, modelling the evolution of the population, the set of points, found in the first stage. The results of profit optimization for an industrial case study are presented.

1 Optimization over an implicitly defined region

In some applications, there occur optimization problems with implicitly defined feasible regions and with objective functions not defined outside the feasible regions. An example of such a problem, from the oil industry, is described below and will be used to demonstrate a new optimization procedure suitable for tackling such problems.

We start with a formal statement of the problem. We need to find the minimum value and a minimizer for the problem

$$\min_{X \in A} f(X), \tag{1}$$

where $f(X)$ is not defined for $X \notin A$, $A \subset B \subset R^n$, $B = \{X : b_i^- \leq x_i \leq b_i^+\}$, and A itself is defined by an indicator function $I(X)$

$$I(X) = \begin{cases} 1, X \in A, \\ 0, X \notin A. \end{cases}$$

We assume that the region A is small with respect to B so that even finding feasible points is challenging. In any case, if a feasible point is known, the construction of a descent trajectory is difficult, since construction of reasonable penalty or other merit functions is challenged by discontinuities on the border of the feasible region.

The class of problems we wish to consider pose a further challenge. The objective function may be non-differentiable and even dis-continuous. The case study presented below exhibits discontinuous behaviour and, furthermore, is obtained by means of computer simulation, causing some small noise to be added to the function values. This introduction of noise makes the objective function non-differentiable. Since the regularity (including unimodality) of the objective function can not be guaranteed, a global search technique should be applied. The majority of global optimization techniques available for broad classes of multimodal objective functions assume that the objective function is defined over a hyper-rectangular feasible region [1]. Application of these techniques to problem (1) is difficult since the objective function is not defined over most of B. If the feasible region A were to constitute a large part of B then the obvious extension of the definition of the objective function $f(X) = c$, $X \notin A$, where c is a large constant, might be helpful. In our case A is relatively small, and the extension would transform the original problem into the minimization of a function which is constant almost everywhere over B. It is well known that the solution of such a problem is difficult [1].

Evolution procedures are appealing for the optimization of objective functions with irregularities [2, 3, 4, 5]. In their basic form, evolutionary algorithms are intended for unconstrained problems. A penalty function based approach is normally used to adjust genetic algorithms to constrained problems [2, 6]. However, in our case, the penalty based approach is not applicable since the feasible region is defined implicitly. We will not review the adjustment of evolutionary algorithms for constrained problems without involvement of penalties since such adjustments are mainly problem dependent.

The properties of problem (1) leave few possibilities to define a rational search procedure, not only for obtaining the minimum but also for finding feasible points. Monte Carlo methods for the generation of random points distributed over bounded regions lose their efficiency when the region is relatively small with respect to the hyper-rectangular enclosure and is defined implicitly [7].

Because of these difficulties, the solution of the general case of problem (1), applying known approaches, is hardly possible. Instead, we define a special case of (1), generalising a favourable property of the problem of process design considered below. Physical intuition suggests that the feasible region of this problem is not disjoint. Moreover, it is possible to construct enclosures of A whose extent can be controlled via an input parameter of the model for the process considered. Therefore, we will consider a special case of (1) assuming that the feasible region A is not disjoint, although it may be non-convex. Further, we assume that there exist successive enclosures for A. A θ-enclosure of A, denoted A_θ, is defined by the inclusions $A \subset A_\theta$, $0 < \theta < 1$, and $A_\theta \subset A_\pi$, $\pi < \theta$. Using a sequence of enclosures, we will define an evolutionary algorithm to generate a set of points in A. This set will be used as an initial population for an evolutionary search for the minimum over A.

2 A Case Study: Optimization of an Oil Stabilization Process

The case study presented here is an industrial scale process from the oil industry. An oil stabilization process consists of a series of flash drums, mixers and pressure changing devices to process a raw hydrocarbon feed stock into oil and gas products [8]. The design and optimization of such a process is difficult due to a combination of features inherent in the process and in the models used, both for modelling the physical behaviour of the process and for deriving cost estimates of a given configuration and the set of operating conditions. A sample process is illustrated in Figure 1.

The formulation of the oil stabilization process consists of models for the flash vessels, pressure changing devices (compressors and valves), and product stream specifications, described in detail below. The overall process model is a function of a small number of design parameters: each flash vessel has an operating temperature and pressure changing devices each have a target pressure. The example flowsheet in Figure 1 therefore has 7 variables: 4 operating temperatures for the flash vessels and 3 target pressures for the pressure changing devices. It should be noted that the pressure changing devices just prior to the gas product tank (P2 and P3 in the figure) are often fixed at a specified value.

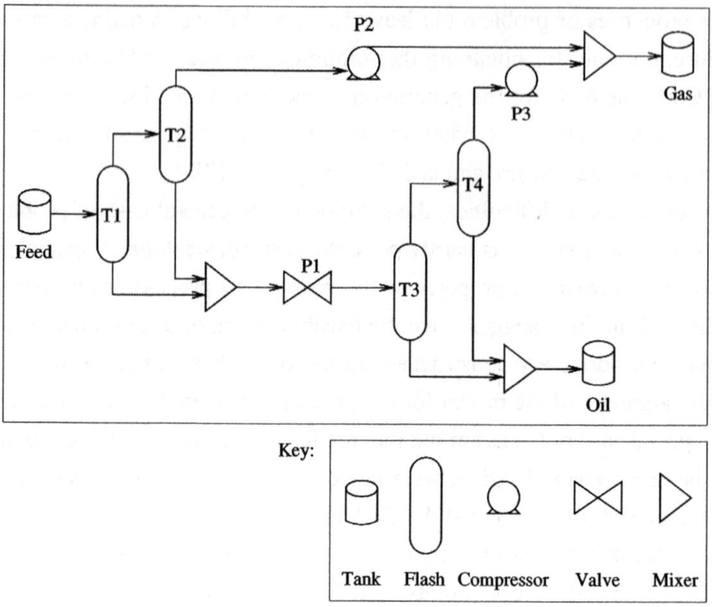

Figure 1: Four flash drum configuration of an oil stabilization process.

2.1 Flash Vessels

An individual flash drum has a single feed, which may be in liquid phase, vapour phase, or as a multi-phase stream containing both liquid and vapour phases. There are two output streams, a vapour stream (indicated by the stream leaving the top of each flash drum in Figure 1) and a liquid stream (bottom stream of each flash drum). A flash drum has one degree of freedom, the operating temperature. This temperature is used to determine, through the use of vapour-liquid equilibrium (VLE) physical property estimation methods, the compositions and flows of the vapour and liquid streams for the given feed stream. VLE models are highly nonlinear.

The specification of the operating temperature is given by a value in [0, 1]. This normalization of the temperature variable maximises the size of the feasible region A relative to B. A 0 value corresponds to the bubble point temperature (the temperature at which a liquid stream, when heated, starts to bubble) of the feed to the flash vessel and 1 the dew point temperature (the temperature at which a vapour stream, when cooled, starts to form droplets of liquid). The bubble and dew point temperatures depend on the composition of the feed stream. Hence, the meaning of the design variable's value depends on the values of the design variables for units up-stream in the process.

Given the output stream specifications, the design and costing of the flash drum requires evaluating a series of nonlinear equations. Although straightforward, there is an implicit binary decision for the orientation of the flash drum. Depending on the relative flows of the vapour and liquid output streams, the drum may be designed to be oriented horizontally or vertically. This implicit binary decision results in a discontinuous function for the cost versus the operating temperature design parameter.

2.2 Pressure Changing Devices

The pressure changing devices include both compressors for increasing the pressure and valves for reducing the pressure. Compressors are chosen from a set of possible types and simple models are used for all types. The type of compressor chosen depends on the amount of pressure change required as well as the flow of the stream involved. Each type has a different cost model and the result again is a discontinuous cost objective function.

2.3 Process Constraints

The overall process model has two types of constraints. The first type is composed of physical constraints, such as the operating limits of each process unit and the availability of utilities at appropriate temperatures and with sufficient capacity to meet the heating and cooling demands of the process. The second type includes the product specifications, which are market specific, and which impose severe restrictions on the feasible space. In particular, the specification of the oil product is tight: the vapour pressure of the oil stream, P_v, must lie between 10 and 12 `psia`,

$$|P_v - 11| \leq 1, \tag{2}$$

equivalent to $P_v \in (69, 83)$ kPa approximately. In practice, we find that the combination of physical and marketing constraints leads to a feasible space, A, whose hyper-volume may be much less than 1% of the domain defined by simple bounds on the design variables, B.

2.4 Model Solution

The case study is formulated as the minimization of a function of 5 variables; the two compressor pressures for the process in Figure 1 are fixed although their design will depend on the upstream unit designs. The physically reasonable intervals are scaled to form the hypercube $[0,1]^5$. The objective function corresponds (multiplied by -1) to the profit of the process. The profit may be calculated only for the products which meet the marketing specifications (2).

The model has been implemented in the *Jacaranda* system for process design and optimization (`http://jacaranda.chemeng.ucl.ac.uk/`). Jacaranda exposes an interface which allows it to be used as a *black box* objective function in other software, such as Octave (`http://www.octave.org/`). The objective function value returned is the negative of the profit for feasible points and 10^{20} for infeasible ones.

This problem has been previously tackled using a targeted genetic algorithm where the genetic operators, mutation and crossover, have been adapted to emphasise the creation of solutions in the feasible region. The feasible region was estimated using a clustering method based on visualization [9]. Although successful, the method required a sufficient number of initial feasible points to provide useful clustering information for the genetic algorithm operators. This need motivates the generation of feasible points in a systematic manner, as described below.

3 Generation of Feasible Points

The feasible region A of the optimization problem (1) is defined implicitly. To construct a rational algorithm for the search for a minimum, a constructive characterization of A is desirable. A set of points uniformly covering A is one of possible characterizations [9]. The generation of such points by brute force is difficult: generating 10^6 random points uniformly distributed over B produces only 107 feasible points, taking approximately 10^4 seconds on a 700 MHz Pentium based computer with 256 MB of RAM. The best profit value found was 3.039×10^8, which is not an acceptable solution since better solutions are known.

The idea of continuation may be applied to construct a more efficient algorithm for the search for feasible points based on the opportunity to use the enclosures of the feasible region. The market specification for the oil product is modified to include the continuation parameter, $0 \leq \theta \leq 1$:

$$|P_v - 11| \leq 1 + 100(1 - \theta).$$

The actual market specification (2) corresponds to $\theta = 1$. The decrease of θ means softening of market requirements and produces enclosures $A_\theta \supset A$.

To gain insight into the size of the enclosures, we start with a random generation of 1500 random trial points with a uniform distribution over the hyper-rectangular region. The percentage of points hitting the largest enclosure A_0 is approximately equal to 4%. As noted before, 10^6 trials produced only 0.01% points hitting the feasible region of (1), $A \equiv A_1$. The sizes of the enclosures for different θ are shown in Figure 2. The

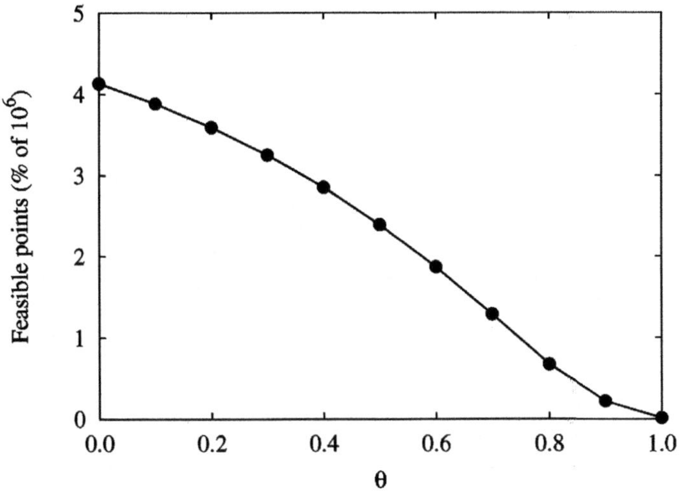

Figure 2: Hypervolume of enclosure depending on θ.

graph for $0.9 \leq \theta \leq 1$ does not illustrate the decrease of the hypervolume of enclosure very precisely because of the scale of the graph; in fact, it behaves as a jump towards zero near $\theta = 1$.

For the construction of an adaptive algorithm rationally generating points in A, the structure of enclosures is important. To understand the structure of the sets in multidimensional space, two dimensional projections of points uniformly distributed over the sets are helpful. The projections of the points hitting A_0 into two dimensional planes of different variables show that the assumption that A_0 is not disjoint is likely to be valid. An example of the visualization is presented in Figure 3, showing the projection of all the trial points in B onto the plane of variables x_1, x_2, where the points in A_0 are specially marked. The hypothesis that the set A_0 is not disjoint is corroborated by the similar projection to the plane defined by two first principal coordinates of the set of points in A_0, as shown in Figure 4. Moreover, the projections show that A_0 is nearly convex. Similar conclusions are made with respect to the other enclosures, analysing the corresponding projections interactively. Projecting a set of points to two dimensions is a heuristic tool which does not guarantee the correctness of any generalization of the properties of the projections for the multidimensional set of interest. However, this visualization supports our engineering intuition with respect to the structure of the enclosures. A correct hypothesis concerning the structure of the enclosures is crucial for the efficiency of the algorithm constructed to generate a desired set of points in A.

The algorithm for generating a set of points in A is a sequential procedure starting with A_0 and finishing with $A = A_1$. The generation of points in A_θ is guided by infor-

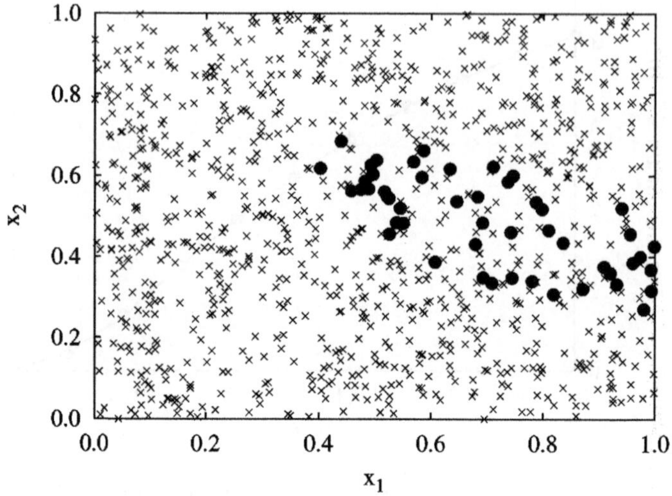

Figure 3: Projection of trial points to the plane of x_1, x_2. Feasible points are indicated with large filled circles and infeasible points with small crosses.

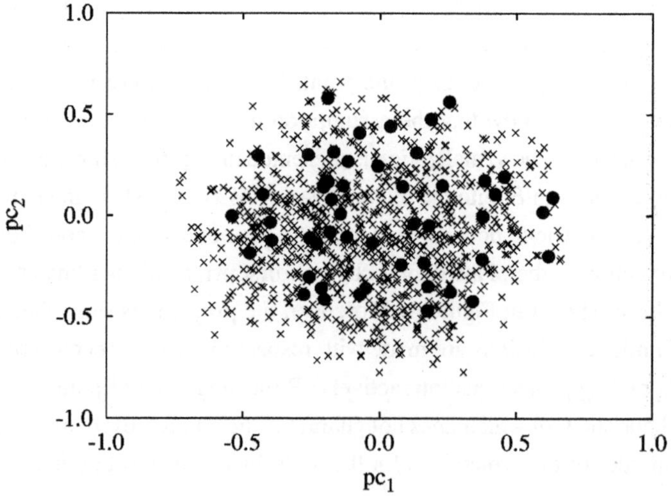

Figure 4: Projection of trial points to the plane of two first principal coordinates. Feasible points are indicated with large filled circles and infeasible points with small crosses.

mation obtained from points in A_π, where $\pi < \theta$. As an increase of the index implies a decrease in the size of the enclosure, the evolution of the set of points is similar to the evolution of a biological population driven by worsening environmental conditions.

Consider a population of size N. The evolution is driven by changing environmental conditions worsening step by step, i.e., $A_\theta \subset A_\pi$, $\theta > \pi$. The generation living in conditions defined by π produces descendants. The new generation is composed of the individuals of the old generation and their descendants who survive in the new conditions, defined by θ. The descendants are produced by multi-parent crossover aiming to ensure diversity in the new generation, where diversity may be interpreted as the uniformity of the distribution of points in A_θ. The parents who would survive under the new conditions are chosen for crossover more frequently than those who would not; however those who would survive normally constitute only a small fraction of the old generation. For a set, J, of j multi-parents, one survivable parent and $j-1$ non-survivable parents are chosen. The choice in corresponding subsets is random with uniform distribution. The crossover is defined by the formula

$$Y = \sum_{i \in J} \beta_i X_i, \qquad \beta_i = \alpha_i + (1-\gamma)(\frac{1}{j} - \alpha_i),\ \alpha_i \geq 0,\ \gamma \geq 1,\ \sum_{i \in J} \alpha_i = 1, \tag{3}$$

where α_i are weights; the weight of the survivable parent is 0.5 and the other weights are generated randomly. The coefficient $\gamma \geq 1$ defines the central extension of the convex hull of points X_i with respect to the centre (average of the points). The crossover (3) is a generalization of the convex (arithmetic) crossover (see [2], [10]) for the multi-parent case. The crossover of survivable and not survivable parents (points satisfying constrains and points not satisfying constrains) is similar to the crossover used in GENOCOP [2].

Experimentally obtained samples generated by the proposed algorithm exhibit a slight concentration of points in some subsets of the enclosures considered, leaving other subsets empty. We have to note, however, that even for samples from theoretically uniform generators, uniformity is guaranteed only asymptotically. For a discussion on the practical generation of low-discrepancy sequences in multidimensional rectangle regions, we refer to [11]. Ensuring low discrepancy in the case of non-rectangular small regions is more difficult; to eliminate the *holes*, we apply a heuristic procedure. Let the size of the population be N. By means of crossover (3), a number of survivable descendants is produced to comprise, together with survivable parents, a generation of size $0.75N$. For these points, a minimum spanning tree is constructed. A number of the longest edges of the minimum spanning tree are selected. Since the sets A_θ are not disjoint, and are nearly convex, the edges normally belong to A_θ. New points in A_θ are generated randomly, with Gaussian distribution centred on the middle of an edge, and the components are non-correlated with standard deviation equal to 0.1 of the average length of the edges of the minimum spanning tree. Several trials may be needed to generate a point in A_θ. In such a way, $0.25 \times N$ points are generated which, together with the previously generated points, gives a new generation in A_θ of size N.

For the case study, the size of population was chosen equal to $N = 10^4$ and six generations with $\theta = 0.2 \times k$, $k = 0,...,5$ were modelled.

4 Minimization

Minimization of a function with a small implicitly defined feasible region is difficult. Moreover, the function itself can be assumed neither smooth nor unimodal. These two factors motivate us to consider an evolutionary approach for minimization.

The minimization is performed by a version of evolution strategies [3]. The starting generation for the evolutionary search for the minimum is found by the evolutionary procedure described in the previous section. It consists of points in the feasible region A. We describe the population at the current step of evolution by the average of these points, and the eigenvectors and eigenvalues of the covariance matrix of these points. The descendants are modelled as realizations of a Gaussian random vector with the average equal to $X_a + \varphi(X_{\min} - X_a)$, where X_a is the average of the parent population and $X_{\min}$ is the best point found. The covariance matrix of the Gaussian distribution is diagonal with respect to the coordinate axes defined by the eigenvectors of the parent generation. The elements of the diagonal are chosen equal to halved eigenvalues of the parent generation. The descendants survive if their function value is smaller than the average function value of the parent generation; the descendants corresponding to the infeasible points do not survive. The production of descendants continues so long as members of the parent population with function values larger than average are substituted by descendants. The formation of a new generation is completed by computing a new average value, new eigenvalues, and new eigenvectors.

This version of evolution strategy was developed experimentally. The performance of the more standard version with the centre of probability density at the point $X_{\min}$ was not satisfactory. While solving the case study, a current vector $X_{\min}$ normally is on the border of the feasible region since the points with the best function values are on the border. But in the case of a Gaussian probability density with the centre on the border, there is a large probability to generate an infeasible point, and the probability to generate a better point than $X_{\min}$ is very small. A probability distribution with non-correlated components and equal variance is frequently used in evolution strategies to model mutations. However, such a model is not suitable for the solution of our case study because the extent of the feasible region in different directions is strongly different. Therefore, the covariance structure of the distribution was tuned to the characteristics of the parent generation which represents the shape of the considered part of the feasible region.

For our case study, ten generations are modelled, each consisting of 10^4 descendants. The minimization was performed 30 times to evaluate the effectiveness of the algorithm. The average minimum value was -3.23×10^8 and the best overall was -3.25×10^8. The average time of one minimization was approximately 2×10^3 s.

5 Conclusions

A two stage procedure has been developed, suitable for handling problems with small, implicitly defined, feasible regions and with a non-smooth objective function. The first stage uses continuation to identify a non-disjoint feasible region. The second stage is an evolutionary algorithm which leads a search in the direction of the best solution available at any point. The two stage procedure has been applied successfully to a difficult industrial case study.

The use of visualization, through a two-dimensional mapping of the higher dimensional data, enables the user of the method to interact with the procedure (see [12] for a discussion on how this type of interaction can be helpful in process design and optimization). The interaction could occur both within the first stage in the generation of sufficient initial feasible points or as a refinement to the initial population passed to the second stage. The evolutionary procedure has the extra advantage that a population of solutions, the majority of which will be feasible, is available to the user at the end. This population can be used to analyse the behaviour of the solution space near the optimum, allowing the user to gain an understanding of the impact of the implicit constraints on the solutions obtained.

The evolutionary procedure presented has characteristics of other evolutionary algorithms, such as genetic algorithms [5] and simulated annealing [13], specifically the concepts of generating new solutions in the neighbourhood of existing good solutions and the use of populations. However, the new procedure differs from a genetic algorithm in having a directed procedure for the evolution of the population. It differs from simulated annealing in only accepting solutions which are better than an average solution. The new procedure has been shown to be effective at solving a complex optimization problem with an implicitly defined feasible region, especially when coupled to a pre-processing stage which identifies an initial population of feasible solutions.

References

[1] A. Törn and A. Žilinskas. Global optimization. *Lecture Notes in Computer Science*, 350:1–255, 1989.

[2] Z. Michalevich. *Genetic Algorithms + Data Structures = Evolution Programs.* Springer, NY, 1996.

[3] H.-P. Schwefel. *Numerical Optimization of Computer Models.* J.Wiley, NY, 1981.

[4] I. Parmee. *Evolutionary and Adaptive Computing in Engineering Design.* Springer, NY, 2001.

[5] J. E. Smith. Genetic algorithms. In Panos M. Pardalos and H. Edwin Romeijn, editors, *Handbook of Global Optimization Volume 2*, Nonconvex optimization and its applications, pages 275–362. Kluwer Academic Publishers, 2002.

[6] K. Miettinen, M. Mäkelä, and J. Toivanen. Numerical comparison of some penalty-based constrainet handling techniques in genetic algorithms. *Journal of Global Optimization*, 27:427–446, 2003.

[7] R. Smith. Efficient Monte Carlo procedures for generating points uniformly distributed over bounded regions. *Operations Research*, 32:1296–1308, 1984.

[8] E. McCarthy, E. S. Fraga, and J. W. Ponton. An automated procedure for multicomponent product separation synthesis. *Computers & Chemical Engineering*, 22(7):S77–S84, 1998.

[9] K. Wang, A. Salhi, and E. S. Fraga. Process design optimisation using embedded hybrid visualisation and data analysis techniques within a genetic algorithm optimisation framework. *Chemical Engineering and Processing*, 43(5):657–669, 2004.

[10] M. Gen and R. Cheng. *Genetic Algorithms and Engineering Optimization.* J.Wiley, NY, 2000.

[11] L. Kocis and W. Whiten. Computational investigation of low-discrepancy sequences. *ACM Trans. on Mathematical Software*, 23:266–294, 1997.

[12] E. S. Fraga, K. Wang, and A. Salhi. Interactivity and automated process design. *Chemical Engineering Technology*, 26(8):823–827, 2003.

[13] P. J. M. van Laarhoven and E. H. L. Aarts. *Simulated annealing: Theory and applications.* Kluwer Academic Publishers (Dordrecht), 1987.

Application of Artificial Neural Networks In Differential Pulse Code Modulation Scheme

H. B. BAHAR
Faculty of Engineering, University of Tabriz, P. O. Box 5166614776, Tabriz Iran
Tel: ++98411 3301781, Fax: ++98411 3344272, gprco_hbb@yahoo.com , hbbahar@tabrizu.ac.ir

Abstract

In this work an Artificial Neural Network with Radial Basis Function (RBF) is employed to model a predictor, utilized in Differential Pulse Code Modulation (DPCM) scheme. The RBF predictor estimates the magnitude of signal incoming to DPCM, the error between incoming signal and estimated one is applied to quantiser unit. The resultant error contains a few bit word length and is transmitted in data format towards a receiver unit. The output of the RBF predictor is added to error at the receiver part of DPCM to produce actual output. In this study application of RBF predictor in DPCM is explained. The potential offered by DPCM scheme using RBF predictor leads to a considerable amount of reduction in word length of filter coefficients.

Key Words: Differential Pulse Code Modulation, Radial Basis Function, Predictor, FIR Digital Filter

1. Introduction

Many schemes for the employment of Artificial Neural Networks (ANNs) in control systems have been proposed and also some have practically been made [1, …, 4]. A neural network is in essence a non-linear mapping device and in this respect, at the present time, most of the reported work describing the use of neural networks in a control environment is concerned solely with the problem of process modeling or system identification [5,6,7].

One of the most popular forms of neural network at the present time is the Radial Basis Function network, which has similar weight training properties to CMAC [4]. The Multi-Layer Perceptoran (MLP) Artificial Neural Network is also among the most popular and versatile forms of ANN classifier; see for example [8] for a dynamic weight estimation.

The current work is to investigate a new filter design method on the apparently novel realization that is the action of the Differential Pulse Code Modulator (DPCM) from the field of communication engineering [9]. Non-recursive FIR digital filters hold a large number of coefficients, where each coefficient contains a number of bits. DPCM technique is applied to realize the Non-recursive FIR transfer function in the form of Recursive FIR digital filter, with filter coefficients of –1, 0 and +1. Thus, the incoming signals to such a filter are simply processed by subtraction (-1), no-operation (0) and add (+1) operations.

2. Differential Pulse Code Modulation Scheme

In general a differential pulse code modulation (DPCM) refers to a scheme with a transmitter and a receiver shown in Figures 1 and 2 respectively [9]. In the feedback path of the modulator in Figure 1, a circuit predicts the next input sample to the modulator. For verification, general and classical arrangement of the digital predictor is illustrated in Figure 3.
The input to the modulator is a signal $h(n)$ where_in this study is the impulse response of the desired FIR filter. (Note that in DPCM the $h(n)$ for convenience is represented as h_n, etc.) This Low-Pass filter with 77 coefficients is denoted by LP77 in this work [10]. The LP77 filter coefficients together with frequency response are illustrated in Figures 4 and 5 respectively.
The LP77 digital filter design procedure by the expression given in Appendix (I). In the usual application of DPCM the input signal would be an infinite, or near infinite, sequence of samples forms, for example, a speech waveform. The predictor creates a prediction of the current input sample based on previous samples [9, 11]. This prediction sample is labeled $\hat{\tilde{h}}_n$ as shown in Figure 1.The difference d_n between the actual next input sample h_n and its prediction $\hat{\tilde{h}}_n$ is passed to the quantiser unit. Obviously if d_n is small due to good prediction, few bits are needed by the quantiser to represent this signal. This is the main advantage of the DPCM scheme as applied to communication systems because the signal can be transmitted using fewer bits in total. In an extreme example the quantiser output can be constrained to have two levels +l and -1, in which the DPCM system becomes a delta modulator.
The original signal h_n may be approximately reconstructed by adding the incoming quantised difference $(h_n - \hat{\tilde{h}}_n)$ to the output of an identical predictor at the receiver, to obtain a discrete estimate $\tilde{h}_n'$, as shown in Figure 2. Later it will be shown that the realized recursive digital filter having the desired FIR response, replaces the receiver function in the DPCM system.

A classical form of the predictor is the non-recursive shape as shown in Figure 3. Here $\hat{\tilde{h}}_n$ is a linear combination of past values and is governed by the following difference equations,

$$\hat{\tilde{h}}_n = \sum_{i=1}^{k} p_i . \tilde{h}'_{n-i} \qquad (1)$$

Where, p_i s are the predictor coefficients and k is the order of the predictor. To provide optimum prediction the predictor coefficients have to be chosen to reduce some measure of the error between actual and estimated input samples, to the smallest possible value.
In this work a new approach is used to design an ANN predictor to perform in DPCM system.

3. Recursive Digital Filter Design Using DPCM Scheme

The motivation here is, to apply the same advantage that DPCM has in transmitting signals with fewer bits, to the design of digital filters which require fewer operations in the filter coefficients. In fact the DPCM can be applied in this way, as is now shown.

We now proceed to consider a typical DPCM scheme shown in Figure 6. Here an impulse is applied to the ideal filter $H(z)$ and the sequence $h(n)$ is produced as before. (Note that in DPCM the $h(n)$ for convenience is represented as h_n, etc). The feedback arrangement of Figure 1 is now replaced by the block with a transfer function, $P(z)/(1-P(z))$ which is easily shown to be mathematically equivalent. Note that in ordinary DPCM system, $P(z)$ is the predictor polynomial,

$$P(z) = p_1 z^{-1} + p_2 z^{-2} + \ldots + p_k z^{-k} \qquad (2)$$

The sequence of difference samples d_n is applied to the quantiser, which produces quantised samples at the output. The associated error sequence is $e_n = \tilde{d}_n - d_n$. the sequence e_n can be represented as the output of some filter having response $H_e(z)$ and excited by an impulse, $\delta(n)$. By analogy with the direct quantisation scheme the sequence $\tilde{d}_n$ could be constructed by means of a similar structure such as Figure 3 and would have a transfer function,

$$\tilde{D}(z) = \tilde{d}_0 + \tilde{d}_1 z^{-1} + \tilde{d}_2 z^{-2} \ldots + \tilde{d}_{N+P} z^{-(N+P)} = \sum_{n=0}^{N+P} \tilde{d}_n z^{-n} \qquad (3)$$

Figure 6 viewed as a two input sampled-data linear system and elementary analysis methods give,

$$\tilde{D}(z) = [l - P(z)].H(z) + [l - P(z)]H_e(z) \qquad (4)$$

Which gives,

$$\tilde{D}(z) \approx (l - p(z)).H(z) \qquad (5)$$

if $H_e(z)$ is small.

The filter response $\tilde{D}(z)$ is not yet in the required form because it does not equate to $H(z)$. The correct form is obtained by re-arranging (4) as follows:

$$H(z) = \frac{\tilde{D}(z)}{l - P(z)} - H_e(z) \approx \frac{\tilde{D}(z)}{l - p(z)} \qquad (6)$$

This clearly shows that $H(z)$ can be approximately realized (to within the error function $H_e(z)$) by the recursive transfer function $\tilde{D}(z)/(1-P(z))$ formed by the quantised sequence $\tilde{d}_n$ and the predictor polynomial formed by predictor coefficients p_i.

4. Radial Basis Function Network For Prediction

Of the many types of neural networks possible, the Multilayer Perception (MLP), despite its problems due to slow weight convergence through non-linear learning, has perhaps received most attention. More recently, radial basis function networks have been gaining popularity as an alternative to multilayer perceptorans . These networks have been shown to have the ability to model any arbitrary non-linear function [3, 6]. However radial basis functions require a large number of basis functions to achieve the required approximation properties. This is similar to the problem of estimating the number of layers and nodes in the multilayer preceptoran [3].

The radial basis function output is merely a linear combination of the hidden layer signals. Such that, given an input vector h_n and a set of network weights w_i the network output is found from,

$$f(h_n) = w_0 + \sum_{i=1}^{N} w_i \phi_i(\| h_n - r_i \|) \qquad (7)$$

Where N is the number of radial basis functions (neurons), w_0 is a bias term, r_i is the set of centers and $\phi_i(.)$ is a function of the distance between the current input point h_n, and the center r_i. It is usual, but not necessary, for the function $\phi_i(.)$ to be a common function type throughout the N neurons, typical examples being:

I. $\phi(t) = t^2 \times \log(t)$ Thin Plate Spline

II. $\phi(t) = \sqrt{(t^2 + c^2)}$ Multiquadratic

III. $\phi(t) = \exp(-t^2 / 2\sigma_i^2)$ Gaussian

IV. $\phi(t) = \dfrac{1}{\sqrt{(t^2 + c^2)}}$ Inverse Multiquadratic

in which σ and c are simply scaling or "width" parameters.

Of the selections possible, and it should be stressed that many other options exist, the Gaussian function is the most intuitive, providing a peak for each function which coincides with the input vector bring equal to the basis function centre, the function trailing off as the input vector moves away from the centre. However, practical experience has shown that the thin plate spline actually works very well in terms of modelling accuracy.

Dynamic, can readily be taken into account in a radial basis function network through the input vector h_n, consisting of elements which are sampled plant or controller, input and output signals. For a system prediction exercise, if $\hat{\tilde{h}}_{n-1}, \hat{\tilde{h}}_{n-2}, \ldots$ indicate the system output at time instant *n-1, n-2,* etc.; and $h_{n-1}, h_{n-2}, \ldots$, indicate the system input at time instant *n-1, n-2,* etc. , from the network, the input vector h_n can be formed from,

$$h_n = f[\hat{\tilde{h}}_{n-1}, \hat{\tilde{h}}_{n-2}, \ldots : h_{n-1}, h_{n-2}, \ldots] \qquad (8)$$

The network output, $\hat{\tilde{h}}_n$ can then be made to model the system output, at time instant, n by an appropriate choice of basis functions, center positions and weightings. For multiple output systems, several RBFs can be combined to form a network. All outputs share the same set of centers but have different coefficients. In a basic form of operation, one basis function type is selected and the basis function centers are fixed. The network weights, w_i, can then be adjusted to satisfy an error criterion such as linear least squares.

The crucial feature of radial basis function networks is the selection of basis function centers, both in terms of their number and respective positions. The choice made is critical to the successful operation of the network, too few centers and the network's approximation properties will be poor, however unfortunately the number of required centers tends to increase exponentially with regard to the input space dimension, thereby making radial basis function networks unsuitable for a use on a highly complex input space.

It is possible to select radial basis function network centers in a number of ways, e.g. a uniform distribution across the input space, or indeed a random distribution, of a set total number of centers. More interestingly a clustering algorithm, such as the Mean-Tracking algorithm can be employed, this giving a statistically best section of Center quantity and position. A brief comparison of the different methods of center selection , in terms of their approximation properties , can be found in [3]. Some of the activation functions require an extra parameter, which determines the " width " of the basis function. Again no rigorous method exists to calculate this parameter. In the case of Gaussian basis functions, a suitable heuristic is to set the width to the mean of the distances from the center to the nearest neighboring centers.

A RBF network to model the predictor at DPCM scheme is illustrated in Figure 7. A pair of Input samples, $\tilde{h}'_{n-1}$ and $\tilde{h}'_{n-2}$ at DPCM scheme creates the current output sample, which is labeled as $\hat{\tilde{h}}_n$ in Figure 1. Note that in Figure 1 the difference between input samples to the predictor, $\tilde{h}'_n$ and input samples to DPCM, h_n is due to quantisation error e_n, thus,

$$e_n = \tilde{h}'_n - h_n \qquad (9)$$

The sequence e_n can be represented as the output of some filter having response $H_e(z)$ and excited by an impulse, $\delta(n)$ as shown in Figure 6. However, utilizing an accurate predictor will reduce the error to a negligible amount hence,

$$\tilde{h}'_n \approx h_n \qquad (10)$$

On subsequent section the importance of reduction in error to satisfy equation (10) will be discussed.

5. Simulated Performance

In this study the RBF Network architecture with an output, $\hat{\tilde{h}}_n$ shown in Figure 7 is used for simulation purposes. The RBF Network was trained with time series samples, $h_{n-1}, h_{n-2}, \ldots$ to emulate the dynamic behavior of the DPCM system. The output of the predictor, $\hat{\tilde{h}}_n$ is an estimated sample, where the difference between, h_n and $\hat{\tilde{h}}_n$ is applied to quantiser unit to produce –1, 0 and +1 digit at the output of the transmitter unit, shown in Figure 1. The proposed RBF Network model shown in DPCM scheme was found to be,

- **Number of input samples $\widetilde{h}'_{n-1}$ and $\widetilde{h}'_{n-2}$ is taken as N=2.**
- **Single output sample $\hat{\widetilde{h}}_n$.**
- **Number of layers is 2, a hidden layer (as input layer) and an output layer.**

A set of 77 patterns is used for RBF Network training process. Each input pattern is composed of 2 sequence samples, $h_{n-1}, h_{n-2}, \ldots$ following the application of impulse to the desired FIR filter, LP77. Input patterns for training were generated by C++ programming language for simulation purposes. The selected network parameters are, spread constant sc=5.90 and sum-squared error goal eg= 90 and sampling interval is $t_s = 1\ ms$.

RBF Network was trained by applying the noise free patterns where theoretical background of training mechanism was as explained in the previous section. The overall performance of the trained RBF predictor in DPCM for given parameters is illustrated in Figure 8.

In Figure 8, the output samples of ANN, $\hat{\widetilde{h}}_n$ are labeled '+' and input samples to DPCM, h_n are labeled '*'. An arithmetic examination on illustrations in Figure 8 reveals that error between $\hat{\widetilde{h}}_n$ and h_n can be disregarded. Therefore, this confirms that ANN is capable of modeling the predictor in DPCM scheme. In the following section the features of ANN modeling and recalling process will be discussed.

6. Further Discussions and Results

The main objective in this study is to realize an ANN predictor to function in DPCM scheme. Initially, a learning process is carried out by the impulse response of non-recursive digital filter with transfer function of $H(z)$, see Figure 6. For recalling purposes, the trained ANN predictor, is utilized in DPCM system to estimate the incoming sample, h_n. In fact, in recalling stage unseen patterns, i.e. sum of quantiser and predictor outputs were fed to ANN predictor to obtain subsequent output sample, $\hat{\widetilde{h}}_n$. Illustrations in Figure 8 indicate that the estimated output for unseen patterns practices the sequence samples of the impulse response of transfer function, $H(z)$.

Alternative way to show the proper and accurate performance of the designed ANN is to study the frequency response of the designed ANN predictor. Illustration in Figure 9 shows the frequency response of ANN predictor, which has quite similarity to frequency response of $H(z)$, shown in Figure 5.

Considerable amount of work has been carried out to make that inputs to quantiser unit lay within a desired range as shown in Figure 10-(a). Here, each input to quantiser unit randomly acquires a magnitude between –1.5 and +1.5. The quantiser output, -1, 0 and +1, is outlined in Figure 10-(b). See Equation (3).

The gain factor, G, plays an important role once the operation of quantiser is predetermined, where referring to Appendix (I) the gain factor is chosen, G=1000.

Considering Equation (6) in section 3, $H(z)$ can be classically realized (within the error function $H_e(z)$) by the recursive transfer function $\widetilde{D}(z)/(1-P(z))$. For constructing the transfer function, $\tilde{D}(z)/(1-P(z))$, the quantised sequence $\tilde{d}_n$ and the ANN predictor polynomial are required. In ANN modeling the nature of the predictor polynomial is parallel computation and quantised sequence, $\tilde{d}_n$ involves in serial computation. Therefore, ANN predictor transfer function, $P(z)$ would be realized by an enormous polynomial to investigate the performance of, $\tilde{D}(z)/(1-P(z))$. In this study indirect approach is used to observe the dynamic behavior of the ANN predictor. The output sequence samples of ANN predictor in DPCM, $\tilde{\hat{h}}_n$ is considered as the impulse response of ANN predictor transfer function. The sketch of impulse response samples, $\tilde{\hat{h}}_n$ and corresponding frequency response of the ANN predictor are illustrated in Figures 8 and 9 respectively. Similarities between frequency responses in Figures 5 and 9, confirm that ANN predictor satisfies the performance of DPCM scheme in parallel processing.

Finally, a further point that needs to be dealt with is selecting RBF training parameters, spread constant, sc and sum-squared error goal, eg. The typical recalling performances of models for various, sc with unchanged, eg were demonstrated in Figures 11-(a), (b), (c). Study on the models behavior concludes that the optimal ANN model with minimum root mean square (rms) error, rms=0.0282 is illustrated in Figure 11-(c), where both trained and recalled ANN samples within negligible error are equal in magnitude.

7. Conclusion

The purpose concerned here is that of wanting to apply the enhancement of ANN predictor in DPCM scheme. The advantage that DPCM has in transmitting signals with fewer bits, is used to the design of ANN based FIR digital filter, which require fewer operations in the filter coefficients. The ANN based recursive FIR digital filter embodies the coefficients, -1, 0 and +1, at the numerator with RBF predictor at its denominator.

At the recalling step unseen patterns in DPCM scheme were fed to ANN predictor. The simulated performance shows that in DPCM system the actual estimated output samples, $\tilde{\hat{h}}_n$ practice the desired input samples, h_n.

Appendix (I)

Non-recursive Digital Filter Design Example

The commonly used lowpass filter is taken for the main test example and is designed using the impulse invariant technique. The procedure is now described after first presenting the relevant basic theory. The non-recursive filter transfer function,

$$H(z) = \sum_{i=o}^{N} a_i \; z - i \tag{1}$$

and the corresponding difference equation,

$$y(n) = \sum_{i=0}^{N} a_i \; x(n-i) \tag{2}$$

are obtained from equations 1.

It can be proved that the coefficients a_i are the impulse response of the non - recursive digital filter. Thus the coefficients of this filter are equal to the impulse response. The transfer function (1) is:

$$H(z) = \sum_{n=0}^{N} \hat{h}(n).z^{-n} \qquad (3)$$

From this property arises the basic idea of the "impulse - response - invariant design technique" for FIR filters. That is first to find a finite impulse response, which approximates that of the desired filter, and then to equate it to the filter coefficients.

Deriving the Fourier series of the repetitive ideal lowpass response, H(z), gives $\hat{h}(n)$ as the Fourier series coefficients in the time domain. The result is:

$$\hat{h}(n) = Gf_c.T . \frac{Sin\ (2\pi n f_c T)}{2\pi n f_c T} \qquad (4)$$

where f_c is the filter cut - off frequency and G is the gain factor.

For negative values of n, samples are shifted in the positive direction, which does not affect the frequency response of the filter. This is done by replacing the argument n in equation (4) by the argument $n-(\frac{N}{2}+1)$. Thus the final form windowed lowpass impulse response is

$$h(n) = W(n).\hat{h}(n)$$

$$= \frac{1}{2}(1 + Cos\pi(n-(\frac{N}{2}+1)).\ (Gf_c T \frac{Sin\ 2\pi f_c T(n-(\frac{N}{2}+1))}{2\pi f_c T(n-(\frac{N}{2}+1))} \qquad (5)$$

where, n = 0, 1, 2, … , N. Note the central 'lobe' of $h(n)$ occurs at $n=\frac{N}{2}+1$ and equation (5) shows that: $h(\frac{N}{2}+1) = 2Gf_c T$. For a test example, specific parameter values in equation (5) must be chosen. In practical non-recursive filters the number of samples usually ranges from a few tens to several hundreds. Any value of N in this range would serve for a useful example and somewhat arbitrarily, the value of N=76 are chosen, this also being used in work of Lockhart [10] (note that as equation 1 shows, the degree of the filter example is also N = 76, but the number of coefficients is N + 1 = 77). Also G = 1000 and $f_c T = f_c / f_s$ = .038 (in line with reference [10]), are chosen.

References

1. Hunt K.J., Irwin G.R., Warwick K "*Neural Network Engineering in Dynamic Control Systems*", 2nd printing, Springer, (1996).
2. Pandya, A. S. and Macy, R. B, *Pattern Recognition with Neural Networks in C++. CRC Press Inc*, (1996).
3. Haykin, S, *Neural Networks, A Comprehensive Foundation.* Macmillan College Publishing Company, (1994).

4. Ananthraman, A. and Garg, D. P, Training Back propagation and CMAC Neural Networks for Control of a SCARA Robot. *Journal of Engineering Applications of Artificial Intelligence*, Vol. 6, No. 2, (1993), 105-115.
5. Pham, D. T. and Liu, X, *Neural Networks for Identification, Prediction and Control.* 2nd Edition, Springer-Verlag, (1995).
6. Bahar H.B, "*Dynamic System Identification Using Artificial Neural Network*", the Fifth International Conference on Adaptive Computing in Design and Manufacture, ACDM, (2002).
7. Bahar H. B., " Engineering Application of Correlation on ANN Estimated Mass", *International Journal of Engineering, Transaction B: Applications*, Vol. 15, No. 2, July 2002, 157-166.
8. Bahar, H. B. and Horrocks, D. H, Dynamic weight Estimation using Artificial Neural Network. *Artificial Intelligence In Engineering*, Vol. 12, (1998),135-139.
9. Jayant N. S, "*Digital Coding of Speech Waveforms PCM, DPCM*", Proc. IEEE, Vol. 62, (May 1974), 611 – 632.
10. Lockhart G. B, "*Binary transversal filters with quantised coefficients",* Electronics Letters, Vol. 7,No. 11, (June 1971).
11. Gustafsson.F, "*Adaptive Filtering and Change Detection*", John Wiley & Sons Ltd, (2001).

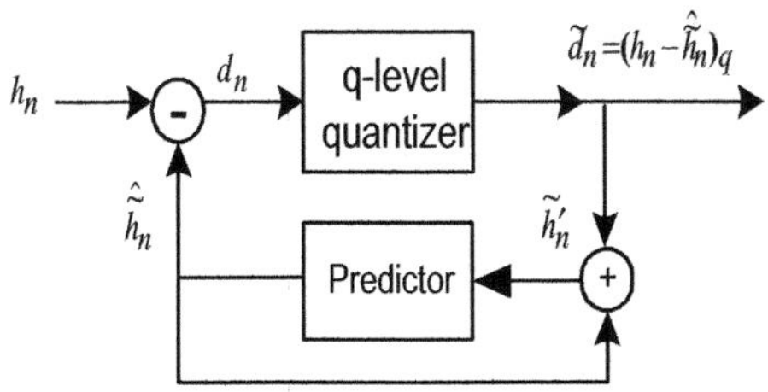

Figure 1 A typical scheme of DPCM transmitter.

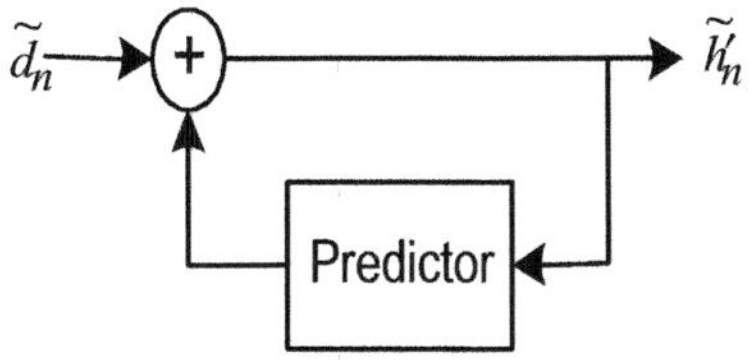

Figure 2 A typical scheme of DPCM receiver.

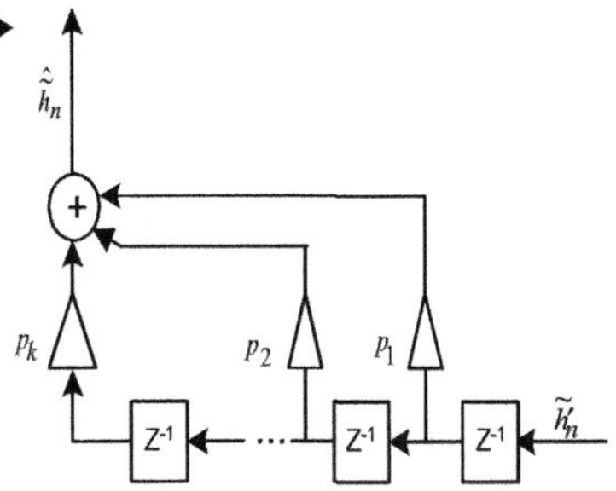

Figure 3 A typical predictor. used in DPCM.

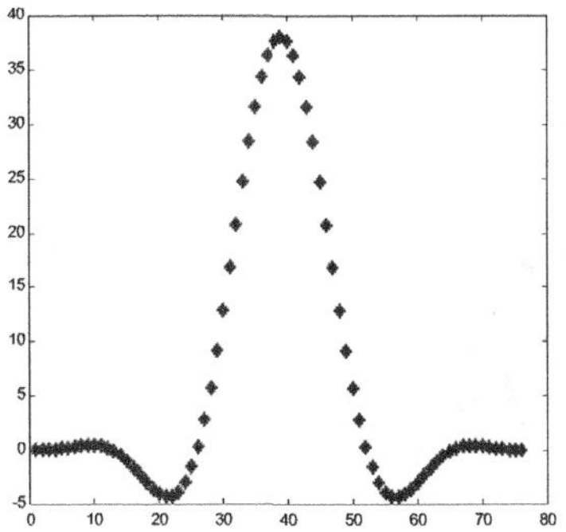

Figure 4 Impulse response of FIR digital filter, LP77.

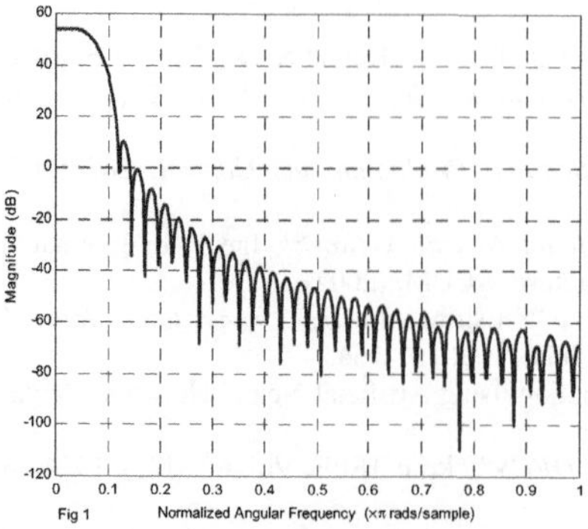

Figure 5 Frequency response of Non-recursive FIR digital filter, LP77.

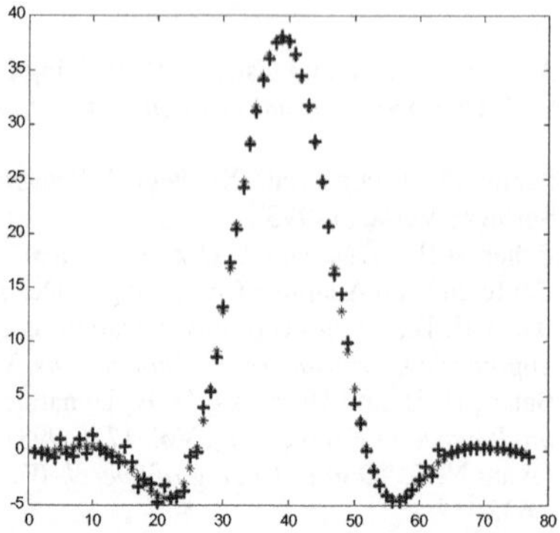

Figure 8 ANN predictor output labeled'+' and input to DPCM labeled '*'.

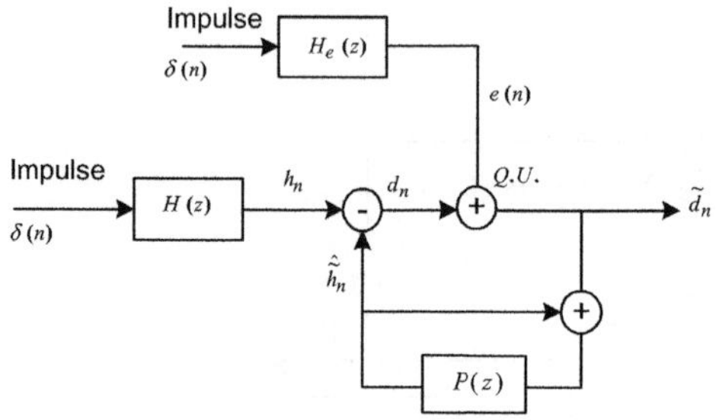

Figure 6. DPCM coefficient quantisation scheme.

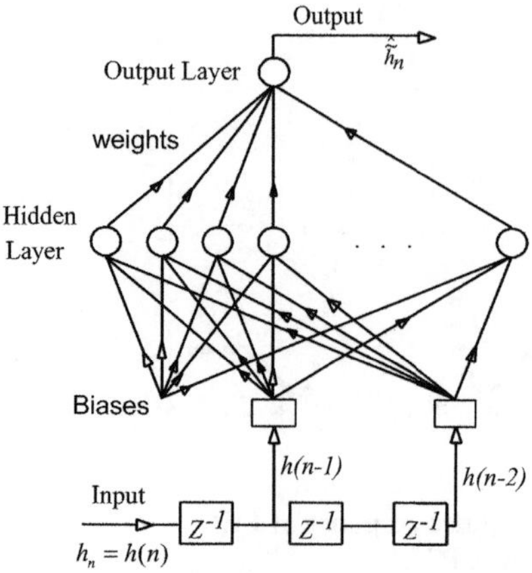

Figure 7 A RBF network to model the predictor at DPCM scheme.

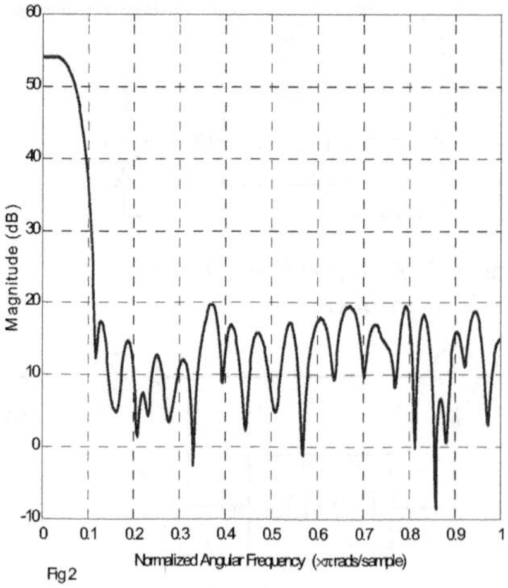

Figure 9 Frequency response of ANN predictor.

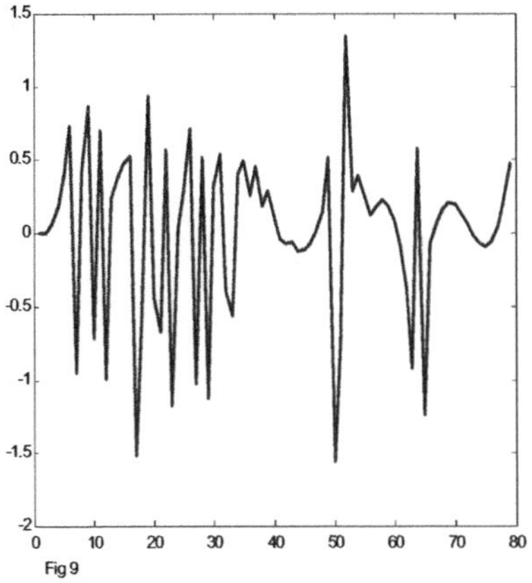

(a) Error input samples.to quatiser unit

0	0	0	1	-1	-1	1	0	-1	1	-1	0
0	1	-1	1	-1	0	1	0	-1	0	0	0
0	0	0	0	0	0	0	0	0	0	0	0
1	-1	-1	1	0	0	0	0	0	0	0	0
0	0	-1	1	-1	0	0	0	0	0	0	0
0	0	0	0	0	0	0					

(b) quantised output samples, -1, 0 and +1.

Figure 10 Quantiser input and output samples, (a) error input samples, (b) quantised output samples, -1, 0 and +1.

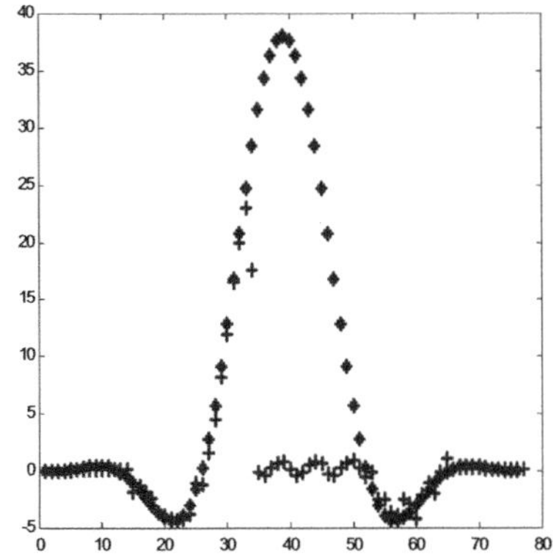

(a)

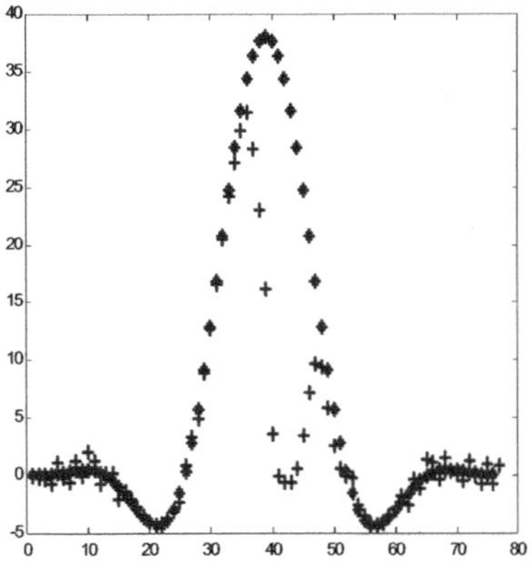

(b)

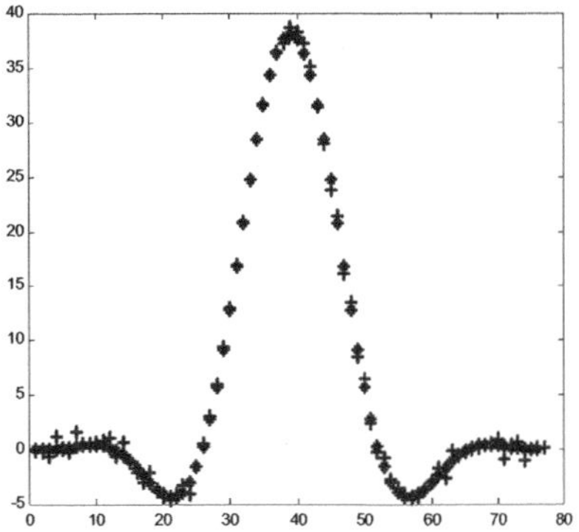

(c)

Figure 11 Recalled samples labeled'+' and training samples labeled '*',G=1000 and eg=.07, (a).sc=.90 and error, rms =1.47, (b) sc=2.012 and error, rms=1.05, (c) sc=2.89 and error, rms=0.029.

Chapter 3

Manufacturing Processes

Job Shop Scheduling using the Clonal Selection Principle

Carlos A. Coello Coello[1], Daniel Cortés Rivera[2] and Nareli Cruz Cortés[3]
CINVESTAV-IPN (Evolutionary Computation Group)
Depto. de Ingeniería Eléctrica, Sección de Computación
Av. IPN No. 2508, Col. San Pedro Zacatenco
México, D. F. 07300, MEXICO
[1]ccoello@cs.cinvestav.mx
[2]dcortes@computacion.cs.cinvestav.mx
[3]nareli@computacion.cs.cinvestav.mx

Abstract

In this paper, we propose an algorithm based on an artificial immune system to solve job shop scheduling problems. The approach uses clonal selection, hypermutations and a mechanism that explores the vicinity of a reference solution. It also uses a decoding strategy based on a search that tries to eliminate gaps in a schedule as to improve the solutions found so far. The proposed approach is compared with respect to three other heuristics using a standard benchmark available in the specialized literature. The results indicate that the proposed approach is very competitive with respect to the others against which it was compared. Our approach not only improves the overall results obtained by the other heuristics, but it also significantly reduces the CPU time required by at least one of them.

Introduction

The purpose of scheduling is to allocate a set of (limited) resources to tasks over time [1]. Scheduling has been a very active research area during several years, both in the operations research and in the computer science literature [2,3] with applications in several disciplines. Research on scheduling basically focuses on finding ways of assigning tasks (or jobs) to machines (i.e., the resources) such that certain criteria are met and certain objective (or objectives) function is optimized.

A wide variety of scheduling problems (e.g., job shop, flowshop, production, etc.) have been tackled with diverse heuristics such as evolutionary algorithms [3,4,5], tabu search [6], and simulated annealing [7], among others. Note, however, that the use of artificial immune systems for the solution of scheduling problems of any type has been scarce (see for example [8,9]).

This paper extends our previous proposal of a new approach based on an artificial immune system (basically on the clonal selection principle) to solve job scheduling problems, which was introduced in [10]. Three are the main changes with respect to our previous proposal are the following: (1) we no longer use a library of antibodies, (2) we introduced two new domain-specific mutation operators, and (3) we use a new backtracking mechanism. As we will see later on, these changes introduce important improvements in our algorithm with respect to the original version. The proposed approach is compared with respect to GRASP (Greedy Randomized Adaptive Search Procedure), a Hybrid Genetic Algorithm (in which local search is used), a Parallel Genetic Algorithm and our previous AIS [10] in several test problems taken from the specialized literature. Our results indicate that the proposed approach is a viable alternative for solving efficiently job shop scheduling problems and it also improves on our previous version reported in [10].

Statement of the Problem

In this paper, we will be dealing with the Job Shop Scheduling Problem (JSSP), in which the general objective is to minimize the time taken to finish the last job available (makespan). In other words, the goal is to find a schedule that has the minimum duration required to complete all the jobs [2]. More formally, we can say that in the JSSP, we have a set of n jobs $\{J_j\}_{1\le j\le n}$, that have to be processed by a set of m machines $\{M_r\}_{1\le r\le m}$. Each job has a sequence that depends on the existing precedence constraints. The processing of a job J_j in a machine M_r is called operation O_{jr}. The operation O_{jr} requires the exclusive use of M_r for an uninterrupted period of time p_{jr} (this is the processing time). A schedule is then a set of duration times for each operation $\{c_{jr}\}_{1\le j\le n, 1\le r\le m}$ that satisfies the previously indicated conditions. The total duration time required to complete all the jobs (makespan) will be called L. The goal is then to minimize L.

Garey and Johnson [11] showed that the JSSP is an **NP-hard** problem and within its class it is indeed one of the least tractable problems [3]. Several enumerative algorithms based on *Branch & Bound* have been applied to JSSP. However, due to the high computational cost of these enumerative algorithms, some approximation approaches have also been developed. The most popular practical algorithm to date is the one based on *priority rules* and *active schedule generation* [12]. However, other algorithms, such as an approach called *shifting bottleneck* (SB) have been found to be very effective in practice [13]. The only other attempt to solve the JSSP using an artificial immune system that we have found in the literature is the proposal presented in [8,9] and our previous version of the algorithm presented here [10] (whose differences with our current proposal have been previously indicated). In [8,9], the authors use an artificial immune system in which an antibody indirectly represents a schedule, and an antigen describes a set of expected arrival dates for each job in the shop. The schedules are considered to be dynamic in the sense that sudden changes in the environment require the generation of new schedules. The proposed approach compared favorably with respect to a genetic algorithm using problems taken from [14]. However, the authors do not provide the problems used nor their results.

Description of our Approach

As indicated in [17], an artificial immune system is an adaptive system, inspired on our immune system (its observed functions, principles and models), and intended to be used as a problem-solving tool. Our approach is based on the clonal selection principle, and can be seen as a variation of an specific artificial immune system called CLONALG, which is has been successfully used for optimization [15]. CLONALG uses two populations: one of antigens and another one of antibodies. When used for optimization, the main idea of CLONALG is to reproduce individuals with a high affinity, then apply mutation (or blind variation) and select the improved maturated progenies produced. Note that "affinity" in this case, is defined in terms of better objective function values rather than in terms of genotypic similarities (as, for example, in pattern recognition tasks), and the number of clones is the same for each antibody. This implies that CLONALG does not really use antigens when solving optimization problems, but, instead, the closeness of each antibody to the global optimum (measured in relative terms with respect to the set of solutions produced so far) defines the rate of hypermutation to be used. It should also be noted that CLONALG does not use a mechanism that allows a change of the reference solution as done with the approach reported in this paper. In order to apply an artificial immune system to the JSSP, it is necessary to use a special representation. In our case, each individual represents the sequence of jobs processed by each of the machines. An antibody is then a string with the job sequence processed by each of the machines (of length $m \times n$). An antigen is represented in the same way as an antibody. The representation adopted in this work is the so-called *permutations with repetitions* proposed in [16] (see an example in Table 1).

Job	**machine(time)**			
1	1(2)	2(2)	3(2)	4(2)
2	4(2)	3(2)	2(2)	1(2)
3	2(2)	1(2)	4(2)	3(2)
4	3(2)	4(2)	1(2)	2(2)
5	1(2)	2(2)	3(4)	4(1)
6	4(3)	2(3)	1(1)	3(1)

Table 1: A problem of size 6 x 4

Input data include the information regarding the machine in which each job must be processed and the duration of this job in each machine. Gantt diagrams are a convenient tool to visualize the solutions obtained for a JSSP. An example of a Gantt diagram representing a solution to the 6×4 problem previously indicated is shown in Step 1 of Figure 1 also requires some further explanation:

- The string at the bottom of Figure 1 corresponds to the solution that we are going to decode.
- **Step 1**: This shows the decoding before reaching the second operation of job 2.
- **Step 2**: This shows the way in which job 2 would be placed if a normal decoding was adopted. Note that job 2 (J_2) is shown to the extreme right of machine 3 (M_3).
- **Step 3**: Our approach performs a local search to try to find gaps in the current schedule. Such gaps should comply with the precedence

constraints imposed by the problem. In this case, the figure shows job 2 placed on one of these gaps for machine 3.

- **Step 4**: In this case, we apply the same local search procedure (i.e., finding available gaps) for the other machines. This step shows the optimum solution for this scheduling problem.

Our approach extends the algorithm (based on clonal selection theory) proposed in [17] using a local search mechanism that consists of placing jobs in each of the machines using the available time slots.

Requiere: Input file (in the format adopted in [18]).
  Input parameters: *#antigens*, *mutation rate*, *random seed* (optional), *degree of freedom*
    p - number of iterations
    i - counter

  Retrieval of problem (read file) and algorithm's parameters.
  Generate (randomly) an *antigen* (i.e., a sequence of jobs) and decode it.
  Generate (randomly) an *antibody*.

  repeat
    Decode the *antibody*.
    if (the (*antibody - degree of freedom*) is better than the *antigen1*) ***then***
      Make the *antigen1* the same as the *antibody*
      if (the *antibody* is better than the *antigen2*) ***then***
        Make the *antigen2* the same as the *antibody*
      end if
    end if
    Generate a *clone* of the *antibody*
    Mutate the *clone* generated
    Select the best *antibody*
  until $i > p$
  Report the best solution found, stored in *antigen2*

Algorithm 1: Our AIS for job shop scheduling

Our approach is described in Algorithm 1. First, we generate the initial population. What we do is to randomly generate an antibody and an antigen (it is important to keep in mind that we use a special representation and that both the antibody and the antigen have the same structure). To generate these two elements (antibody and antigen), we adopt a string of length $m \times n$, which is filled with m values ranging from 0 to $n-1$. Once we fill in the array, we perform a set of random permutations in order to obtain the individual to start the search. At the next stage, the main cycle of the algorithm is executed. Within this cycle, we first decode the antibody (note that the antigen was decoded at a previous step).

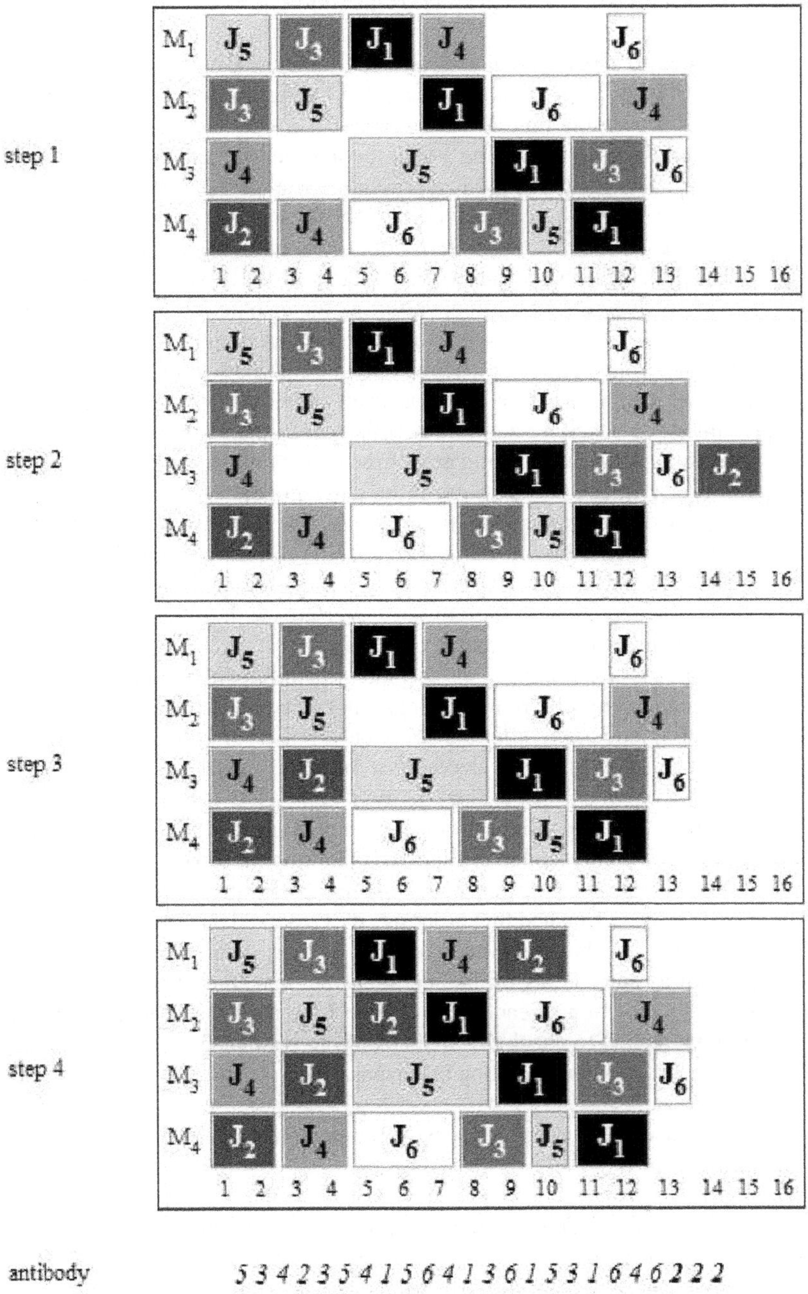

Figure 1: The graphical representation of a solution to the 6×4 problem shown in Table 1 using a Gantt diagram. The string at the bottom of the figure indicates the antibody that we are going to decode. See the text for an explanation of the different steps included.

The process required to decode an individual as to determine its fitness (i.e., the makespan of the corresponding schedule) is the following:

1. We need to have in a matrix all the problem's data (i.e., the processing order of each of the jobs to be handled by the machines available as well as their processing times).
2. We read the string encoding a solution in order to identify each of the jobs contained within (each job is represented by a number between 0 and m).
3. Once we know the corresponding job number, we keep a count of the order of occurrence of each of the numbers as to identify the corresponding job operations (i.e., if this is the first occurrence, then it corresponds to the first job operation). We also determine the machine in which each job is processed using the corresponding input matrix.
4. The following step is to place the operation in the schedule. In order to do this, we use a structure that has been previously initialized and which contains the schedule with the necessary information to accommodate the operations without violating any of the constraints of the problem.
5. In order to place an operation in its corresponding place in the schedule, we provide an example in Figure 1. In this figure we can see that the operation is first placed in its corresponding machine (based on the input matrix). After that, we try to locate a gap in the schedule in which we can place this operation, avoiding to interfere with other operations and avoiding to violate the existing constraints.
6. This process of finding gaps to place operations may cause that several strings encoding different orderings can be decoded to the same solution.
7. The next step is to reorder the string encoding a solution such that the next time that such string is decoded it becomes unnecessary to apply the strategy previously described to find gaps. The ordering performed is based on the order of appearance of each operation and considering each machine from the first to the last. By adopting these criteria, we minimize the amount of possible gaps available in the next iteration.
8. Once we have finished this ordering, we create a data structure that is very important for the mutation operator. Such a data structure consists of generating an ordering of the operations per machine such that it is easy to know the position of each operation and the machine to which it belongs without having to check this in an exhaustive manner.
9. We report the corresponding makespan.

The decoding process is the most expensive (computationally speaking) part of our algorithm.

In the next stage of the algorithm, we compare the antigen with respect to the antibody. Note that we do not adopt a phenotypic similarity metric as the affinity measure. Instead, we use the makespan value as our affinity measure. During this process, we use the best solution found so far as a reference for further search (this is called *antigen1* in Algorithm 1). Each time a better solution is found, it is used as a new reference. In the original version of our algorithm [10], we used a single antigen as a reference. However, we decided to keep a second antigen to allow good (but not the best) solutions to be used as references as well (this is called *antigen2* in Algorithm 1). What we do is to keep a second solution that is one or two units away (in terms of makespan value) from the best solution found so far and we also use it as a reference. This second antigen serves as some sort of backtracking mechanism of the algorithm that allows it to escape from local optima. By using this second antigen, we were able to obtain significant gains in terms of computational time. Once we finish the verification stage in which we check if any of our two antigens (or both) have been improved, the following stage is the cloning of the antibody. This cloning stage consists of copying the antibody a certain number of times without doing any changes to its structure. The number of clones to be produced was varied from 1 to 10 depending on the complexity of the problem tackled. Note however, that if many clones are adopted, the improvement gained is only marginal and the high computational cost increase makes this option unattractive. Thus, we adopted values of either 1 or close to 1 for the number of clones to be produced. Once the clones are available, each of them is mutated in such a way that they suffer a slight variation in in their structure. The algorithm has two types of mutation operators available, and the one to be used is selected with a 50% probability. The similarities and differences between these two mutation operators (which we will call Mutation-A and Mutation-B) are the following:

- In both cases, the mutation operator is applied by using $flip(pm)$ at each string location.[1]
- In both cases, the mutation rate is a function of the antibodies length and it is defined such that 1 mutation takes place for each string (i.e., antibody).
- In both cases, the operator locates an operation, then finds another operation of another job and then swaps the positions of the 2 operations.
- In order to have a quick indexing of the positions of each operation, we use the data structure previously created for the current antibody.
- The only difference between the two mutation operators is that in the case of Mutation-A, we find the first operation and then locate the other operation with which it will swap places. However, if there are other operations of the same job before the current operation, we traverse them all. As a consequence, we not only change those operations, but we also produce more changes to the schedule.
- In the case of Mutation-B, we only locate two operations and swap their locations without any further exploration.

These are all the processes performed by our algorithm. Once it reaches its convergence criterion (a maximum number of iterations), the algorithm reports the best solution found during the process, which is stored in one of the two reference antigens (in *antigen2*). The algorithm then reports the full schedule with all the detailed information regarding the ordering of the machines and the initial and termination times for each of the available operations.

---

[1] The function $flip(pm)$ returns TRUE with a probability pm.

Comparison of Results

We compare our Artificial Immune System (AIS) with respect to 3 different approaches: a Hybrid Genetic Algorithm (HGA) reported in [19], a GRASP approach [20], and a Parallel Genetic Algorithm (PGA) [21]. We chose these references for two main reasons: (1) they provide enough information (e.g., numerical results) as to allow a comparison; (2) these algorithms have been found to be very powerful in the job shop scheduling problem studied in this paper. Note that the test problems adopted were taken from the OR-Library [18]. Additionally, we also compared results with respect to our previous AIS [10]. All our tests were performed on a PC with an Intel Pentium 4 running at 2.6 GHz with 512 MB of RAM and using Red Hat Linux 9.0. Our approach was implemented in C++ and was compiled using the GNU g++ compiler.

	deviation	**Deviation AIS**	**Improvement**
HGA	0.42%	0.18%	0.23%
GRASP	0.47%	0.18%	0.28%
PGA	0.93%	0.18%	0.74%

Table 2: Comparison of results between our Artificial Immune System (AIS) and three other algorithms: Greedy Randomized Adaptive Search Procedure (GRASP) [20], the Hybrid Genetic Algorithm (HGA) [19], and the Parallel Genetic Algorithm (GA) [21].

Table 2 shows the overall comparison of results. In the first column, we show the algorithm with respect to which we are comparing our results. In the second column, we show the average deviation of the best results obtained by each algorithm with respect to the best known solution for the 43 test problems adopted in our study. In the third column, we show the average deviation of our AIS with respect to the best known solution for the 43 test problems adopted in our study. The last column indicates the improvement achieved by our AIS with respect to each of the other algorithms compared. From Table 2, we can see that our approach was able to improve on the overall results produced by the 3 other techniques. The most remarkable improvement produced was with respect to the PGA [21].

	AIS		
	Win	**Tie**	**Lose**
HGA	3	32	8
GRASP	3	30	10
PGA	0	23	17

Table 3: Overall performance of our AIS with respect to the 3 other algorithms against which it was compared. The column labeled **Win** shows the number of problems in which each algorithm beat our AIS. The column labeled **Tie** indicates ties between our AIS and the other algorithms. Finally, the column labeled **Lose** indicates the number of problems in which each algorithm lost with respect to our AIS.

In Table 3, we show the overall performance of our AIS with respect to the 3 other algorithms against which it was compared. Results indicate that the HGA beat our

AIS in 3 problems and it lost in 8. In the remainder (32 problems), they tied. GRASP beat our AIS in 3 problems and lost in 10. The worst contender was the PGA, which was not able to beat our AIS in any problem and lost in 17 problems.

Table 4 summarizes the results obtained by each of the 4 approaches compared in the 43 test problems taken from the OR-Library [18]. We use **boldface** to indicate both the best known results and when an algorithm reached such result. Note that the number of evaluations performed is only reported for our two AIS and for GRASP. The reason is that we only found such information available for GRASP. We can clearly see that our AIS obtained competitive results with respect to the other approaches compared. Furthermore, the number of evaluations performed by our AIS was significant lower than those performed by GRASP.[2] Some remarkable examples are the following:

- **FT10**: In this problem, GRASP found the best known solution, but it required 2.5 million evaluations. Our AIS found a solution which is only 1% away from the best known solution and it only required 250,000 evaluations. Note that in our original AIS produced a poorer solution than the new version, and it required 20 million evaluations.
- **LA28**: In this problem, the best solution found by GRASP is slightly worse than the best solution found by our AIS (1225 vs. 1216). However, our AIS required 1 million evaluations whereas GRASP required 20 million iterations. Note that the original version of our AIS found a worse solution using 5 million evaluations.
- **LA16**: Both GRASP and our AIS reached the best known solution. However, GRASP required 1.3 million evaluations and our approach required only 10,000 evaluations. Note that the original version of our AIS required 2 million evaluations to reach this solution.

As can be seen in Table 4, the new version of the algorithm presents a significant improvement with respect to the original version reported in [10], both in terms of the quality of the solutions obtained as in terms of the computational efforts required to obtain them.

Instance	Size (m x n)	BKS	HGA	AIS	Evals AIS	oAIS	Evals oAIS	GRASP	Iters. GRASP	PGA
FT06	6 x 6	**55**	**55**	**55**	0.0001	**55**	0.001	**55**	0.00001	**55**
FT10	10 x 10	**930**	**930**	936	0.25	941	20	**930**	2.5	936
FT20	20 x 5	**1165**	**1165**	**1165**	0.5	-	-	**1165**	4.5	1177
LA01	10 x 5	**666**	**666**	**666**	0.001	**666**	0.01	**666**	0.0001	**666**
LA02	10 x 5	**655**	**655**	**655**	0.01	**655**	0.01	**655**	0.004	666

[2] In fact, what we report as the number of evaluations for GRASP is actually the number of iterations performed by the algorithm. Since at each iteration, GRASP performs several evaluations of the objective function, the real number of evaluations is much higher than those reported in Table 4.

LA03	10 x 5	**597**	**597**	**597**	0.01	**597**	10	**597**	0.01	**597**
LA04	10 x 5	**590**	**590**	**590**	0.001	**590**	0.01	**590**	0.001	**590**
LA05	10 x 5	**593**	**593**	**593**	0.001	**593**	0.01	**593**	0.0001	**593**
LA06	15 x 5	**926**	**926**	**926**	0.001	**926**	0.01	**926**	0.0001	**926**
LA07	15 x 5	**890**	**890**	**890**	0.001	**890**	0.01	**890**	0.0001	**890**
LA08	15 x 5	**863**	**863**	**863**	0.001	**863**	0.01	**863**	0.0003	**863**
LA09	15 x 5	**951**	**951**	**951**	0.001	**951**	0.01	**951**	0.0001	**951**
LA10	15 x 5	**958**	**958**	**958**	0.001	**958**	2	**958**	0.0001	**958**
LA11	20 x 5	**1222**	**1222**	**1222**	0.001	-	-	**1222**	0.0001	**1222**
LA12	20 x 5	**1039**	**1039**	**1039**	0.001	-	-	**1039**	0.0001	**1039**
LA13	20 x 5	**1150**	**1150**	**1150**	0.001	-	-	**1150**	0.0001	**1150**
LA14	20 x 5	**1292**	**1292**	**1292**	0.001	-	-	**1292**	0.0001	**1292**
LA15	20 x 5	**1207**	**1207**	**1207**	0.001	-	-	**1207**	0.0002	**1207**
LA16	10 x 10	**945**	**945**	**945**	0.01	**945**	2	**945**	1.3	977
LA17	10 x 10	**784**	**784**	**784**	0.01	785	2	**784**	0.02	787
LA18	10 x 10	**848**	**848**	**848**	0.01	**848**	2	**848**	0.05	**848**
LA19	10 x 10	**842**	**842**	**842**	0.01	848	10	**842**	0.02	857
LA20	10 x 10	**902**	907	907	0.25	907	5	**902**	17	910
LA21	15 x 10	**1046**	**1046**	**1046**	0.25	-	-	1057	100	1047
LA22	15 x 10	**927**	935	**927**	0.25	-	-	**927**	26	936
LA23	15 x 10	**1032**	**1032**	**1032**	0.25	-	-	**1032**	0.01	**1032**
LA24	15 x 10	**935**	953	**935**	0.25	-	-	954	125	955
LA25	15 x 10	**977**	986	979	0.25	1022	5	984	32	1004
LA26	20 x 10	**1218**	**1218**	**1218**	0.2	-	-	**1218**	3.5	**1218**
LA27	20 x 10	**1235**	1256	1240	0.5	-	-	1269	10.5	1260
LA28	20 x 10	**1216**	1232	**1216**	1	1277	5	1225	20	1241
LA29	20 x 10	**1157**	1196	1170	5	1248	6.4	1203	50	1190
LA30	20 x 10	**1355**	**1355**	**1355**	0.1	-	-	**1355**	3	1356
LA31	30 x 10	**1784**	**1784**	**1784**	0.005	-	-	**1784**	0.01	**1784**
LA32	30 x 10	**1850**	**1850**	**1850**	0.025	-	-	**1850**	0.0001	**1850**
LA33	30 x 10	**1719**	**1719**	**1719**	0.025	-	-	**1719**	0.001	**1719**
LA34	30 x 10	**1721**	**1721**	**1721**	0.01	-	-	**1721**	0.05	1730
LA35	30 x 10	**1888**	**1888**	**1888**	0.05	1903	5	**1888**	0.01	**1888**
LA36	15 x 15	**1268**	1279	1281	0.5	1323	6.4	1287	51	1305
LA37	15 x 15	**1397**	1408	1408	0.5	-	-	1410	20	1441
LA38	15 x 15	**1196**	1219	1204	0.5	1274	6.4	1218	20	1248
LA39	15 x 15	**1233**	1246	1249	0.5	1270	6.4	1248	6	1264
LA40	15 x 15	**1222**	1241	1228	2.5	1258	6.4	1244	2	1252

Table 4: Comparison of results between our artificial immune system (AIS), GRASP (Greedy Randomized Adaptive Search Procedure) [20], HGA (Hybrid Genetic Algorithm) [19], and PGA (Parallel Genetic Algorithm) [21]. oAIS refers to our original AIS, reported in [10] and is included only to have a rough idea of the improvements achieved with the new

version. The number of evaluations reported is in millions. Only the number of evaluations of GRASP and our two AIS versions are reported because this value was not available for the other approaches. We show in **boldface** both the best known solution and the cases in which an algorithm reached such value.

Conclusions and Future Work

We have introduced a new approach based on an artificial immune system to solve job shop scheduling problems. The approach uses concepts from clonal selection theory (extending ideas from CLONALG [15]), and adopts a permutation representation that allows repetitions. The comparison of results indicated that the proposed approach is highly competitive with respect to other heuristics, even improving on their results in some cases. It also improves in the previous version of the algorithm reported in [10]. In terms of computational efficiency, our approach performs a number of evaluations that is considerably lower than those performed by GRASP [21] while producing similar results.

As part of our future work, we intend to add a mechanism that avoids the generation of duplicates (something that we do not have in the current version of our algorithm). It is also desirable to find a set of parameters that can be fixed for a larger family of problems as to eliminate the empirical fine-tuning that we currently perform. Finally, we also plan to work on a multiobjective version of job shop scheduling in which 3 objectives would be considered [3]: 1) makespan, 2) mean flowtime and 3) mean tardiness. This would allow us to generate trade-offs that the user could evaluate in order to decide what solution to choose.

Acknowledgments

The first author acknowledges support from CONACyT project No. 34201-A. The second and third authors acknowledge support from CONACyT through a scholarship to pursue graduate studies in Computer Science at the Sección de Computación of the Electrical Engineering Department at CINVESTAV-IPN.

References

1. M. Pinedo. *Scheduling---Theory, Algorithms, and Systems.* Prentice Hall, Englewood Cliffs, 1995.
2. Kenneth R. Baker. *Introduction to Sequencing and Scheduling.* John Wiley & Sons, New York, 1974.
3. Tapan P. Bagchi. *Multiobjective Scheduling by Genetic Algorithms.* Kluwer Academic Publishers, New York, September 1999.
4. R. Cheng, M. Gen, and Y. Tsujimura. A tutorial survey of job-shop scheduling problems using genetic algorithms: I. Representation. *Computers and Industrial Engineering*, 30:983--997, 1996.
5. R. Cheng, M. Gen, and Y. Tsujimura. A tutorial survey of job-shop scheduling problems using genetic algorithms: II. Hybrid genetic search strategies. *Computers and Industrial Engineering*, 36(2):343--364, 1999.

6. J.W. Barnes and J.B. Chambers. Solving the Job Shop Scheduling Problem using Taboo Search. *IIE Transactions*, 27(2):257--263, 1995.
7. Olivier Catoni. Solving Scheduling Problems by Simulated Annealing. *SIAM Journal on Control and Optimization*, 36(5):1539--1575, September 1998.
8. Emma Hart, Peter Ross, and J. Nelson. Producing robust schedules via an artificial immune system. In *Proceedings of ICEC'98*, pp. 464--469, Anchorage, Alaska, 1998. IEEE Press.
9. Emma Hart and Peter Ross. The Evolution and Analysis of a Potential Antibody Library for Use in Job-Shop Scheduling. In David Corne et al., eds, *New Ideas in Optimization*, pp. 185—202, McGraw-Hill, 1999.
10. Carlos A. Coello Coello, Daniel Cortés Rivera, and Nareli Cruz Cortés. Use of an Artificial Immune System for Job Shop Scheduling. In Jon Timmis et al., eds, *Proceedings of ICARIS'2003*, pp. 1--10, September 2003. Springer-Verlag. Lecture Notes in Computer Science Vol. 2787.
11. David S. Johnson Michael R. Garey. *Computers and Intractability: A Guide to the Theory of NP-Completeness.* W. H. Freeman & Co., 1979.
12. Albert Jones and Luis C. Rabelo. *Survey of Job Shop Scheduling Techniques*. National Institute of Standards and Technology, 1998.
13. J. Adams E. Balas and D. Zawack. The shifting bottleneck procedure for job shop scheduling. *Management Science*, 34(3):391-401, 1988.
14. Thomas E. Morton and David W. Pentico. *Heuristic Scheduling Systems: With Applications to Production Systems and Project Management.* John Wiley & Sons, 1993.
15. Leandro Nunes de Castro and Fernando José Von Zuben. Learning and Optimization Using the Clonal Selection Principle. *IEEE Transactions on Evolutionary Computation*, 6(3):239--251, 2002.
16. Takeshi Yamada and Ryohei Nakano. Job-shop scheduling. In A.M.S. Zalzala and P.J. Fleming, editors, *Genetic Algorithms in Engineering Systems*, pp. 134--160. The Institution of Electrical Engineers, 1997.
17. Leandro Nunes de Castro and Jonathan Timmis. *Artificial Immune System: A New Computational Intelligence Approach.* Springer Verlag, Great Britain, September 2002. ISBN 1-8523-594-7.
18. J. E. Beasley. OR-Library: Distributing Test Problems by Electronic Mail. *Journal of the Operations Research Society*, 41(11):1069--1072, 1990.
19. José Fernando Goncalves, Jorge José Mendes, and Mauricio G.C. Resende. *A Hybrid Genetic Algorithm for the Job Shop Scheduling Problem.* Technical Report TD-5EAL6J, AT&T Labs Research, 180 Park Avenue, Florham Park, NJ 07932 USA, September 2002.
20. Renata M. Aiex, S. Binato, and Mauricio G.C. Resende. Parallel GRASP with path-relinking for job shop scheduling. *Parallel Computing*, 29(4):393--430, 2003.
21. José Fernando Goncalves and N.C.Beirao. Um algoritmo genético baseado em chaves aleatórias para sequenciamiento de operacoes. *Revista Associacão Portuguesa de Desenvolvimento e Investigacão Operacional*, 19:123 -- 137, 1999. (in Portuguese).

Genetic Optimisation of Machine Tool Paths

M.K.A. Mohd Ariffin [1], N.D.Sims [2], K.Worden [3]

Department of Mechanical Engineering, Faculty of Engineering,
University of Sheffield S1 3JD, United Kingdom.
[1] mep02mm@shef.ac.uk;
[2] n.sims@shef.ac.uk;
[3] k.worden@shef.ac.uk

Abstract

Many aircraft components are designed from monolithic structures to reduce manufacturing cost, reduce weight, and increase stiffness. To enable machining of such structures it is necessary to perform high-speed machining that can remove a large amount of material within a shorter time. However, the performance of high speed-machining operations is limited by the static and dynamic stiffness of the tool and part, which can cause problems such as regenerative chatter and push-off.

The tool path plays a key role in avoiding these problems as it helps to minimise the workpiece vibration during machining. This work aims to optimise the tool path by simulating the removal of material in a finite element environment, which is controlled by a Genetic Algorithm. To simulate the physical removal of material during machining, a finite element model is designed to represent a thin walled workpiece. The target was to reduce the deflection after each element was removed, according to the sequence suggested by the Genetic Algorithm.

As a first step, a cantilever beam is created, meshed and numbered. There are 6 elements to remove, representing the tool path sequence for a physical machining process. A Genetic Algorithm was used to find the element removal sequence that gave the greatest workpiece stiffness in the cutting region. It is concluded that tool path optimisation can be performed successfully by the proposed technique.

1 Introduction

The design change of aircraft structures from assembly techniques towards monolithic structures has required high-speed machining processes capable of producing thin walled structures with substantial material removal rates. As a result manufacturing cost, machining time and manufacturing lead times have been reduced. However, workpiece dynamics is a major limiting factor that prevents full utilisation of the capabilities of high-speed machining. This is due to issues such as the static deflection of the workpiece during machining, and unstable self-excited vibrations known as chatter [1]. Chatter occurs when the cutting forces are modulated, changing the uncut chip thickness, which in turn results in greater variation of uncut chip thickness for the next tooth [2] .

Several approaches have been employed to tackle this machining problem. One of the most popular methods is to vary the cutting process by choosing the appropriate cutting conditions (e.g. spindle speed and depth of cut) so as to avoid chatter [3]. An alternative approach is to optimise the way to remove the excess material during machining, i.e the tool path [4, 5].

Nowadays, the tool path sequence is invariably designed using CAD/CAM systems capable of automatically generating the CNC programme. Typically, the software will work from existing 3D models of the workpiece and raw material, and the decision over the sequence of regions of material to remove lies with the software user. This is generally a subjective process, which relies on prior experience and considers factors such as the length of tool path, workpiece fixturing, and workpiece static and dynamic behaviour.

As an example, consider the machining problem depicted in Figure 1. Here, a total of 12 pockets must be removed from each side of the workpiece, and the programmer must choose the sequence of these pocketing operations. As material is removed the workpiece becomes more flexible and is more susceptible to chatter. Within each pocket, the tool path can usually be programmed automatically by the CAD/CAM software. The pocketing sequence is currently an entirely subjective process that relies on operator experience. The tool path for each pocket can be based on a number of factors [6], but current technology does not consider the workpiece static and dynamic stiffness [4].

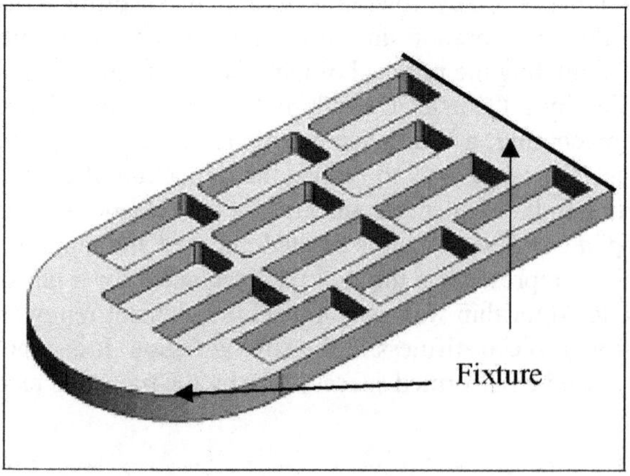

Figure 1: Typical machining scenario

The aim of this work is to develop an alternative approach to these decision-making processes with particular emphasis on the role of the workpiece stiffness. The method requires an understanding of chatter, and optimisation tools based upon genetic algorithms. For this work a simplified problem, based upon the machining of a cantilever beam, is considered. The paper begins with the theory of workpiece dynamics and genetic algorithms. A methodology is then described where a Genetic Algorithm (GA) controls the removal of elements from a Finite Element (FE) model of a workpiece, so as to find the optimum sequence of removal of material. Then a numerical study is presented with corresponding results. Finally some conclusions are drawn and suggestions made for future work.

2 Theory

2.1 Workpiece Vibration and Stiffness

Chatter is generated by two types of vibration, either forced vibration or self-excited vibration. However, self-excited vibration is the most common factor that creates surface waviness on the machine surface [1]. When a milling cutter vibrates during cutting, it creates a 'wavy' surface. When the next tooth engages in the workpiece, it will experience a variable chip thickness as a result as shown in Figure 2 [1]. Referring to Figure 2, the relative vibration between the tool and the workpiece results in surface waviness (A). In milling the next tooth pass experiences an instantaneous chip thickness, h, is given by:

$$h = h_m + Y_o - Y \qquad (1)$$

Where h_m is the mean chip thickness, Y_o is the relative deflection between the tool and the workpiece for the old surface, and Y is the equivalent relative deflection for the new surface. Assuming that the tool is rigid, the cutting force, F is related to the displacement, Y, by:

$$Y = FG(\omega) \qquad (2)$$

Where $G(\omega)$ is a workpiece frequency response function. The system can therefore be represented as the block diagram shown in Figure 3, where b is the depth of cut and K_s is the cutting stiffness that relates the cutting area (bh) to the cutting force F, in the direction normal to the cut surface.

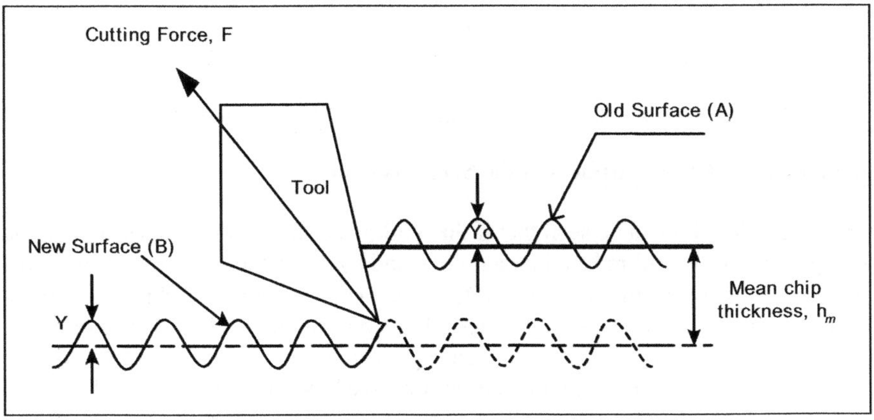

Figure 2: Schematic diagram of the cutting process

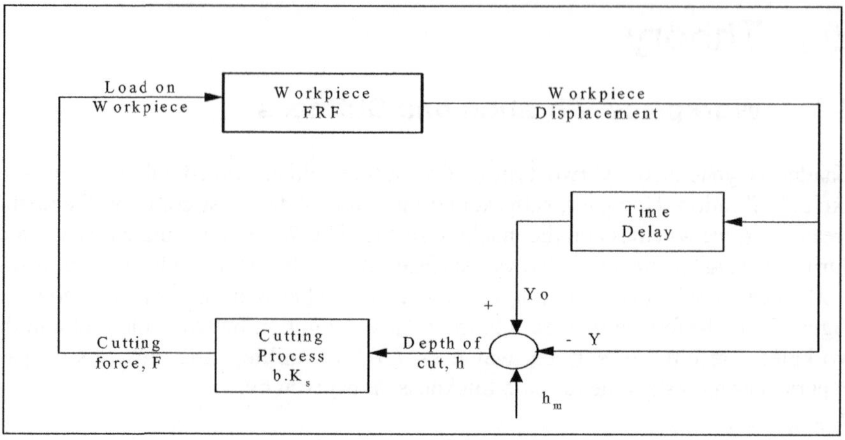

Figure 3: Simplified Block diagram

The stability of the process can be determined by the Nyquist stability criterion as [1]:

$$b_{\lim} = \frac{-1}{(2K_s \operatorname{Re}[G(\omega)])} \qquad 3$$

Where $b_{\lim}$ is the limiting depth of cut. Only the negative real part of the frequency response $G(\omega)$ will cause chatter.

Considering a single degree-of-freedom (SDOF) mass-spring-damper system as an example, the real part of the frequency response function is given by [7]:

$$\operatorname{Re}[G(\omega)] = \frac{1}{k}\left[\frac{1-r^2}{(1-r^2)^2+(2r\xi)^2}\right] \qquad 4$$

Here, ξ is the damping ratio, and r is the ratio of ω/ω_n, where ω_n is the natural frequency and *k* is the stiffness of the SDOF system.

From Equation 4 it can be seen that if the workpiece stiffness is increased, then the real part of G(ω) will reduce in magnitude and so the stable depth of cut, b_{lim}, will increase. This means that it is desirable to maintain maximum workpiece stiffness at the cutting location. This is also desirable from the point of view of workpiece 'push-off' due to the static forces during machining [8]. The stiffness of the workpiece at the cutting location can be improved by judicious choice of the tool path sequence.

2.2 Genetic Algorithm

A genetic algorithm is an optimisation search method based on a random selection that uses stochastic global search to find a near optimum solution. The initialisation of the GA begins when the initial population is generated using a random selection of the initial parents, or from a potential solution for the problems. The GA uses the survival of fittest principle as a means of producing hopefully better and better approximations to a solution. Goldberg [9] defines Genetic algorithm as search algorithm based on the mechanics of natural selection and natural genetics, and they are widely used for many applications. In this project the GA is incorporated as a type of travelling salesman problem to find the tool path or route that gives the lowest possible workpiece deflection during machining.

There are 4 important stages that should be included for the simple genetic algorithm: evaluation, selection, crossover and mutation [10]. Evaluation is a process of finding the fitness values of the population according to the specific objective function. Selection is a genetic operator that is used to select the fittest string inside the population. Crossover occurs when the two parents cross their genotype with each other and finally the mutation operation makes random changes of the genotype within the same parent. In order to use the Travelling Salesman Problem this study has used the ordinal method [11] to represent the permutation sequence of the material to remove.

3 Method

Consider the machining scenario depicted in Figure 4a and 4b, for a machining process and a finished product respectively. The aim is to find the sequence of removal of material that produces the lowest workpiece deflection at the location of the cutter. This deflection is evaluated numerically by a finite element solver, where the FE model represents the current workpiece geometry and a static load is applied at the cutter location. As additional material is removed, so the corresponding elements are removed from the FE model, and the deflection re-evaluated.

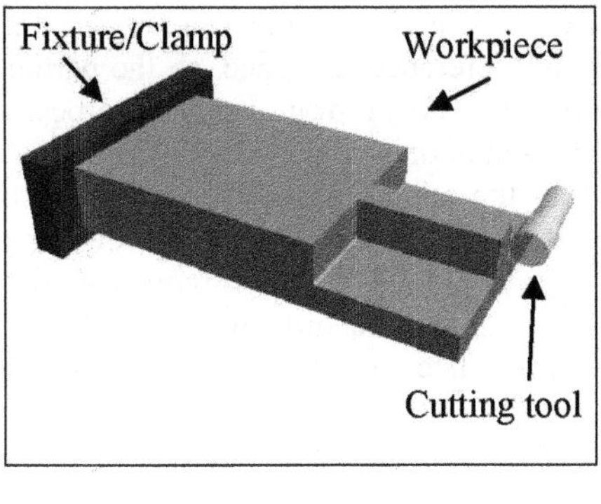

Figure 4a: Machining Process

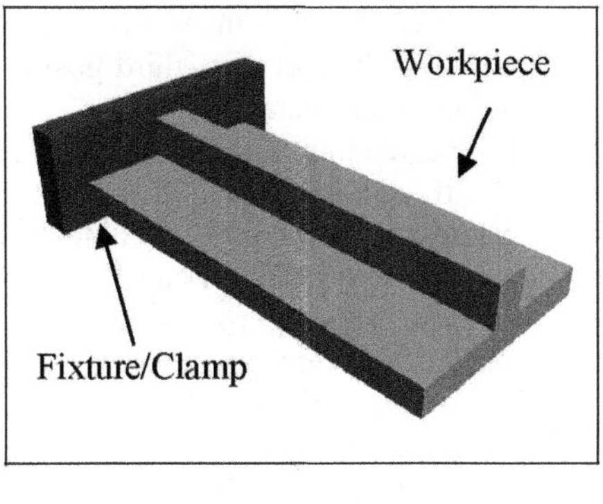

Figure 4b: Finished Product

The simulation undergoes several processes: initialisation, analysis, genetic process and finally the result. With reference to Figure 5, the processes are as follows:

1. Initialisation is the process of setting up the initial value for the genetic parameters. The process selects a random population and creates the initial FE input file with respect to the number of nodes in the design. This is necessary to specify which element to remove and at which node to give the load.
2. Before the Genetic algorithm commences, the FE program evaluates the fitness function. In this operation the TSP permutation value gives the fitness value according to the deflection analysis. The lowest deflection is where the highest fitness value will be, and so the calculation for the fitness function is 1/deflection.
3. Genetic operator process. This operation integrates the genetic operators of selection, crossover and mutation to select a population that contains the chosen chromosomes from the previous population. In this operation the permutation has to change to ordinal representation [11] before the crossover takes place. In the ordinal representation, there is a path order that acts as a Reference Sequence for each sequence in ordinal representation. Consider the following example, for a Reference Sequence: A = (1 2 3 4 5 6):

1	2	3	4	5	6	Sequence 1
1	1	1	1	1	1	Ordinal representation
4	5	6	3	1	2	Sequence 2
4	4	4	3	1	1	Ordinal representation

   With reference to the illustration of Sequence 2, the ordinal representation is determined as follows: The first number in Sequence 2 is 4, which corresponds to fourth place in A, and it is represented as number 4 in ordinal representation. A partial ordinal representation is 4. Having matched number 4, it is removed from the Reference Sequence A. The next number in Sequence 2 is 5, and it is in fourth place in A, so the next number of the ordinal representation is 4. The partial ordinal representation becomes '44'. Number 5 is removed from A. The next number in Sequence 2 is 6, corresponding to fourth place in A, so the partial ordinal representation becomes '444'. Number 6 is removed from A. The next number in Sequence 2 is 3, which represents third position in the reference sequence, so the partial ordinal representation is '4443'. Number 3 is removed from A and the process is repeated until the complete ordinal representation is obtained: '444311'.

   If sequence 1 and 2 cross together at the middle point, then sequence 1 would have number 3 two times, which is physically impossible. However the ordinal representation can be crossed to give '111311', which represents the physical removal of elements of 123645. Using the ordinal representation ensures that each number can only appear once in the sequence, after the crossover operation.

   In order to stop the genetic algorithm a termination criteria needs to be included, such as how many generations the genetic algorithm should run for, or if the objective function is achieved.
4. Store the result.

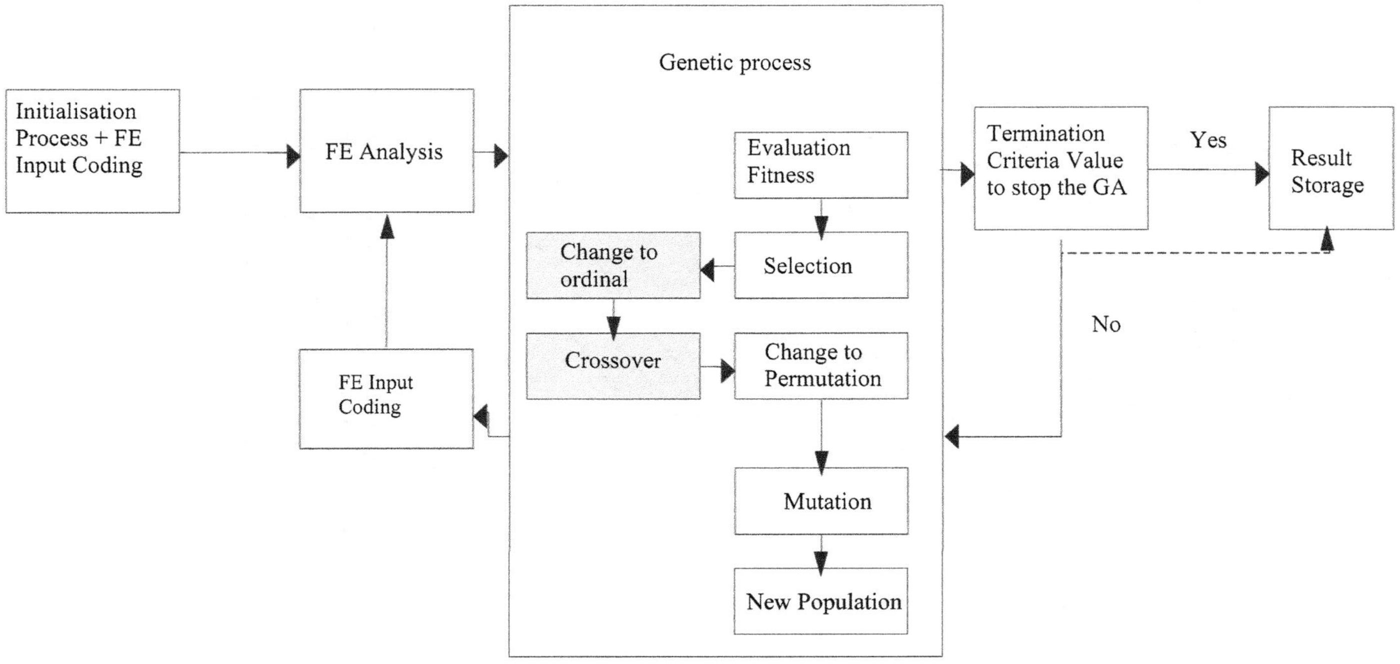

Figure 5: Process flow chart

4 Numerical Analysis

To demonstrate the process the cutting problem has been simplified to produce the FE model shown in Figure 6. There are six elements to remove in total. It is obvious that the lowest deflection is the sequence 1,4,2,5,3 and 6 that in theory the process of removing these elements should start from the element situated far from the boundary condition and finish with the elements nearest to the boundary condition. The aim is to assess the performance of the proposed GA/FE approach in terms of finding this optimal solution.

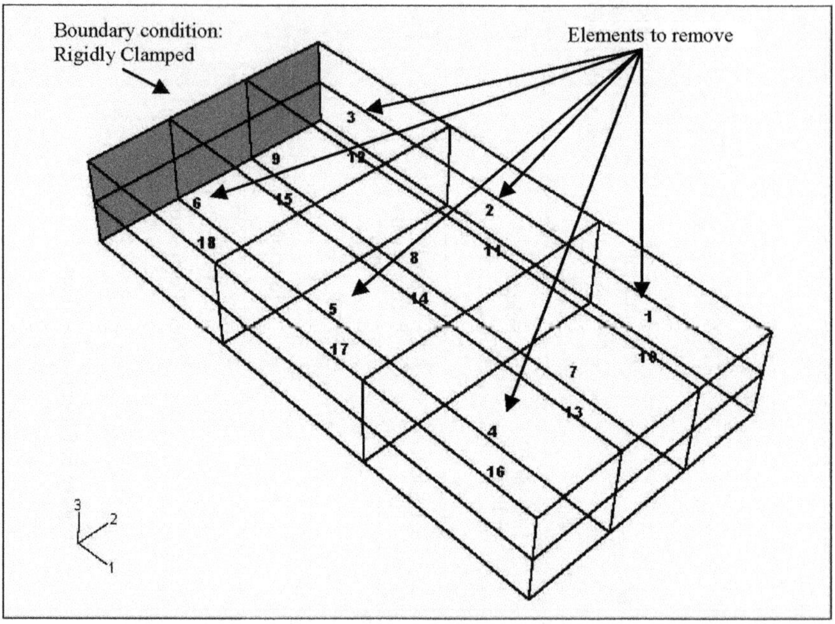

Figure 6: Model of the Beam fix point at grey area.

The effectiveness of the genetic algorithm depends on parameter settings such as the probability of crossover, probability of mutation and population size. In general, there is no agreement between the researchers what value is the best for those parameters [12]. To assess the performance of the GA under difference conditions, the optimisation process was repeated using the parameters shown in Table 1.

Table1: Typical parameters setting for GA's operations [12]

Parameters	Value
Population Size	20 – 30
Probability crossover	0.7 – 0.9
Probability mutation	0.005 – 0.01

5 Result

For each set of parameters, the simulation was run five times with different random initial populations. The results were plotted for the number of the generations against the fitness values. It should be noted that the highest fitness value is 1499. The simulation was performed with a range of parameter settings as shown in Table 1. Typical results for a probability of crossover of 0.07 and probability of mutation of 0.01 with population size number of 10 are shown in Figure 7. In the first generation where the population was randomly created the maximum fitness value was 1376 corresponding to a sequence of 241563. With reference to Figure 6, the element removal sequence starts at the middle, which increases the workpiece flexibility for the material removal at elements 1 and 4. This is clearly undesirable. After 6 generations or 60 sequences the highest fitness value was 1499. The sequence that gave the highest fitness value was 145236, which is obviously optimal because it starts from the less supported area and works toward to the uncut portion of the workpiece. In an exhaustive search there are 720 sequences and any combination of the numbers ‘14’, ‘25’, and ‘36’, will give optimal solutions as for all 8 of these cases element removal starts from the less supported area and works toward the supported area.

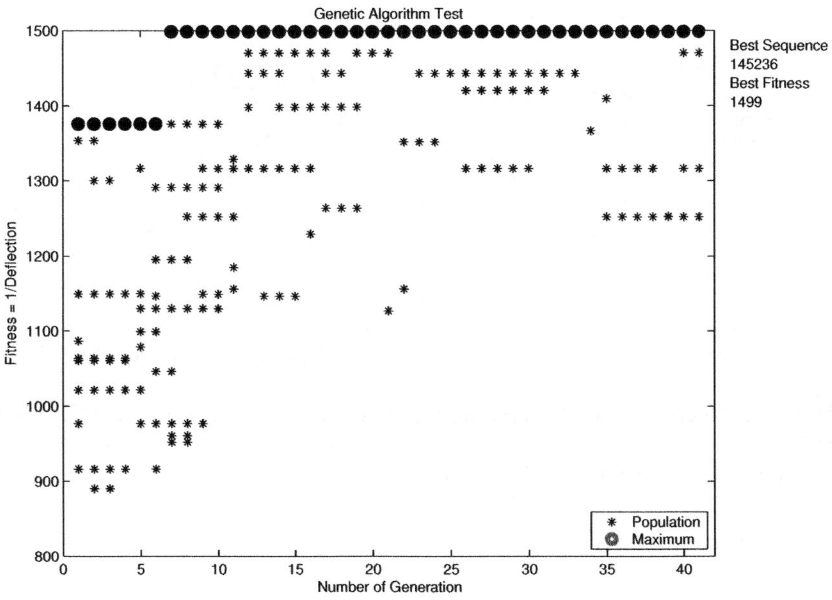

Figure 7: Typical result of genetic optimisation

5.1 Discussion

The size of the initial population determines the success of the genetic algorithm, as a larger initial population will initiate more possible solutions and thus the search is said to be in a global search area. If the initial population size is large however then this tends towards an exhaustive search, where all possible solutions are considered resulting in a significant loss of time. The maximum generation number is a time constraint, where after this value the time taken to find a solution is considered excessive. There is possibly a more convenient solution such as changing the probability of crossover and mutation that is suitable with the specific problems. In this study, a number of tests were performed using different parameters (as shown in Table 1), and the results agreed with published findings concerning the effect of the GA parameters [12].

Many tool path strategies currently focus on optimising the distance travelled by the cutting tool, in an effort to reduce machining time. In future work, it is envisaged that this aspect could be included in the GA/FE concept. For example, with reference to figure 8a and 8b, if the tool travelling time is taken into account there are only two optimum sequences, which are '145236' and '412563'. This could be included in the GA algorithm by introducing a neighbourhood constraint into this simulation process.

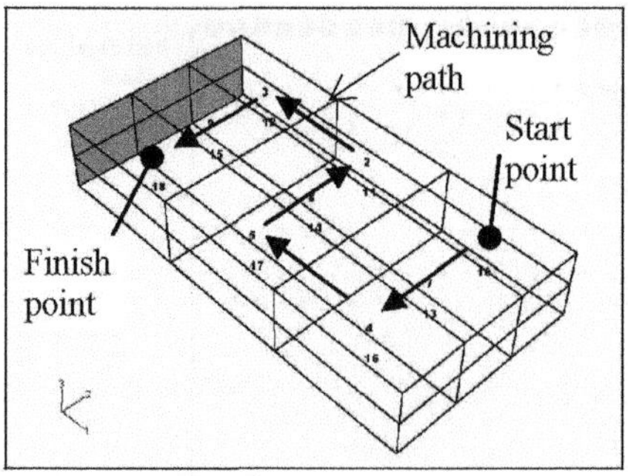

Figure 8a: Sequence '145236'

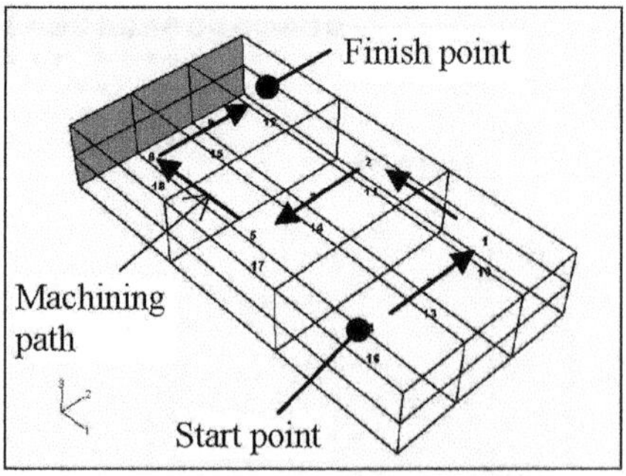

Figure 8b: Sequence '412563'

The time taken to find the optimum solution depends on how complex is the problem. In this project the problem size is kept into minimum, so that the result can be compared with the result obtained from the exhaustive search.

6 Conclusion

Tool path optimisation is one of the approaches to avoid problems such as workpiece chatter and static deflection during machining. This problem can be solved using the genetic algorithm Travelling Salesman Problem coupled to a Finite Element analysis of the workpiece. In this article a simple problem has been studied based upon a cantilever beam, and the results compared with engineering intuition and an exhaustive search. The proposed method was successful in converging upon the optimum solution in terms of the fitness criterion.

A large number of simulations have been carried out to investigate the performance of GA using different operating conditions such as lowering the probability of crossover, varying the initial population size, and changing the probability of mutation.

The Genetic Algorithm may not give the optimum solution for a general problem, but the method can give a near optimum solution, which is often adequate for most practical problems. Further work will extend the approach to more complex (and realistic) workpieces and investigate the effectiveness of the algorithms in more detail.

References

1 J. Tlusty, "Manufacturing processes and equipment." Upper Saddle River, NJ: Prentice Hall, 2000, pp. 559 - 604.

2 D.Montgomery and Y.Atlintas, "Mechanism of cutting force and surface generation in dynamic milling," *Journal of Engineering for Industry*, vol. 113, pp. 160-168, 1991.

3 J.Tlusty, "Dynamics of High-Speed Milling," *Journal of Engineering for Industry*, pp. 59 - 67, 1986.

4 S. Smith and D. Dvorak, "Tool path strategies for high speed milling aluminum workpieces with thin webs," *Mechatronics*, vol. 8, pp. 291-300, 1998.

5 M. Monreal and C. A. Rodriguez, "Influence of tool path strategy on the cycle time of high-speed milling," *Computer-Aided Design*, vol. 35, pp. 395-401, 2003.

6 V. Tandon, H. El-Mounayri, and H. Kishawy, "NC end milling optimization using evolutionary computation," *International Journal of Machine Tools and Manufacture*, vol. 42, pp. 595-605, 2002.

7 S. S. Rao, "Mechanical vibrations." Upper Saddle River, New Jersey: Pearson Education, 2004, pp. 236.

8 S.Smith and J.Tlusty, "An overview of modelling and simulation of the milling process," *Journal of Engineering for Industry*, vol. 113, pp. 169-175, 1991.

9 E. Goldberg David, *Genetic algorithms in search, optimization, and machine learning*. Reading: Addison-Wesley Pub. Co, 1989.

10 J. E. Baker, "Reducing bias and inefficiency in the selection algorithm," *Proceedings of the second international conference on Genetic Algorithms*, pp. 14 - 21, 1987.

11 Z. Michalewicz, "Genetic algorithms + data structures = evolution programs." Berlin: Springer-Verlag, 1992, pp. 165 - 192.

12 M. Mitchell, *An introduction to genetic algorithms*. Cambridge, London: MIT Press, 1996.

Automating the Analysis of Wafer Data Using Adaptive Resonance Theory Networks

Emilio Miguelañez[1], Ali M S Zalzala[1] and Paul Tabor[2]

[1]Department of Electrical, Electronic and Computer Engineering
School of Engineering and Physical Sciences
Heriot-Watt University, Edinburgh, EH14 4AS, UK
e.miguelanez@hw.ac.uk, a.zalzala@hw.ac.uk

[2]Test Advantage, Inc, 1525 West 10th Place,
Tempe, Arizona 85281, USA
paul.tabor@testadvantage.com

Abstract

In semiconductor manufacturing, finding needles in haystacks is easy compared with finding sub micron defects in modern ICs like complex microprocessors. The problem is likely to grow much worse as the relative complexity of chips (number of transistors and total wiring length) increases as the size of the smallest defects that can cause failures decreases. The use of unsupervised learning is a promising strategy towards the development of fully automated classification tools. This research intends to develop an automatic defect classification system for electrical test analysis of semiconductor wafer using an adaptive resonance theory network as a classifier. As a primary input source to the network, the system employs e-binmaps obtained from the test stage of the manufacturing process. To accomplish this task, a filtering algorithm is also implemented able to discard those wafermaps without pattern. This paper reports satisfactory results showing that the proposed system can recognised defect spatial patterns with a 82% correct e-binmap classification rate.

1 Introduction to Automated Analysis

Automated analysis for semiconductor wafer defect data has become increasingly important over the past several years as a means of quickly understanding and controlling contamination sources and process faults which impact product yield. It has been estimated that up to 80% of the yield loss in the mature production of high volume, very large-scale integrated (VLSI) circuits can be attributed to random visual particle and pattern defects [1]. Larger semiconductor wafer and smaller defect size have caused an exponential increase in the volume of visual and parametric defect data that must be analysed, evaluated and maintained. The ability of a manufacturer to readily access this data and to quickly extract useful process knowledge is required as market trends cycle drive the cycle time down [2].

The 2001 NTRS Roadmap [3] describes the need to develop automated intelligent analysis and reduction algorithms that correlate facility, design, process, test and work-in-process (WIP) data. Automated data/image mining and reduction must be acquired to source defects from multiple data sources to enable rapid root cause analysis of yield limiting conditions. Nowadays, the semiconductor fabricator has a large variety of tools available to provide wafer measurements ranging from in-line whole-wafer optical inspection for defect detection, to off-line scanning electron microscopy (SEM), energy dispersive x-ray spectroscopy (EDX), and focussed ion beam (FIB) for defect analysis and fault isolation. The wafer is also electrically tested to isolated bit fails for memory cache, and to provide binmapping for die sort. The result is the collection of a wide variety of data types that must be correlated to quickly source manufacturing problems to errant tools and processes.

Methods such as automatic defect classification (ADC), spatial signature analysis (SSA), statistical process control (SPC), and datamining are but a few of the technologies being developed and tested in the fabrication environment today as a means of adding informational content to the raw measured data.

Automatic Defect Classification (ADC) has been developed to provide automation of the tedious manual inspection processes associated with defect detection and review. ADC was initially developed in the early '90s to automate the manual classification of defects during off-line optical microscopy review [4, 5]. Since this time, ADC technologies have been extended to include optical in-line defect analysis [6] and SEM off-line review. For in-line, ADC, a defect may be classified 'on-the-fly', i.e. during the initial wafer inspection tool, or during a re-visit of the defect after the initial wafer scan. During in-line detection the defect is segmented from the image using a die-to-die comparison or a 'golden template' method [7]. These methods are also used during off-line review, when the defect is re-detected using the specified electronic wafermap coordinates.

Spatial Signature Analysis (SSA) is an automated procedure developed by researchers at the Oak Ridge National Laboratory to address the need to intelligently group, or cluster, wafermap defects together into spatial signatures that can be uniquely assigned to specific manufacturing processes [2, 8, 9, 10]. The SSA method is built on two primary technologies: advanced clustering which segments the defects of a given signature into distinct groups; and signature classification which assigns a user-defined class label to each unique group. For this work, a pair-wise fuzzy k-Nearest Neighbour (kNN) approach was adapted. Initially, SSA was developed for detection and classification of optical defect spatial signatures, but has also been extended to analyse electrical test binmap data to recognise process-dependant patterns [11, 13]. SSA and ADC technologies have been merged to enable intelligent wafer sub-sampling for efficient off-line review and improved ADC classifier performance [12].

Automatic defect and signature classification in semiconductor manufacturing has been shown to have an average performance of 75%, with performance being define as the number of correct class assignments divided by the total number of samples tested based on comparison with a human expert [13, 14].

Among datamining techniques, few studies report the application of neural networks as main classifier in an automatic defect classification system. Fei-Long Chen *et al.* [15] presented a neural network approach to automatically identify the classes of patterns in the data set and assign each wafer to one of the classes. In their work, they proposed this approach comparing two architectures, namely Adaptive Resonance Theory Network 1 (ART1) [16] and Kohonen's Self Organising Features Maps (SOFM) [22, 23]. Based on the analysis of simulated and real wafers, they conclude that ART1 network classifies better and more efficiently compared to the SOFM architecture. However, the main source of data used in this experiment was essentially visual, therefore it is totally dependent on different inspection machines detecting visual defects. And then, before analysis of the collected data can be performed, transformation of the data coordinates is necessary because different machines produce different original coordinates. And K. Kameyama *et al.* [21] reported an ADC system for visual inspection of semiconductor wafers using a neural network classifier. They proposed a Hyperellipsoid Clustering Network employing a Radial Basis function in the hidden layer. The network was trained with additional penalty conditions for recognising unfamiliar inputs as originating from an unknown defect class.

Looking at the constraints of the above methods, this research intends to develop a novel intelligent classification system able to recognise spatial patterns of clustered defects in electrical test wafermaps (e-test wafermaps). In order to obtain a better classification rate, our study is based on a neural network architecture named adaptive resonance theory 1 (ART1).

The paper is organised as follows. In section 2, the mechanisms of the classification system is explained including the definition of the input vector, the description of the implemented pre-processing tool and the explanation of ART1 as the main classifier. Results from the classification using real and simulated data is treated in Section 3, whereas the classification obtained in this research is compared with other techniques is discussed in Section 4. Finally, in Section 5 we draw some conclusions and indicate some directions for future research.

2 Classification System: ART Miner

2.1 The input vector

Electrical Wafer Sorting (EWS) is the step of the process where all the devices are tested. The wafer-under-test is moved to each site specified in a pre-programmed map, and electrical characterisation tests are made. Test data is stored and reported in output formats, called electrical test wafermaps (e-test wafermaps). In other words, mapping the results of these electrical tests onto a 2-D space creates the e-test wafermap. E-test wafermaps contain in many cases characteristics that give insight into the health of the semiconductor manufacturing process. Electrical failures are the consequence of many different process causes: human mistakes, wrong recipes, equipment or process out of control, equipment malfunctions, too narrow process windows, scratches due to human and machine causes, particles.

In this study, e-test wafermaps have been comprised of bin data and the electrical test binmap is obtained (e-binmaps). Bin data describes the results of electrical testing of individual die on a wafer. A bin can be viewed as a 'bucket' classification into which all die that meet that classification fall. The e-binmaps are the input vectors employed in the defect classification system. Examples of input vectors can be observed in the following images.

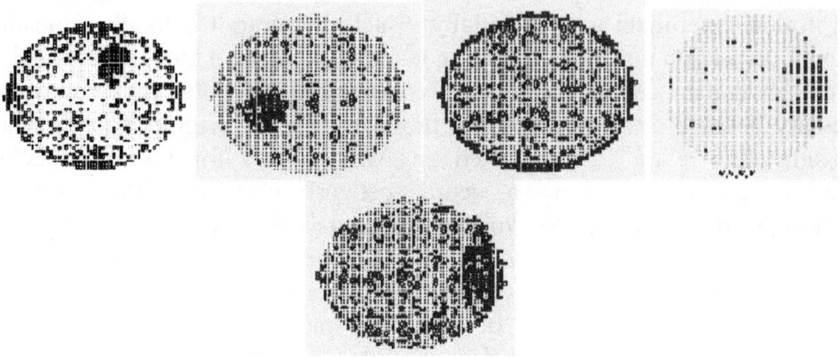

Fig 1. Examples of E-Binmaps

2.2 Pattern Miner

To assist ART Miner, a pre-processing algorithm has been implemented to perform some operations on the e-test wafermap. This pre-processing algorithm, called Pattern Miner, is able to detect if there is any pattern in the e-binmap without losing any information from the original e-binmaps. (see figure 2). This tool looks across the 2D e-binmap and performs median filtering of the matrix. Median filtering is a nonlinear operation often used in image processing to reduce "salt and pepper" noise (outliers). Median filtering is more effective than convolution when the goal is to simultaneously reduce noise and preserve edges [17].

Firstly, the identification of an appropriate mask is needed. This mask will identify which devices in the e-binmap are selected to perform the median filter. The mask should be similar to the pattern. A test engineer's knowledge is essential in order to create the appropriate mask, although – in absence of this data – any simulated theoretical pattern could be used too.

Around each value that has been selected by the mask in the original e-binmap, the median is calculated considering a neighbourhood of a size n-by-n. Due to the dimensionality of the data and the 'salt and pepper' noise in the image, a severe smoothing is not needed. This fact implies that the size of the window must be small. Different experiments were performed with different sizes, and the best performance was obtained with the window of size 3x3.

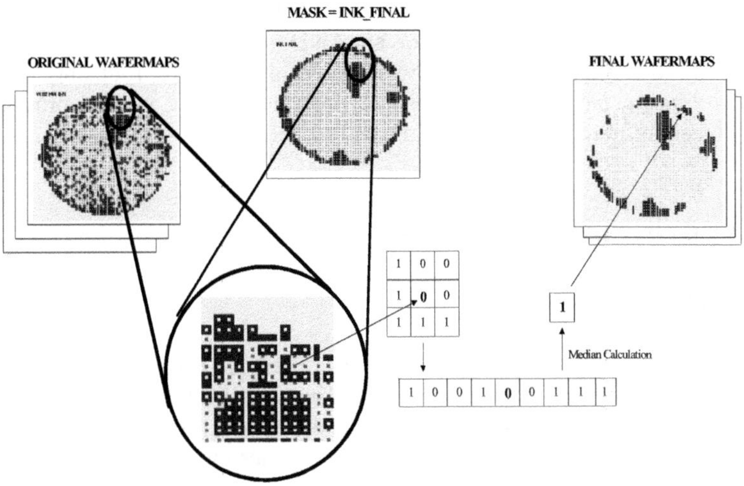

Fig 2. Pattern Miner algorithm

2.3 ART1 Network Model

Adaptive resonance theory (ART) is best introduced in the article of Carpenter and Grossberg [16] which describes the ART 1 model for classifying binary (0 or 1 to each node) inputs. An input vector is classified into one of a number of categories depending on similarity with the previous patterns. If the similarity criterion is not satisfied, a new category is created. Therefore, the ART network solves the dilemma between plasticity and stability by enabling the learning to occur only in a resonance state. In this way the network learns new information without destroying the old one. The ART NN works with binary input patterns that are compared to prototype vectors in order of decreasing similarity until either an adequate match is found or a new category is created.

Anyone who has ever used an ART network will be very familiar with the vigilance parameter, which regulates the system and provides a balance between memorising as many patterns as possible and generalising patterns into broader classes. By adjusting a suitable vigilance parameter from to 1 to near 0, it is possible to reduce category proliferation while at the same time improving the performance of the network - a problem known as the bias variance dilemma. A vigilance parameter $\rho = 0.7$ means that an input vector x with 10 ones will resonate with a winner vector that has at least 7 ones of the same position as the input x. If the vigilance test is not satisfied for all categories and the input pattern cannot be classified to neither of the existing categories, a new category is created. Methods have been proposed to automatically adjust the vigilance parameter common in Adaptive Resonance Theory networks [18]. In this approach study, the vigilance parameter was found optimal at the value 0.3 after several experiments were performed with different values, and it stays fixed for all simulation runs.

It is well known that using ART1 architecture it is difficult to obtain a number of classes close to the true one. It seems that the method has difficulties in handling a large number of wafers. This may be due to the fact that the ART1 network identifies the classes on the basis of an AND logic. Therefore, to solve this problem and to avoid decreasing the vigilance parameter and producing too many classes, reference patterns of classes are presented to the system providing a visually meaningful representation of the wafers belonging to the class.

3. Experimental Results

3.1 Source Data

The source data is stored in three different data files. ART Miner performs pattern mining on all e-binmaps in every data file. The classification has been tested on these files separately, because they are from different customers testing different products. The number of devices per wafer determines the size of the input vector. These three files are:

Test_file_1.dat: Contains real data from 37 wafers with 665 devices in each one. The lot was processed in two separate batches during its front-end steps. There is one primary 'Edge Effect' pattern in this lot, which is determined by 16 wafers.

Test_file_2.dat: Contains real data from 6 wafers with 1708 devices in each one. Every wafer contributes to two primary patterns in one: 'Ring Effect – Hot Spot' pattern.

Test_file_3.dat: Contains simulated data from 45 wafers with 1710 devices in each one. There are three different defect patterns within this lot: Ring Effect, Hot Spot and Edge Effect.

Looking at the first two test files, only one pattern can be found for each lot: 'Edge Effect' pattern in the first file and, 'Ring Effect - Hot-Spot' pattern in the second. In the real world, different patterns could be found in the same wafer or in different wafers for the same lot. Therefore, an extra file has been created containing 39 simulated e-binmaps (1710 devices) with three different patterns: Hot Spot, Ring Effect and Edge Effect. These binmaps were created using theoretical patterns as a base. Then, random noise was added to this base in order to obtain a pseudo-real e-binmap. Considering this case, three different masks must be applied with the pattern miner tool. These masks are theoretical patterns.

3.2 ART Miner Classification

The experiment begins by applying pattern miner, as a pre-processing tool, to the original e-binmaps to discard noise. This noise could the responsible for the diversity of pattern categories due to the existence of outliers in the e-binmap, confusing the algorithm in order to match the pattern among the data. The pattern miner is also able to detect which e-binmap contains a possible pattern. From the original 88 e-binmaps, pattern miner successfully discriminates 18 e-binmaps from the lot because they do not have any possible pattern.

The output from pattern miner is then considered as main input to ART 1 network. The classification results produced by ART Miner on the three test files have been summarised in Table 1. The numbers in the table represent the identification number of the wafer in the lot.

Category	E-binmap($\rho = 0.3$)	(Test_file_1.dat)
Ring – Hot Spot	1,2,3,4,5,6	
Residual 1		
		(Test_file_2.dat)
Edge Effect 1	11, 12, 20,	
Edge Effect 2	22, 23, 24, 25, 26, 27, 28, 29, 30, 32, 34, 35, 36,	
Residual 1	1, 2, 3, 4, 5, 6, 7, 8, 9, 10, 13, 14, 15, 16, 17, 18, 19, 21, 31, 37,	
		(Test_file_3.dat)
Hot Spot	1, 2, 3, 4, 5, 6, 7, 8, 9,	
Ring Effect	14, 15, 16, 17, 18, 19, 20, 21, 22, 23,	
Edge Effect	27, 28, 29, 30, 31, 32, 33, 34, 35, 36, 37, 38, 39,	
Residual 1	10, 11, 12, 13,	
Residual 2	24, 25,	
Residual 3	26,	

Table 1. Results from ART miner

In the top section of the table, it can be observed that Art Miner discriminates with high success the test data file containing the 'Ring Effect – Hot Spot' pattern. Classifying test_file_2, the network classifies in detail the e-binmaps that have not pattern (residual category) and the ones with pattern. With a vigilance test equal to 0.3, these e-binmaps have been also classified into different categories: Edge Effect 1 and Edge Effect 2. Decreasing this parameter to 0.2, all e-binmaps will be classified in the same category. Looking test_file_3, each e-binmap has a particular pattern. Therefore, an ideal classification would give no elements in the residual category. But in this experiment, the network classifies correctly 32 out of 39 e-binmaps, i.e. a performance of 82%. Looking in detail each category, it could be observed that the network categorised successfully each e-binmap with an 'Edge Effect' pattern. In the 'Hot Spot' category, 9 out of 13 e-binmaps are properly allocated, while the e-binmaps no. 10, 11 12 and 13 are classified in the residual category. And a better performance can be appreciated in the 'Ring Effect' category, where 10 out of 13 e-binmaps are distributed correctly. The e-binmaps no. 24, 25 and 26 have be allocated in the residual category, although they should be classified in the 'Ring Effect' category. Looking at the e-binmaps in the residual category, they have a pattern that does not match the theoretical pattern. ART miner is able is created a new category to store these unknown patterns. Now the test engineer has to decide if these e-binmaps should be considered part of a new class or they are not too different of already known class.

3.3 ART Miner Performance

The easiest way to asses the performance rate is to choose a test set independent of the training set (and validation set if used), to classify its examples, count the examples that have been correctly classified and divide by the size of the test set. The proportion of test-set examples that are classified correctly estimates the performance rate. These measures are clearly unbiased estimated under all circumstances, but they can be highly variable, and having to use a test set may waste data which could otherwise have been used for training. The idea of a test set is sometimes called the *hold-out* (HO) method [19]. The HO performance is also known as the expected performance and it is an estimate of how well the trained classifier will perform in the field when exposed to new data.

Table 2 gives a summary of ART Miner classifier HO performance for test file 3. To describe the statistical performance of the classifier, the information regarding the classification of patterns is presented in a simple matrix format known as confusion matrix [20]. As shown in the table, rows represent the actual class of the data, i.e., as classified by the user. The columns represent the ART Miner assigned class. By summing across a row, the actual number of samples of a particular class is ascertained. The diagonal elements of the matrix contain the number of patterns for each class that have been correctly classified. Therefore, the ratio of the diagonal element to the sum of the row gives an estimate of the performance of the classifier for a particular class.

	Hot Spot	Ring Effect	Edge Effect	Unknown	Rate
Hot Spot	9	0	0	4	69.2%
Ring Effect	0	10	0	3	76.9%
Edge Effect	0	0	13	0	100%
			Total unweighted performance		**82.03%**
			Total weighted performance		**84.12%**

Table 2.Confusion matrix describing the statistical HO performance of the classifier.

There are two values of performance reported in the table; 'total weighted performance' and 'total unweighted performance'. The first of these measures is weighted by the number of actual samples collected for each class. This metric allows the pattern of highest occurrence to carry the most weight in the performance estimate. The later measure assumes weight for each class. This is more indicative of performance assuming that every defined signature type is equal importance. The metric that is most representative of the system performance depends on the point of view of the reader.

Another indicator of the efficacy of the method is the purity of a classification. Purity is determined for a class by taking the ration of the diagonal element to the sum of the column. While performance is a measure of how often a given class was identify correctly, purity is a measure of how often the identification of a given class was correct. Looking at table 2, the purity of each class for ART miner classifier was 1, i.e., all the identifications were 100 % correct.

4. ART Miner: Comparisons

Fei-Long Chen *et al.* [15] also presented a neural network approach to automatically identify defect spatial patterns in semiconductor fabrication. The neural network architecture applied was also ART1. The following table shows the difference between this approach and ART Miner.

	F. Chen work	ART Miner
Data Source	Imagery	Parametric/Binmap
Performance (unweighted)	**71.42%**	**82.03%**
Vigilance Parameter	0.11	0.30
No. of patterns	7	3
No. of training sample per pattern	5	2
No. of wafers	14	39
No. of die per wafer	294	1710

Table 3. Comparison between F. Chen *et al.* approach and ART Miner

The main source of data in F. Chen and Liu work was made of only 14 real wafers, each of which containing 294 die. They visually judge that the real wafers included twelve ring-type patterns and two random-type patterns. With that limited data set, it seems very difficult to draw definitive conclusions on the performance of unsupervised algorithms. Table 3 shows that ART Miner classifies better and more efficiently using a more extensive data set of simulated and real wafers, with a higher classification rate and vigilance parameter. Partly, this due to the fact that the data source is based on patterns formed by electrical failures.

Previously in this document, the SSA method has been described as a technique for detection and classification of optical defect spatial signature. This approach has been adapted to analyse also electrical test data [13]. Many of the features used in the optical SSA techniques are also used in the binmap signature analysis, and the fuzzy k-NN is identical to that used for optical defect SSA. Table 4 reproduces a summary of SSA E-test classifier performance for the three participating sites determined during the validation exercise and compares those results with ART Miner performance.

	SITE 1	SITE 2	SITE 3	ART Miner
Total Weighted Performance	**79%**	**84%**	**72%**	**84.12%**
Total Unweighted Performance	**74%**	**81%**	**64%**	**82.03%**
Average Purity	69%	80%	66%	100%
No. of Defined Classes	9	15	11	3
No. of Signatures (Bins) Analysed	2004	2105	1511	39
Average no. of Bins per Wafer	7	30	7	1
Approximates no. of Wafers	286	70	196	39

Table 4. Comparison between SSA E-test classifier and ART Miner.

They reported an average overall system performance of 75% (70%), which is lower that the overall performance obtained by ART Miner. Although both techniques attain classification on the same type of parametric/binmap data, it has to be said that the amount of data used by SSA is considerably bigger than the one classified by ART Miner.

5. Conclusions and Further Research

This research developed a novel intelligent system, ART Miner that can recognise spatial patterns of clustered defects in electrical test wafermaps. Using adaptive resonance theory 1 network as the main classifier has shown the advantages of unsupervised learning where no subjective classification is needed, and, being completely automatic, the method can identify unsuspected patterns that could have been overlooked by a human operator. To complete the system, a pre-processing tool, called pattern miner, is implemented to assist the classification. The aim of this tool is to discard noise that could be responsible for the diversity of pattern categories due to the existence of outliers in the e-binmap. This noise effect could confuse the algorithm in order to match the pattern among the data. The pattern miner has not only the ability to discard noise but also is able to detect which e-binmap contains a possible pattern. This tool has proved its importance in the system by successfully discriminating 18 e-binmaps from the original 88 e-binmaps lot. The consequence of this operation is the increment of the vigilance parameter to 0.3, therefore more classes can be classified closer to the true one. The experimental results show that this approach provides not only the automated classification of known patterns but also the detection of new unknown patterns (residual category). The classification rate, 82% (84%) on randomly sampled binmaps, can be considered remarkable considering the shortage of training data and variety of techniques that have been developed by other research groups.

Considering future investigation, an interesting direction of research is the classification of wafer taking into account that the electrical testing does not simply classifies dies as failures or successes but also specifies the type of failure. Along this direction, an alternative type of architecture may be worth exploring to classify non-binary wafers.

6. References

1 T. Hattori, Detection and Identification of Particles on silicon Surfaces, Particles on Surfaces, Detection, Adhesion, and Removal, Edited by K.L. Mittal, Marcel Dekker, Inc., New York.

2 K.W. Tobin, S.S. Gleason, F. Lakhami, and M.H. Bennet, (1997) Automated Analysis for Rapid Defect Sourcing and Yield Learning. Future Fab International, Vol. 4.

3 The National Technology Roadmap for Semiconductors, Semiconductor Industry Association, San Jose, 2001.

4 M.M. Slama, M.H. Bennett, and P.W. Fletcher, (1992) Advanced In-Line Process Control of Defects. Integrated Circuit Metrology, Inspection, and Process Control VI, SPIE vol. 1673, p. 496.

5 B. Trafas, M.H. Bennet, M. Godwin, (1995) Meeting Advanced Pattern Inspection System Requirements for 0.25-μm Technology and Beyond. Microelectronic Manufacturing Yield Reliability, and Failure Analysis, SPIE vol. 2635, p. 50.

6 R. Sherman, E. Tirosh, Z. Smilansky, (1993) Automatic Defect Classification System for Semiconductor Wafers. Machine Vision Applications in Industrial Inspections, SPIE Vol. 1907, p.72.

7 P.B. Chou, A.R. Rao, M.C. Sturzenbecker, F.Y. Wu, V.H. Brecher, (1997) Automatic Defect Classification for Semiconductor Manufacturing. Machine Vision and Applications, Vol. 9(4), p. 201.

8 K.W. Tobin, S. S.Gleason, T.P. Karnowski, and S.L. Cohen, (1997) Feature Analysis and Classification of Manufacturing Signatures on Semiconductor Wafers. SPIE 9th Annual Symposium on Electronic Imaging: Science and Technology, San Jose, CA.

9 K. W. Tobin, S. S. Gleason, T. P. Karnowski, S. L. Cohen and F. Lakhani, (1997) Automatic Classification of Spatial Signatures on Semiconductor Wafermaps. SPIE 22nd Annual International Symposium on Microlithography, Santa Clara Convention Center, Santa Clara, CA.

10 K. W. Tobin, S. S. Gleason, T. P. Karnowski and M. H. Bennett, (1996) An Image Paradigm for Semiconductor Defect Data Reduction, SPIE's 1996 International Symposium on Microlithography, Santa Clara, CA.

11 S.S. Gleason, K.W. Tobin, T.P. Karnowski, and Fred Lakhani, (1999) Rapid Yield Learning through Optical Defect and Electrical Test Analysis. SPIE's Metrology, Inspection, and Process Control for Microlithography XII, Santa Clara Convention Center, Santa Clara, CA.

12 S. S. Gleason, K. W. Tobin and T. P. Karnowski, (1997) An Integrated Spatial Signature Analysis and Automatic Defect Classification System. 191st Meeting of The Electrochemical Society, Inc.

13 T.P. Karnowski, K.W. Tobin, S.S. Gleason, Fred Lakhani, (1999) The Application of Spatial Signature Analysis to Electrical Test Data: Validation Study. SPIE's 24th Annual International Symposium on Metrology, Inspection and Process Control for Microlithography XIII, Santa Clara Convention Center, Santa Clara, CA.

14 Bennett, M.H., Tobin, K.W., Gleason, S.S., (1995) Automatic Defect Classification: Status and Industry Trends. Integrated Circuit Metrology, Inspection, and Process Control IX, Proc. SPIE, Vol. 2439, p. 210-220.

15 F. Chen and S. Liu, (2000) A Neural-Network Approach To Recognize Defect Spatial Pattern In Semiconductor Fabrication. IEEE Transactions on Semiconductor Manufacturing, Vol. 13 (3).

16 Carpenter, G. A. and S. Grossberg, (1998) The ART of adaptive pattern recognition by self-organization neural network. Computer, vol. 21, no. 3, pp. 77-88.

17 Lim, Jae S. (1990 Two-Dimensional Signal and Image Processing. Englewood Cliffs, NJ: Prentice Hall, pp. 469-476.

18 A. Canuto, M. Fairhurst and G. Howells, (2001) Improving ARTMap learning through variable vigilance. International Journal of Neural Systems, Vol. 11, No. 6, pp. 509-522

19 W.H. Highleyman, (1962) The Design and analysis of pattern recognition experiments. Bell Systems Technical Journal, vol. 41, p. 723-744, 1962.

20 B.D. Ripley, (1996) Pattern Recognition and Neural Networks, Cambridge University Press, p. 75.

21 K. Kameyana, and Y. Kosugi, (1999) Semiconductor Defect Classification using Hyperellipsoid Clustering Neural Networks and Model Switching. International Joint Conference on Neural Networks, Vol. 5, p. 3505 - 3510.

22 Kohonen, T., (1998) Self organization and associative memory. 3rd Ed, New York Springer-Verlag.

23 Kohonen T., (1990) The self-organizing map. Proceedings of the IEEE 78, pp. 1464-1480.

Job Shop Scheduling using Multi-Agent Systems

A. M. Booth

School of Computer Science
University of Birmingham
Birmingham
B15 2TT

Abstract

This paper investigates heuristics to improve the intelligence of the head of the supply chain in a contract net framework for a multi-agent system used to solve a job shop scheduling problem. The heuristics prioritise the selection of the next operation to be scheduled and allow the head of the supply chain to intelligently influence the scheduled time for the operation. It was found that the policy of scheduling the longer operations first at the earliest time possible, then scheduling the shorter operations allowing them to be scheduled in the gaps left by the longer operations, leads to the improvement of the quality of schedule produced with respect to the objectives of meeting customer delivery deadline, minimising average batch makespan and maximising machine utilisation.

1. Introduction

The job scheduling problem is a naturally distributed problem both from a physical and logical perspective, since manufacturing involves several resources and tasks that can be performed simultaneously [1]. It is also suggested by [2] that centralised control of systems is no longer practical or efficient in a competitive market and that distributed localised information and cooperative decision making is now needed. As a result of these factors distributed architectures and increased use of intelligence have now been proposed for scheduling systems.

The distributed system proposed by the artificial intelligence community is the multi-agent system. A definition, as used by [3], of an agent is an entity able to act on itself and its surrounding environment. It produces behaviour to achieve its objectives using its observations, knowledge and interactions with other agents and its environment. A multi-agent system is an artificial system composed of a population of agents that cooperate to achieve a common goal whilst pursuing individual objectives. Agents can be used to solve a complex problem beyond their individual capabilities by cooperation in a multi-agent system.

Rather than using traditional methods to provide scheduling capabilities, multi-agent systems use communication to produce overall schedules, rather than searching for viable solutions. Agents are used to represent resources or jobs and each agent produces a local schedule for the object it represents. Agents may be competitive and bid against each other, to provide their service or to be scheduled, in an auction mechanism. Agents may also be cooperative, negotiating to find mutually acceptable schedule assignments. An effective negotiation or auction mechanism used in a multi-agent system must provide the coherence and knowledge of the scheduling problem found in a centralised system whilst providing dynamic and reactive scheduling. The contract net protocol [4] is a commonly used auction mechanism for distributed problem solving and has been successfully used for job shop scheduling [5-8].

To improve the scheduling ability of a multi-agent system, agent intelligence can be investigated. A typical representation for multi-agent system using a contract net protocol consists of agents representing batches of jobs to be completed and machine agents to process the various operations. Enhancement to agent intelligence has been concentrated at a low level, where the intelligence of the machines, when forming bids and creating their local schedules, is used to improve the overall scheduling performance of the system. However, intelligence could also be improved at a higher level to enhance the scheduling ability of a multi-agent system. This paper investigates heuristics to improve the intelligence of the head of the supply chain, in a contract net framework for a multi-agent system in development at an engineering company. Two types of heuristics are identified, operation announcement heuristics and utility profile heuristics. The heuristics aim to prioritise the selection of the next operation to be scheduled and to allow the head of the supply chain to intelligently influence the scheduled time for the operation. The remainder of the paper is organised as follows. Section 2 describes the multi-agent systems and contract net protocol used to test the developed heuristics, which are detailed in Section 3. Section 4 describes the experimentation carried out. Finally Section 5 provides a discussion of the results and the conclusions of the investigation.

2. Multi-Agent System

Currently a multi-agent system with a contract net framework is being developed at a company to attempt to solve a job shop scheduling problem. Batches of components are introduced into the system dynamically and have earliest start dates, due dates and a value of importance given by the customer. Each batch requires several operations, where each operation requires a machine and a specific tool to process it. There are three types of agents, Batch agents, Tool agents and Machine Agents. Each agent maintains a diary of its activities. The batch agent selects an operation at random and creates an initial utility profile for the operation. A utility profile is a representation of the window of time available to schedule the operation and shows the time periods when the operation cannot be scheduled. Figure 1 shows an example of a utility profile where the earlier a time slot is, the more desirable it is. The flat zero regions of the profile indicate times where the operation cannot be scheduled. For the other time slots, the desirability of scheduling the operation at that time is shown by its value.

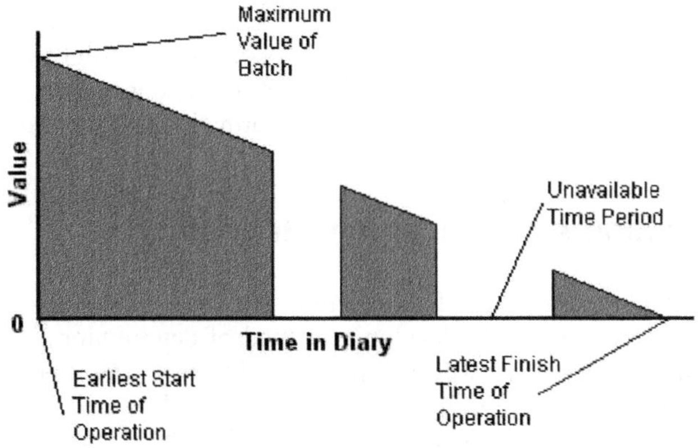

Figure 1: Example Utility Profile

The batch agent then announces the operation and its initial utility profile to the tool agents that are capable of processing that operation type. The tool agents consult their diaries and modify the utility profile by setting the value of time periods where they cannot work on the operation to zero. They then announce the operation to the machine agents, who again modify the utility profile taking into account when they can work on the operation. The utility profile is therefore a method of ensuring that bids are valid. The most desirable time slot left then forms the machines bid for the work.

A test engine was also designed and implemented using JADE. A test engine is used initially rather than the real system, as many features of the real system are not needed to test the performance of the heuristics, such as reacting to resource failure and rescheduling ability. Implementing a simpler test engine than the real system enables easier incorporation and implementation of the heuristics and prevents unnecessary changes to the system to enable the testing of the heuristics. Results from experimentation using a test engine can then be used to guide experimentation within the real system.

3. Heuristics

To improve the scheduling ability of the current system, improving the intelligence of the batch agent, when selecting the next operation for announcement and for creating the operation's initial utility profile, can be investigated. The order in which operations are announced can greatly affect scheduling ability. Five different scheduling methods were compared in [9]. For each of the five methods, they tried three different batch arrival processes and found that each method had large effects on the scheduling abilities of the scheduling methods. A fuzzy rule base was used by [6] to determine which operation should be announced next, providing a means of formalising uncertain expert knowledge and allowing human influence to be incorporated into the system. In the system described in [1] the task manager maintains a list of tasks waiting to be announced. The task manager selects the task with the highest priority using heuristics. A rating system is used in [10] to determine the scheduling priority of tasks using a most-tightly-constrained-first opportunistic heuristic. Tasks are also selected for announcement according to their priority in [11]. However, they provide no information on how they determined the priority.

3.1. Operation Announcement Heuristics

When developing a method for selecting the next operation to announce for scheduling, it is necessary to develop a method of determining the priority of that operation, as seen above. However, sophisticated methods are no use if they take too long to compute [9]. So therefore it is important to use knowledge easily available to the batch agent when determining operation priority. Due to the distributed nature of multi-agent systems, the only information available to the batch agent is its local knowledge. This includes a list of operations still to be scheduled, start times of scheduled operations, precedence relations of operations and the value of whole batch as set by the customer.

The dispatching rule, shortest processing time first, is intended to reduce the makespan of the schedule, which schedules operations in order of the length of their processing time, starting with the shortest. However, the operations with the longer processing times act as bottlenecks in the system. Since they require the most resource time and where they are scheduled has a greatest effect on the ability of the batch to meet its customer deadline. Batch agents will have to compete for their operations to be scheduled with the tools and machines. So if all batches announce their operations in the same order, they will often be competing over the same tool agents at the same time. Since only one batch agent can be successful the others will have wasted their scheduling efforts. This could be reduced if the batch agents announce their operations in a random order. The operation announcement heuristics to be implemented will be:

1. Select an operation at random
2. Select the operation with the shortest processing time
3. Select the operation with the longest processing time

3.2. Utility Profile Heuristics

When the batch agent creates an operation's initial utility profile, setting the values assigned to the available time slots is the batch agent's opportunity to "advise" the tool and machine agents where it thinks the operation should be scheduled. Where the maximum value it can assign is the value of the job specified by the customer and the time periods where its operations are already scheduled are set values of zero. There are many possible policies for determining the desirability. In this investigation only piecewise linear profiles will be considered. Heuristics will only consider the free time slots as a whole, rather than individual times within slots and only those large enough to fit the operation are take into account. Table 1details the heuristics that will be used for creating the initial utility profiles and the reasoning behind there choice.

Table 1: **Utility Profile Heuristics**

Utility Profile Heuristic	**Description**	**Reasoning**
Earliest	The earlier a time slot the greater its value.	Scheduling objective is to minimise the batch makespan and to meet the customer delivery date.
Smallest	The smaller the time slot the greater its value. If tied the earlier slot is more desirable.	Find time slot that is the 'closest fit' for operation and discourage shorter operations using larger slots, therefore preventing longer operations from using them.
Largest	The larger the time slot the greater its value. If tied the earlier slot is more desirable.	Encourage operations to be scheduled in large slots so it is more likely there will room left for other operations. Also to avoid small unusable gaps being left in the diary.
Considers	Ranks the slots according to largest number of unscheduled operations that could still fit in the slot if the operation was scheduled in it. Again if tied the earliest slot is best.	Take into account the operations still left to schedule by trying to ensure there is still room for them to be scheduled when deciding where to schedule the current operation.
Earliest unless exact	The earliest time slot has the greatest value unless there are later slots that are of the exact size for the operation to fit in or are of a size such that no other unscheduled operation could fit in them.	To try and ensure that operations are scheduled as early as possible whilst taking into account the other tasks that are still to be scheduled and trying to prevent unusable slots being created in the diary.
Earliest unless other fit	As above but if earliest slot is the only slot an unscheduled operation can fit in its desirability is decreased.	As above but also trying to prevent shorter tasks using slots that larger tasks can fit in, leaving gaps in the diary.

The heuristics define methods of ranking the slots according to their desirability. They do not specify how the magnitude of the value for each slot should be determined. This is done by firstly considering the utility profile that would be created if the first time in the profile was given the maximum value and the last time was given a zero value. The profile would be a straight line with decreasing gradient. The value of the start time of each slot is then taken from this, so a list of possible values is created. The start of each slot is then reassigned a value according to its ranking given by the heuristic in use. Within a slot the values for each time decrease according to the original gradient. So within a slot the earliest time is always the best. However, the value must never be negative, so if using the current gradient, values would become negative, the gradient for that particular slot is recalculated by fixing the end point of the slot to be zero. Figure 2 shows how a utility profile may look before and after the reassignment of start values for the free slots.

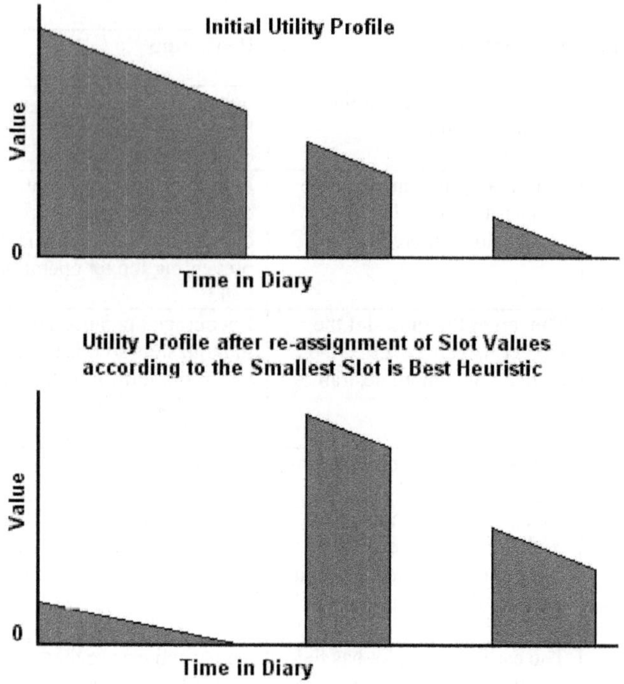

Figure 2: Example Utility Profile before and after re-assignment of slot values produced by the Smallest utility profile heuristic

4. Experimentation

4.1. Initial Experimentation using the Test Engine

Initial testing was carried out using the test engine for each operation announcement and utility profile heuristic combination on a job shop problem provided by the company. The problem involves the production of batches, where each batch has 24 components. 10 operations, requiring various processing times, need to be completed on each component. The operations have the following precedence constraints – operation 1 must be completed first, then operation 2, operations 3-9 may then be completed in any order and then finally operation 10 can be completed. There are four machines and one tool to process each type of operation.

To give an indication of the heuristics ability to meet customer delivery demands, the makespan for each batch is compared. As the heuristics are intended to improve scheduling when the batch agent has a choice between operations, the makespan for operations 3-9 of each batch is also compared. The results for each heuristic are averaged over all test runs and can be seen in Table 2. The results in Table 2 show that although Random operation announcement has the lowest batch makespan, it has the largest makespan for operations 3-9. SPT has only a slightly greater makespan for whole batches, however it has a significantly better makespan for operations 3-9. LPT has the largest makespan for a batch although not by a large amount and although it does not perform as well as SPT when considering operations 3-9 it is significantly better then random operation announcement. Table 2 shows that the utility profile heuristic Smallest performed best when considering the entire batch makespan and the Considers and Earliest Unless Exact heuristics also performed well. When looking at the makespan of operations 3-9, Considers, Earliest Unless Exact and Earliest all performed well, all about 40 minutes greater than the minimum completion time of 888 minutes. These results suggest that the policy of announcing the shortest operations first and scheduling them as early as possible, can lead to shorter completion times.

Table 2: Table showing the Results for each Heuristic tested in the Test Engine

Operation Announcement Heuristic	**Average Batch Makespan (Minutes)**	**Average Makespan for Operations 3-9 (Minutes)**
Random	4773.8	1000.4
SPT	4778.8	925.5
LPT	4823.6	989.4
Utility Profile Heuristic		
Earliest	4845.0	929.0
Smallest	4737.6	945.9
Largest	4809.1	994.4
Considers	4754.1	923.1
Earliest Unless Exact	4759.7	927.7
Earliest Unless Other Fit	4847.1	1110.4

However, looking at specific combinations of heuristics, SPT combined with Earliest Unless Other Fit utility profile heuristic performed the best and resulted in the optimum average of 888 minutes for the completion of the operations 3-9, for batches completed within a week. SPT combined with Considers and Earliest also produce good results, with average makespan of 890 minutes and 899 minutes respectively. This suggests that to minimise completion time, short operations should be announced first and scheduled as early as possible unless they are going to "use up" a slot suitable for a larger operation or there is a later time slot only the short operation can fit in. Larger operations are then announced and are either scheduled in slots left for them or as early as possible after the completion of the shorter operations.

4.2. Test Runs using the Engineering Company System

After analysing the results from the initial testing carried out using the test engine, the heuristics were then tested using the company system. The role of the heuristics is to determine which operation to announce when there is a choice, i.e. when scheduling operations 3 – 9, in the test problem described above. So a different test problem was created with batches containing nine operations of various lengths that could be processed in any order. A tool agent for each type of operation and five machines were the resources allowed. Seven batches were introduced simultaneously, with the earliest start time of each batch spaced at four-hour intervals. This was to represent a more realistic situation of orders arriving at a factory. This test problem was the largest that could be efficiently run, given the need for a statistically significant number of runs in a reasonable timescale.

Table 3: Table showing the Results for each Heuristic when tested in the Company System

Operation Announcement Heuristic	Average Makespan (Minutes)	Average Machine Utilisation %
Random	1656.7	71.9
SPT	1825.4	71.6
LPT	1650.2	75.0
Utility Profile Heuristics		
Earliest	1672.2	74.1
Smallest	1714.6	72.6
Largest	1751.2	72.4
Considers	1776.4	72.7
Earliest Unless Exact	1673.4	72.9
Earliest Unless Other Fit	1676.5	72.3

Table 3 shows the results of the average batch makespan over all test runs for each heuristic. The results show that LPT and Random operation announcement both performed significantly better than SPT. This suggests that the longer operations act as bottleneck in the schedule and to reduce the makespan of the batch should be scheduled first. Using random operation announcement reduces the competition between batches over a tool at any one time, therefore making it more likely that a batch can schedule the operation at a time to fit in with its current schedule. The results show this leads to a shorter makespan for the batch. For the utility profile heuristics, Earliest, Earliest Unless Exact and Earliest Unless Other Fit were the best performers, all with similar average makespan. This suggests that scheduling operations as early as possible can minimise the makespan of the batch.

Machine utilisation is another objective to be considered when comparing the performance of the heuristics, since it is desirable for the machines to be getting maximum use. This considers the percentage of time, from when a machine begins the processing of its first operation to when it finishes its last operation, which the machine is in use. Table 3 shows the average machine utilisation between the five machines across all test runs for each of the heuristics. The results show that LPT announcement heuristic has the best average machine utilisation at 75.0%. Random and SPT perform similarly with lower averages of 71.9% and 71.6%. Comparing the performances of the utility profile heuristics, Earliest has the highest utilisation of 74.1%. The remaining heuristics all perform similarly with values from 72.3% - 72.9%. When comparing specific heuristic combinations, it could be seen that LPT combined with either Earliest, Considers or Earliest Unless Exact produced a significantly higher machine utilisation than the other combinations with averages of

77.2%, 76.3% and 74.9% respectively. This shows that scheduling the largest operations as early as possible, in slots that are an exact fit or in slots that allow shorter operations to be scheduled in the gaps left, increases machine utilisation.

A completion interval is the time between the completion of one batch to the completion of the next batch. In the test problem, the batch's earliest start times are staged at intervals of 4 hours. So a new batch becomes available for processing every 4 hours. Ideally the completion intervals should be the same as the introduction interval with a low standard deviation, showing that the batches are being produced at even intervals. However, experimentation in the company system found that all heuristic combinations lead to batches being completed at a wide variety of intervals. Work-in-progress is another objective used to access the quality of schedule produced. However, testing using the company system did not show very significant differences in performance for each heuristic.

4.3. Further Experimentation using the Test Engine

The Utility Profile heuristics use the concept of free slots and according to various conditions determine which slots are the best to schedule the operation. So to further test the performance of the heuristics a new test problem was created. This test problem involved the initial scheduling of three operations with earliest start times that ensure empty time slots would instantly be created. The batch then contained a further seven operations to be scheduled in any order in the time period. Seven batches were introduced sequentially into the system and all the operation selection and utility profile heuristic combinations were tested.

The average completion time for the 7 batches introduced to the test engine when each heuristic is used over all test runs is compared, as it is an indicator to which heuristics are most likely to produce schedules that meet customer deadlines. The results in Table 4 show that LPT has a significantly lower average completion time. SPT has the largest average completion time, over 2 hours longer than that of LPT. Comparing the average completion times for the utility profile heuristics, shows that Earliest has a significantly lower average than the rest. Earliest Unless Exact and Earliest Unless Other Fit also perform well. The remaining heuristics have considerably longer average completion times. This suggests that the longer operations have the greatest effect on the ability to meet customer deadline and therefore should be scheduled first and at the earliest time possible. So when the shorter operations are then scheduled, in as early positions as possible, they can then fill any gaps left by the scheduling of the larger operations so as not to increase the completion time greatly.

Table 4 also shows the results of the average batch makespan over all test runs for each heuristic. The results show that LPT produces schedule with significantly lower average batch makespan. It has an average time of over 2 hours shorter than SPT and over 3 hours shorter than Random. The results shows that the Smallest utility profile has the lowest average makespan with Earliest only slightly longer. This suggests that the policy of selecting the slot that is the closest fit for the operation can lead to a low batch makespan. Comparing the results for specific heuristic combinations, shows that the combinations of LPT and Earliest, Earliest

Unless Other or Earliest Unless Exact perform the best. Again suggesting a good policy is to schedule the longer task as early as possible.

Table 4: Table showing the Results for each Heuristic tested in the Test Engine on the new test problem

Operation Announcement Heuristic	Average Completion Time for 7 Batches (Minutes)	Average Batch Makespan (Minutes)	Average Machine Utilisation %
Random	3185	2142.4	83.5
SPT	3238	2072.5	83.4
LPT	3103	1951.8	83.3
Utility Profile Heuristics			
Earliest	3016	1973.0	87.4
Smallest	3336	1955.9	78.9
Largest	3238	2189.0	83.4
Considers	3238	2042.8	82.4
Earliest Unless Exact	3112	2084.6	84.0
Earliest Unless Other Fit	3112	2088.1	84.4

In this test problem four machines were used and Table 4 shows a comparison of the average machine utilisation for the machines averaged over all test runs for each of the heuristics. The results suggest that the operation announcement heuristics have no effect on the machine utilisation, as they all have very similar percentages. The Earliest utility profile heuristic produces significantly higher machine utilisation percentage than the other heuristics of 87.4%. Earliest Unless Exact and Earliest Unless Other are the next best and perform similarly. The results suggest that to maximise machine utilisation operations should be scheduled as early as possible.

5. Conclusion

Heuristics to improve the intelligence of the head of the supply chain in the contract net framework have been developed and implemented. The designed heuristics are methods in which the batch agent can select the next operation to schedule and to guide the tool and machine agents to the time at which the operation should be scheduled. So the heuristics are divided in to two types, operation announcement and utility profile heuristics. Experimentation was then carried out into the effects of the heuristics on the quality of schedule produced, when considering various objectives both within the developed test engine and the company system. All results were then compare and analysed.

From the initial experiments carried out using the test engine, it was found that the combination of the SPT operation announcement heuristic and the Earliest Unless Other Fit utility profile heuristic, produced the optimum makespan for the operations 3-9. However, further experimentation using the test engine and within the company system, found that the LPT heuristic resulted in the lowest average batch makespan. This suggests that if the period in the batch diary where the operations under consideration can be scheduled is initially empty the SPT operation announcement heuristic should be used. However, if there are already operations scheduled in the time period or there is greater competition between batches for resources, as is the case of the company system, the LPT operation announcement heuristic should be used to minimise the average batch makespan. The possibility of changing the operation announcement heuristic, according to the number of free slots in the diary, during scheduling, should therefore be investigated.

The results from all the experiments show that scheduling an operation as early as possible reduces the average batch makespan. Further investigation is needed into the extent of the effects produced by the additional conditions in the Earliest Unless Exact and Earliest Unless Other Fit heuristics. However, results suggest that Earliest Unless Exact and Earliest Unless Other Fit heuristics could lead to greater improvements in reducing average batch makespan and meeting customer deadlines, particularly in more complex test problems that produce a greater number of slots in the diary.

The experimentation using the test engine found the operation announcement heuristics to have no effect on the machine utilisation. However, testing in the company system found that the LPT heuristic increased the machine utilisation. This may be due to the differing scheduling abilities of the machine agents in the test engine and in the company system. It may also be due to the differing nature of the test problems. All results showed that scheduling operations as early as possible maximises the machine utilisation.

Experimentation in the company system found that all heuristic combinations lead to batches being completed at a wide variety of intervals, rather the desired regular interval of the same size as the batch introduction intervals. Therefore further research is needed into developing methods to create schedules that have regular batch completion intervals. The results for work-in-progress did not show very significant differences in performance and further research into methods of reducing the work in progress of the system is required.

The paper has demonstrated that heuristics to increase the intelligence of the batch agents can be used to improve the quality of schedule produced with respect to the objectives of meeting customer deadline, minimising average batch makespan and maximising machine utilisation. For overall good performance the operation announcement heuristic longest processing time first and utility profile heuristic earliest time slot is best should be used. Larger test problems than those that could realistically be attempted in this study would reveal more effectively the different capabilities of the proposed utility-profile heuristics, due to the larger number of free slots that would be present in the diaries. Although the second problem used in the test engine insured that slots were initially created in the diary, further testing is required to investigate fully the performance improvements the additional conditions used in the Earliest Unless Exact and Earliest Unless Other Fit utility profile heuristics could provide on larger more realistic test problems. Since both heuristics performed as well as the Earliest heuristic on simpler problems. Also observing the scheduling behaviour of the batch agents in the company system, certain batches appeared to dominate or get left out of scheduling. Batch agents dominate if they always succeed in getting the bids they select from the tool agent processed. Batch agents get left behind if the bids they select become invalid because another batch has secured the resource first. These effects appear to have a large influence in the scheduling. This suggests that investigation into methods of preventing dominating agent behaviour in negotiations and the contract net should be carried out. This could be done by creating a priority level for each batch, adjusted dynamically according to how many successful negotiations it has had in its recent attempts.

References

1. Sousa P, Ramos C (1998) A dynamic scheduling holon for manufacturing orders. Journal of Intelligent Manufacturing, 9(2): 107-112.
2. Gou L, Luh PB, Kyoya Y (1998) Holonic manufacturing scheduling: architecture, cooperation mechanism, and implementation. Computers in Industry, 37(3): 213-231.
3. Kouiss K, Pierreval H, Mebarki N (1997) Using multi-agent architecture in FMS for dynamic scheduling. Journal of Intelligent Manufacturing, 8(1), pp. 41-47.
4. Smith RG (1980) The contract net protocol: High-level communication and control in a distributed problem solver. IEEE Transactions on Computers, c-29(12): 1104-1113.
5. Shen W, Norrie DH (1998) Combining mediation and bidding mechanisms for agent-based manufacturing scheduling. Proceedings of the International Conference on Autonomous Agents, 469-470.
6. Vojdani N (1996) Distributed manufacturing control using fuzzy contract net. IEEE International Conference on Fuzzy Systems, 3: 1655-1659.
7. Saad A, Kawamura K, Biswas G (1996) Performance evaluation of contract net-based heterarchical scheduling for flexible manufacturing systems. International Journal of Intelligent Automation and Soft Computing: Special Issue on Intelligent Manufacturing Planning and Shop Floor Control.
8. Kim KH, Song, JY, Wang KH (1997) Negotiation based scheduling for items with flexible process plans. Computers and Industrial Engineering, 33(3-4): 785-788.
9. Adacher L, Agnetis A, Meloni C (2000) Autonomous agents architectures and algorithms in flexible manufacturing systems. IIE Transactions, 32(10): 941-951.
10. Neiman D, Hildum D, Lesser V, Sandholm T (1994) Exploiting meta-level information in a distributed scheduling system. Proceedings of the Twelfth National Conference on Artificial Intelligence, 394-400.
11. Zhang Z, Cheng T, Wu B, Yang S (2003) Price-based negotiation for task assignment in a distributed network manufacturing mode environment. International Journal of Advanced Manufacturing Technology, 21(2): 145-156.

Tool Wear Prediction Approach for Turning Operations Based on General Regression Neural Network (GRNN) Technique

E.A. Almeshaiei[1], S.E. Oraby[1], M. A. Mahmoud[1]

[1]College of Technological Studies, PAAET, P.O.
Box 42325 Shuwaikh 70654, Kuwait.
(eam@paaet.edu.kw)

ABSTRACT

Detection of tool failure is very important in automated manufacturing. Recent trends, being towards mostly unmanned automated machining systems and consistent system operations, need reliable on-line monitoring processes. A proper on-line cutting tool condition monitoring system is essential for deciding when to change the tool. Many methods have been attempted in this connection. Recently, artificial neural networks have been tried for this purpose because of their inherent simplicity and reasonably quick data-processing capability. The present work investigates the feasibility of using general regression neural networks (GRNN) for estimating the level of the nose wear on the cutting edge. Experimental data of different force components, as well as corresponding nose wear values and the three controlling cutting parameters (speed, feed, and depth of cut) are used to train the neural networks. The technique shows close matching of estimation of nose wear and directly measured wear value. Results indicated the need for inclusion of six input parameters (speed, feed, depth, and three force components) in order to get better prediction capability. However, some parameters, such as feeding and radial force components, have shown higher impact than others (power force component). Therefore, it is possible that the trained neural networks can accurately assess tool wear on-line using an appropriate system.

1 Introduction

The need for accurate and reliable prediction of the tool wear in machining has been a long-term goal of tool manufacturers and users especially after the introduction of advanced fully automated and smart systems. Since, in-process situations, tool wear is an immeasurable parameter, indirect methods are usually used to relate its variations to visible detectable variables such as cutting forces, tool vibration, acoustic emission, cutting temperature, etc. To be a reliable and practical tool, any proposed strategy should consider the complexity and diversity that usually characterize the nature of the machining environment. One of the methods that has been recently acknowledged to monitor tool state and condition is the intelligent discrimination of the tool abnormalities [1]. Many articles have proposed implementation of neural networks (NNs) as a means to monitor, assess and predict tool wear [2-11]. The actual measured sensor signals, cutting forces for instance, can be compared with predicted ones to determine the state of tool wear, and thus prevent damage to the tool and work piece during the course of manufacturing.

In this paper, the feasibility of using neural networks (NN) to predict different tool deformation modes in turning operation is investigated. Special attention is focused on nose wear for many reasons. It is the first failure mode to develop on the cutting edge, then it rapidly triggers the other failure modes such as flank and notch wear. Besides, it is the most controlling factor regarding the dimensional accuracy and the surface finish of the machined part.

The NN used in the analysis is of the general regression neural network (GRNN) type since this is appropriate for metal machining environment where involved variables are of continuous nature. In addition to that it is noticed that among the relevant published literature [1], very little attention was devoted to the application of prediction of continuous variables, and also since it appears that very little attention was given to the general regression neural network (GRNN). Cutting forces; feeding component F_x, radial component F_z and power component F_y, are used individually or, in different mathematical forms and ratios, as measurable sensing variables.

An artificial neural network is usually defined as a network composed of a large number of processors (neurons) that are massively interconnected, operate in parallel, and learn from experience (examples).

2 NEURAL NETWORK ALGORITHM

The general regression neural network (GRNN) is a one-pass learning algorithm that can be used for estimation of continuous variables, and converges to the underlying regression surface. The principal advantages of the GRNN are its quick learning and fast convergence to optimal regression surface as the number of samples becomes large. The overall block diagram of the GRNN in its adaptive form is shown in Fig. 1. to show a feed forward network that can be used to estimate a vector *Y* from a measured vector *X*. The input units are merely distribution units, which provide the (scaled) measured variables *X* to all of the neurons on the second layer, which contains the pattern units. Each pattern unit (neuron) is dedicated to one exemplar (pattern) or one cluster center. When a new vector *X* is entered into the network, it is subtracted from the stored vector representing each cluster center. The squares of the differences are summed and fed into a nonlinear activation function. The activation function used herein is logistic in the form: $f(x) = \frac{1}{1-e^{-x}}$, where x is the input. This function is the most popular and has been found useful for most network applications [12]. The pattern units' output is passed on to the summation units. Details of the GRNN paradigm were provided by Specht [13].

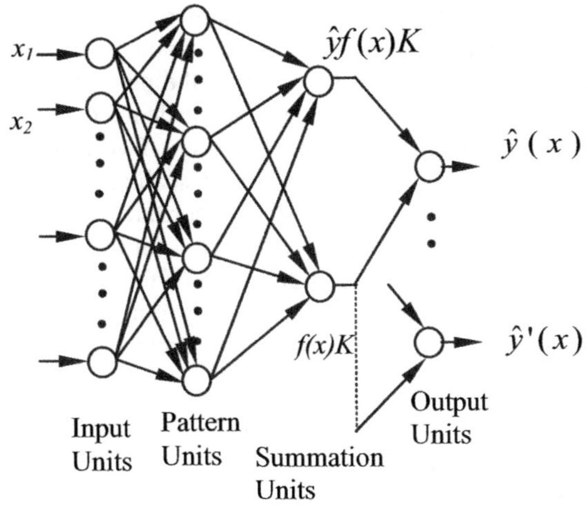

Figure 1: Block diagram of a general regression NN

The network "learns" by adjusting the interconnection weights between layers. The answers the network is producing are repeatedly compared with the correct answers, and each time the connecting weights are adjusted slightly in the direction of the correct answers. Eventually, if the problem is learned, a stable set of weights adaptively evolves which will provide good answers for all of the sample predictions. The real test of neural networks occurs when the trained network is able to produce good results for new data.

In this study, over-training was prevented using a so-called Net-Perfect algorithm [14]. This algorithm optimizes the network by applying the current network to an independent test set during training. The algorithm finds the optimum network for the data in the test set (which means that the network is able to generalize well and give good results on new data); the algorithm optimizes the smoothing factor based upon the values in the test set. It does this by trying different smoothing factors and choosing the one that minimizes the mean squared error between the actual and predicted answers.

General regression neural networks (GRNN) work by measuring how far a given sample pattern is from patterns in the training set in N dimensional space, where N is the number of inputs in the problem. In this study, the method of measuring the distance between patterns was the so-called City Block distance metric, which is the sum of the absolute values of the differences in all dimensions between the pattern and the weight vector for that neuron [13].

The GRNN used in this study was genetic adaptive, i.e. it uses a genetic algorithm to find input smoothing factor adjustment. This is used to adapt the overall smoothing factor to provide a new value for each input. Genetic algorithms use a "fitness" measure to determine which of the individuals in the population survive and reproduce [15]. The measure of fitness for GRNN is the mean squared error of the outputs for the entire data set. The genetic adaptive algorithm seeks to minimize this mean squared error.

When variables are loaded into a neural network, they must be scaled from their numeric range into the numeric range that the neural network deals with efficiently. There are two main numeric ranges the networks commonly operate in: zero to one denoted [0, 1], and minus one to one denoted [–1, 1]. One choice is the use of linear scaling functions for this purpose. Two possible alternatives to these linear scaling functions are two non-linear scaling functions: logistic and tanh. The logistic function scales data to (0,1) according to the following formula: $f(x) = 1/(1+\exp(-x_m/s))$ where x_m is the average of all of the values of that variable in the pattern file, and s is the standard deviation of those values. The hyperbolic tangent function (tanh) scales data to (–1, 1) according to: $f(x)=\tanh(x_m/s)$. As detailed later, a parametric study was conducted to select the best scaling function for the present application.

3 EXPERIMENTAL SETUP

In order to use neural networks, comprehensive experimental testing procedures were conducted to record nose wear variations in relation to the employed cutting conditions (speed V, feed F and depth of cut D). The corresponding force signals are in-process simultaneously recorded in three directions: feeding component F_x parallel to work part axis, radial component F_z normal to work part axis and power (vertical) component F_y tangential to work rotation.

Central Composite Design (CCD) experimental design technique was used to arrange five levels of each of the operating parameters: lowest, low, moderate, high and highest. This led to 24 experiments listed in Table 1.

Table 1 CCD of the experimental procedures

Test	Speed, V (m/min)	Feed, F (mm/rev)	Depth, D (mm)	Test	Speed, V (m/min)	Feed, F (mm/rev)	Depth, D (mm)
1	72	0.12	2	13	206	0.2	2.25
2	145	0.3	2	14	50	0.2	2.25
3	145	0.12	2.5	15	104	0.6	2.25
4	72	0.3	2.5	16	104	0.06	2.25
5	104	0.2	2.25	17	104	0.2	3
6	104	0.2	2.25	18	104	0.2	1.5
7	145	0.12	2	19	206	0.2	2.25
8	72	0.3	2	20	50	0.2	2.25
9	72	0.12	2.5	21	104	0.6	2.25
10	145	0.3	2.5	22	104	0.06	2.25
11	104	0.2	2.25	23	104	0.2	3
12	104	0.2	2.25	24	104	0.2	1.5

Each experiment consisted of two-minute subtest intervals until nose wear on the cutting edge has reached a criterion level of 0.3 mm or, whenever the edge has catastrophically failed due to plastic deformation. This led to 669 total number of patterns used in the NN analysis.

4 RESULTS AND DISCUSSION

Out of the 669 total number of patterns used in the NN analysis, the data used for training and testing the NNs (the Pattern set) were 564 patterns. The patterns in this database were divided into two sets. The first set consisted of 452 patterns and was used for training the networks. The second set consisted of the remainder 112 patterns, selected randomly, and was used for testing the trained networks. To evaluate the robustness of the networks, the trained networks are applied to a "production" data set (105 patterns) that the networks have never "seen" before.

The input layer to the neural network consisted of 6 neurons to which the speed V, the feed F, the depth of cut D, the feeding component F_x, the power component F_y, the radial component F_z were fed. The hidden layer must contain a minimum of one neuron for each data pattern; the number was set to 452. The number of neurons in the output layer is 1, which corresponds to the output (the nose wear). The statistical indicator used to evaluate the closeness of fit is the coefficient of multiple determinations, R that can be defined as [14]: $R^2 = 1 - \frac{\sum (y - y_p)^2}{\sum (y - y_m)^2}$ where y is the actual value, y_p is the predicted value of y, and y_m is the mean of the y values. The coefficient of multiple determinations, R compares the accuracy of the model to the accuracy of a trivial benchmark model wherein the prediction is simply the mean of all of the samples. A perfect fit would result in an R value of 1 and a very good fit near 1. At lower R-values, the quality of fit deteriorates. Table 2 shows that R for the Pattern and Production sets for this network are high, which is a proof of very good fit.

The results of a parametric study of the network design intended to quantitatively determine the general regression neural networks (GRNN) design (using a genetic adaptive algorithm) that best predicts the 'production data' are summarized in Table 2. The variables investigated included:

(a) different scaling functions (linear between [–1,1], linear between [0,1], logistic, and hyperbolic tangent "tanh"), and

(b) two possible ways to measure the distance between patterns, namely the City Block distance metric [13] and the Euclidean distance metric [16].

From the table it is evident that:

(a) in most cases, the City Block distance metric is better than the Euclidean metric;

(b) the tanh scaling function is better with the Euclidean distance metric than with the City Block; and,

(c) the linear scaling functions are better with the City Block distance metric than with the Euclidean.

Based on these findings, the optimum design of the GRNN for the present application is one that uses a genetic adaptive algorithm, the City-Block distance metric, and a linear scaling function between [–1,1] for the input data.
An overall comparison of the NN nose wear prediction with experimental values for the 'Production set' (NN Pro) (105 data patterns) is shown in figure 2. At lower levels of nose wear of 0.02 to 0.18 mm, corresponding NN estimates seem to be over estimated. However, except for few outliers, a reasonable correlation is noticed. This is expected since within that domain tool wear is usually of a stochastic nature. Fortunately, however, that wear level is too low to be considered.

Comparisons between NN estimates and corresponding experimental data for individual tests are found in a good agreement as shown in Figs. 3,4.
Correlation may be discussed in the light of machining performance due to level of operating conditions (V,F,D) used in experiment. While low-to-moderate speed (V) level of (72-104 m/min) was employed in experiments 1,8,9 and 14, Fig. 3, higher level was used in experiments 17,18,23 and 24, Fig. 4. While wider scatter is noticed in the former set, better correlation is shown for the later set. Higher speed usually invokes higher wear rate with more stable machining that is free of edge chipping and built-up edge mechanism. This may explain results in Fig. 2 where, at low wear values, NN estimates are over estimated.
However, at a given speed level, experiments 1,8 and 9, Fig. 3, higher feed (F) value seems to degrade NN predictability. This is thought to be due to the conditions discussed above. Depth of cut (D), however, tends to positively influence correlation where deeper value leads to more stable operation and, accordingly, improves estimation accuracy (compare experiments 1 and 9, Fig. 3). However, it may be concluded that error between experimental and NN estimates is of acceptable level and distribution pattern especially with the well known stochastic nature that characterizes machining operations. Additionally, results showed a better predictability over the conventional multiple or non-linear regression techniques as was previously investigated by one of the authors [17].

Table 2 ***R*** **values for various network designs; inputs: V, F, D, F_x, F_y, F_z**

Distance Metric	Scaling Function	*R* for Pattern set	*R* for Production set
City Block	Linear [0,1]	0.9811	0.9261
City Block	Linear [–1,1]	0.9745	0.9280
City Block	Logistic	0.9774	0.9269
City Block	Tanh	0.9394	0.9066
Euclidean	Linear [0,1]	0.9255	0.8624
Euclidean	Linear [–1,1]	0.9143	0.8390
Euclidean	Logistic	0.9529	0.9159
Euclidean	Tanh	0.9434	0.9142

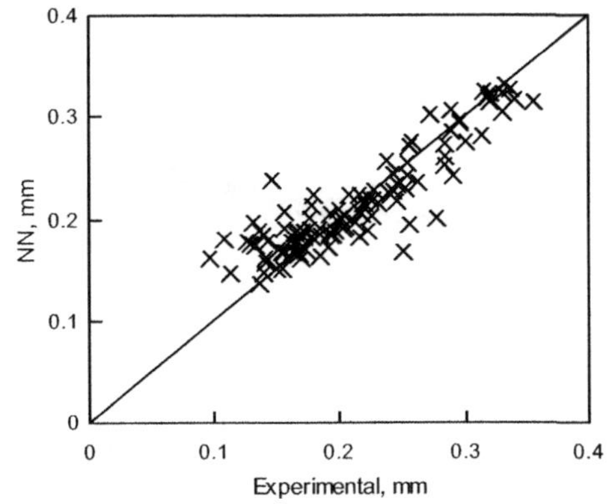

Figure 2: NN nose wear prediction vs experimental values; Production set.

As pointed out in the introduction, there is a need to include (as inputs to the NN) more than one cutting parameter. In the foregoing discussion, the control parameters (S, F, D) and the three cutting forces were considered. An objective of the present study was to quantitatively determine which combination of these parameters gives best wear prediction. Therefore, a total of 14 possibilities and combinations were considered as shown in Table 3. For each case, 8 different NN designs were evaluated in a manner similar to Table 2, and the best design is found. Due to the emerged outcome, it may conclude that:

- for all cases, the City Block distance metric should be used in the NN design rather than the Euclidean,
- a linear scaling function is useful in most cases,
- the use of all three force components, or their ratios in association with the cutting control parameters (S, F, D) give best wear predictions,
- the use of only cutting force components, without the cutting control parameters, dramatically deteriorates the NN performance as pointed out by case 12,
- the two most significant force components to trace the nose wear are F_x and F_z as pointed out by data of case 4. This agrees with the technical fact that tool waer, especially nose scars, is well correlated to frictional force components (Feeding F_x and Radial F_z) rather that to the power (vertical) component F_y, and
- among the three force components, feeding component F_x shows most correlation (case 6) while the least is found for the power component F_y (cases 13).

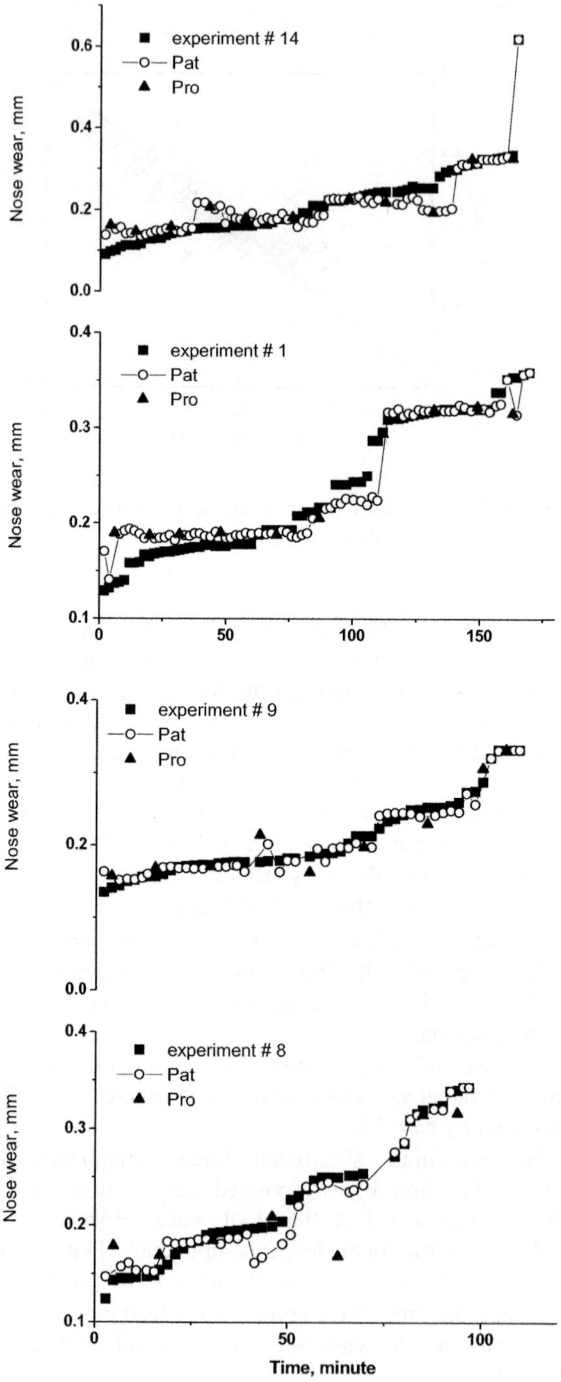

Figure 3: Comparison of Neural and Experimental estimates.

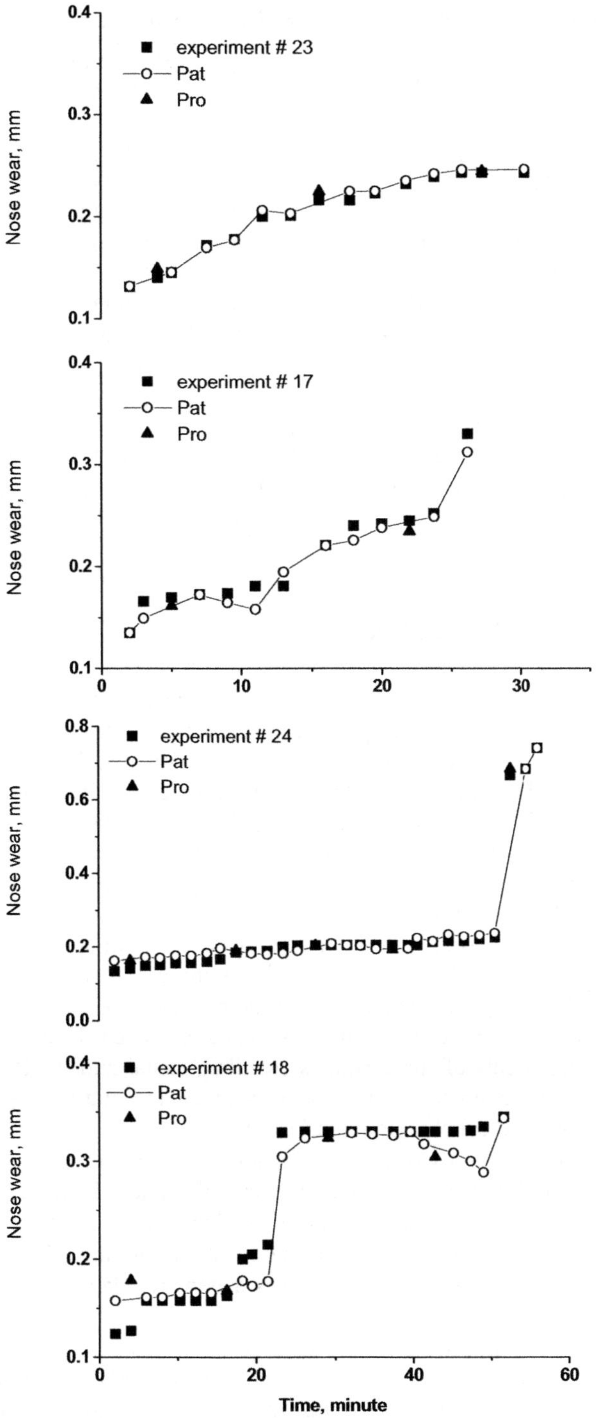

Figure 4: Comparison of Neural and Experimental estimates.

Table 3 ***R*** **for various networks and best design** ($F_{xz}= \sqrt{(F_x^2 + F_z^2)}$, $Px = S\ F_x$)

	Input Variables	Pattern Data	Production Data	Best Network Design
1	F_x F_y F_z V F D	0.9745	0.9280	City Block, -1,1
2	F_x/F_y F_z/F_y V F D	0.9666	0.9243	City Block, -1,1
3	Px Py Pz V F D	0.9580	0.9233	City Block, Logistic
4	F_x F_z V F D	0.9652	0.9222	City Block, Logistic
5	P_x P_y P_z F D	0.9728	0.9200	City Block, -1,1
6	F_x V F D	0.9141	0.9043	City Block, -1,1
7	F_{xz} V F D	0.9553	0.9023	City Block, -1,1
8	P_x/P_y P_z/P_y F D	0.9666	0.9008	City Block, 0,1
9	F_z V F D	0.9521	0.8953	City Block, -1,1
10	F_{xz} F_y V F D	0.9637	0.8780	City Block, -1,1
11	F_x/F_z F_y F D	0.9575	0.8756	City Block, -1,1
12	F_x F_y F_z Only	0.9738	0.8697	City Block, -1,1
13	F_y V F D	0.9236	0.7792	City Block, 0,1
14	F_x/F_z V F D	0.8953	0.7556	City Block, -1,1

5 CONCLUSIONS

Signals of different cutting force components usually show a remarkable functional correlation with the developed wear scars on the nose of the cutting tool when rough turning is performed. This was investigated by many researches using various mathematical and formulation techniques. Preference of one technique over others is always related to its capability to be used in establishing an online monitoring of the operation especially cutting tool performance. Actual in-process collected force signals from the machining system are fed to the established trained NN model and, technical decision is taken accordingly, Fig. 5.

In this work, neural network are used as an applicable intelligent system toward the establishment of a reliable cutting tool condition monitoring using cutting force signals. Different neural routines are investigated. Inputs to the system are the controlling cutting parameters (Speed, feed, and depth of cut) and three components of force signals (feeding, radial and vertical). Output from the system is an estimate level of nose wear on the cutting edge. Among emerged conclusions and concluding remarks are:

- Nose wear is given a special attention in the study and in tool condition monitoring, in general, since it directly affect both the dimensional accuracy and the surface quality of the machined surface.
- Also, It is found that the most appropriate GRNN design for the data involved is the one that uses a genetic adaptive algorithm with a City-Block distance metric and a linear scaling function between [-1,1].
- The use of all the six neurons (parameters) mentioned above gives best prediction of the nose wear. Less performance is found when either set of cutting parameters or set of force components is used independently.

- Among the set of cutting force components, the use of feed and/or radial components showed better predictability than the use of power component.

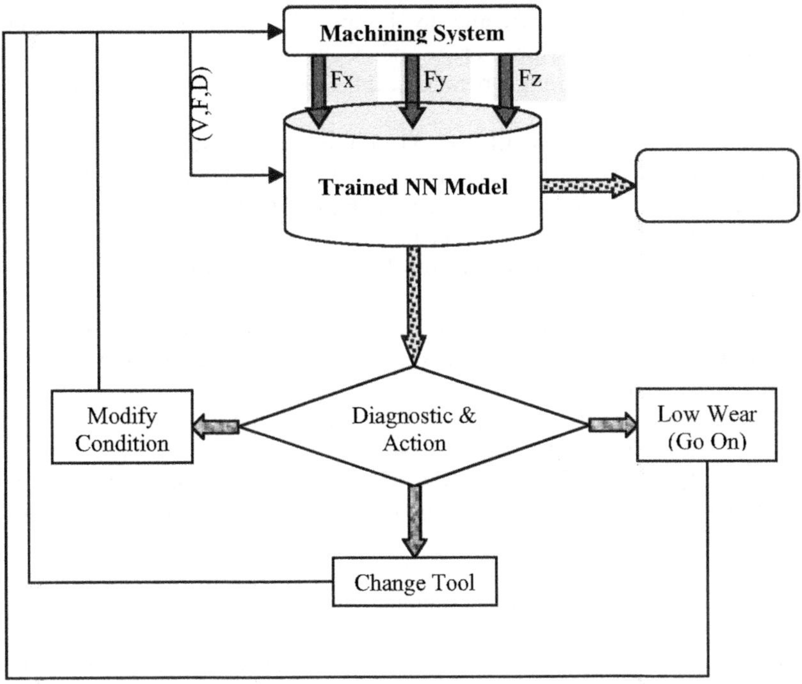

Figure 5: Illustrative diagram of a Tool Wear Monitoring and Control Technique

REFERENCES

1 Dimla JR DE, Lister PM and Leighton NJ, (1997) Neural Network Solutions to the Tool Condition Monitoring Problem in Metal Cutting – A Critical Review of Methods, Int. J. Mach. Tools & Manuf., **30** (29), 1219-1241.

2 Dimla JR DE, Lister PM and Leighton NJ, (1998) Automatic Tool State Identification in a Metal Turning Operation using MLP Neural Networks and Multivariate Process Parameters, Int. J. Mach. Tools & Manuf., **38** (4), 343-352.

3 T. Luo T, Lu W, Krishnamurthy K and McMillan B, (1998) A Neural Network Approach for Force and Contour Error Control in Multi-Dimensional End Milling Operations, Int. J. Mach. Tools Manuf., **38** (10-11), 1343-1359.

4 Stone R and Krishnamurthy K, (1996) A Neural Network Thrust Force Controller to Minimize Delimitation During Drilling of graphic-Epoxy Laminates, Int. J. Mach. Tools Manuf., **36** (9), 985-1003.

5 Yeh Zong-Mu, Tarang Y. S. and Nian C. Y., (1995) A Self-Organizing Neural Fuzzy Logic Controller for Turning Operations, Int. J. Mac. Tools Manuf., **35** (10), 1363-1374.

6 Shang-Liang Chen and Jen YW, (2000) Data Fusion Neural Network for Tool Condition Monitoring in CNC Milling Machining, Int. J. Mach. Tools Manufact., **40** (3), 381-400.

7 Lin CS. and Ting JC., (1996) Drill Wear Monitoring using Neural Networks, Int. J. Mach. Tools Manuf., **36** (4), 465-475.

8 Das S, Roy R, and Chattopadhyay AB, (1996) Evaluation of Wear Turning Carbide Inserts using Neural Networks, Int. J. Mach. Tools Manuf., **36** (7), 789-797.

9 Kim TY, Woo J, Shin D and Kim J, (1999) Indirect Cutting Force Measurement in Multi-Axis Simultaneous NC Milling processes, Int. J. Mach. Tools Manuf., **39** (11), 1717-1731.

10 Raj KH, Sharma RS, Srivastava S and Patvardhar C, (2000) Modeling of Manufacturing Processes with ANNs for Intelligent Manufacturing, Int. J. Mach. Tools Manuf., 40 (6), 851-868.

11 Scheffer C, Kratz H, Heyns PS, and Klocke F, (2003) Development of a Tool Wear-Monitoring System for Hard Turning, Int. J. Mach. Tools Manuf., **43**, 973-985.

12 Hammerstrom D, (July 1993) Working with Neural Networks, IEEE Spectrum, 46-53.

13 Specht DF, (1991) A General Regression Neural Network. IEEE Transactions on Neural Networks, **2** (6), 568-576.

14 Neuroshell 2 manual, (1996) Ward Systems Group Inc., Frederick, Mass.

15 Goldberg DE, (1989) Genetic Algorithms in Search Optimization, and Machine Learning. Addison-Wesley, Reading, MA.

16 Kreyszig E, (1993) Advanced Engineering Mathematics, Seventh Edition, Wiley, New York.

17 Oraby SE and Hayhurst DR, (1991) Development of Models for Tool Wear Force Relationships in Metal Cutting, Int. J. Mech. Sci., **33** (2), 125-138.

An Efficient Genetic Algorithm to Solve the Manufacturing Cell Formation Problem

W. Rojas[1], M. Solar[1], M. Chacón[1], J. Ferland[2]

[1]Computer Engineering Department
University of Santiago of Chile
Santiago of Chile
wrojas@lauca.usach.cl

[2] Département d'informatique et de recherche opérationnelle
Faculté des arts et des sciences
Université de Montréal - Canada

Abstract

A fundamental stage in the design of manufacturing systems is the simultaneous formation of machine cells and families of parts. This problem has been addressed using a number of approaches, but genetic algorithms have had the most success. This paper presents an innovative integer genetic algorithm based on a partial definition of solutions together with a recursive fitness function based on Baldwin effect. The proposed algorithm was tested on a number of problems taken from the literature, and the comparative results are presented.

1 Introduction

Group technology consists of decomposing industrial systems into subsystems. The grouping of parts into families and machines into cells induces the following benefits, among others: reduction of inventory, increase response capacity to changes in demand, and the simplification of programming.

The Manufacturing Cell formation problem is a NP-hard optimization problem. Consequently, exact methods cannot be used to solve practical dimension problems in a reasonable time, while heuristic methods may generate lower quality solutions, albeit in less time. Hence there is a permanent need for methods that can obtain good solutions in a reasonable computational time.

In over thirty years of investigation, a variety of methods for the formation of manufacturing cells have been proposed. King [1] and Chan and Milner [2] present methods based on the reorganizing of rows and columns of the incidence matrix. These methods require visual inspection of the resulting matrix to identify the manufacturing cells. Chu and Tsai [3] compare three methods based on the reorganization of the incidence matrix

McAuley [4], Carrie [5], Seifoddini and Wolfe [6], Yasuda and Yin [7] develop hierarchical clustering algorithm. This approach involves a calculation of similarity (dissimilarity) between pairs of machines/parts. Machines/parts with a greater degree of similarity are included in the same cluster. These algorithms produce a dendogram of potential solutions.

Chandrasekharan and Rajagopalan [8, 9], Srinivasan and Narendran [10], and Srinivasan [11] develop partition cluster algorithms - the set of machines is divided into cells that are homogenous. Each machine belongs to a unique cell. These algorithms use a measure of similarity (distance) to form cells and require an *a priori* specification of the number of cells.

Joines et al. [12], Cheng et al. [13], Onwubolu and Mutingi [14], Dimopoulos and Mort [15], Mark et al. [16], Ribeiro and Lorena [17], Gonçalves and Resende [18] propose different evolutionary strategies. Specifically, Joines et al. [12] proposes a generic algorithm made up of seven genetic operators, to solve an integrated mathematical programming model. Cheng et al. [13] uses an approach based on the traveling salesman problem (TSP), and Whitley's genetic algorithm [19]. Onwubulu and Mutingi [14] propose a genetic algorithm that allows the establishment of manufacturing cells and places limits on the size of the cells. Dimopoulus and Mort [15] develop a genetic programming algorithm to obtain a dendogram of potential solutions in a hierarchical clustering approach. Mark et al. [16] proposes a genetic algorithm that adjusts the genetic parameters during the search process. Each individual is a sequence of $m+n$ integers numbers. The first m numbers correspond to the machines that appear in the rows of the incidence matrix and the remaining n numbers identify the parts that appear in the columns of the matrix. Ribeiro and Lorena [17] proposes a constructive genetic algorithm. Each individual is a sequence $m+n$ symbols. The values of each gene indicate that: (*i*) the machine/part is a median machine/part, (*ii*) the machine/part is assigned to its nearest median - by means of the measure of similarity Jaccard, or (*iii*) the machine/part is not yet assigned to a median. Gonçalves and Resende [18] proposes a genetic algorithm based on the Lamarckian evolution [20] - the result of the improvement in the aptitude of an individual is translated in a change of its genetic structure. Each individual is a sequence of m real numbers more one integer number. The real part of an individual encodes the cell to which a machine is assigned and the integer part of an individual encodes the number of cells of the solution.

Genetic Algorithms (GAs) are stochastic search methods based on the principles of natural selection and genetics, and have been successfully used to solve several combinatorial optimization problems that are characterized by large solution spaces. Goldberg [21] and Michalewicz [22] present a thorough description of GAs. The present work presents a GA for solving the standard manufacturing cell formation problem, which only considers the processing of a set of parts by a set of machines, ignoring other production factors. More complex cases with multiple routing and other production factors are analyzed in [23, 24, 25]. The following section includes a formulation of the manufacturing cell problem. Section 3 introduces the GA and the developed algorithm. Section 4 summarizes the numerical results for a set of problem found in the literature. Finally, section 5 includes our conclusions.

2 The Cell Formation Problem

The incidence matrix $A=[a_{ij}]$ represents a manufacturing system with m machines, $i=1,...,m,$ and n parts, $j=1,...,n$. Each element $a_{ij} \in \{0,1\}$, and $a_{ij}=1$ indicates that machine i is used for processing part j, see Figure 1. The goal is to form manufacturing units composed of machine cells and part families that are as autonomous as possible. Ideally all the parts of one family would be processed in one cell of machines. However, this type of situation is uncommon in real word applications since in general there are parts and machines that are exceptional. An exceptional machine processes more than one family of parts, and an exceptional part requires processing by machines that belonging to more than one cell. Therefore, maximizing autonomy corresponds to minimize the number of exceptional parts and machines.

Machines \ Parts	1	2	3	4	5	6	7	8	9	10	11
1	1	1	0	0	0	1	0	0	0	0	0
2	0	1	0	0	0	1	0	0	1	0	0
3	1	0	1	0	0	0	1	0	0	0	1
4	0	0	1	0	0	0	1	0	0	0	0
5	0	0	1	1	0	0	0	0	0	0	1
6	0	0	0	1	1	0	0	0	0	1	0
7	0	0	0	0	1	0	0	1	0	1	0

(a) Incidence Matrix

Machines \ Parts	4	5	8	10	1	2	6	9	3	7	11
6	1	1	0	1	0	0	0	0	0	0	0
7	0	1	1	1	0	0	0	0	0	0	0
1	0	0	0	0	1	1	1	0	0	0	0
2	0	0	0	0	0	1	1	1	0	0	0
3	0	0	0	0	1	0	0	0	1	1	1
4	0	0	0	0	0	0	0	0	1	0	0
5	1	0	0	0	0	0	0	0	1	1	1

(b) Matrix Solution Figure 1.a

Figure 1.Boctor´s Matrix

Figure 1 (b) illustrates in gray the Boctor's matrix [26] solution composed of machine cells {(6,7), (1,2), (3,4,5)} and of part families {(4,5,8,10), (1,2,6,9), (3,7,11)}. The exceptional elements correspond to entries in the matrix having a value 1, outside of the gray diagonal block. Sarker and Khan [27] carry out a comparative study of the most often used measures of the quality of solutions. In this paper we use the grouping efficacy Γ [28] according to Equation 1, where e denotes the number of entries of the matrix equal to 1, e_1 denotes the number of exceptional elements, and e_v is the number of entries of the gray diagonal blocks that are equal to 0.

$$\Gamma = \frac{e - e_1}{e + e_v} \tag{1}$$

3 Proposed Genetic Algorithm

Genetic Algorithms (GA) are an evolutionary process of a population of individuals representing a set of solutions of an optimization problem. Each individual is made up of *chromosomes.* One *chromosome* is a set of *genes* encoding the problem decision variables. Each member of the population is evaluated to determine its *fitness* or quality.

To evolve towards a better population of individuals, GA uses genetic operators such as selection, crossover and mutation. The selection operator is used to select the pared parents of the population that reproduce themselves. The Crossover operator combines the characteristics of the *parents* - individuals in the population - to generate offsprings - individuals of the next generation. This is achieved by the exchange of gene fragments. The mutation operator alters - with a low probability -

one or more genes of the offsprings. The evolution process continues until some stopping criteria is met (such as maximum number of iterations, convergence of the population). The pseudo-code for GA can be summarized as follows. $P(g)$ denotes the population in generation g and $P'(g)$ an intermediate population.

```
Begin
   g←0;
   Generate(P(g))
   Evaluate (P(g))
   While Continue(P(g))
      P'(g)←Selection(P(g))
      Crossover(P'(g))
      Mutate(P'(g))
      Evaluate(P'(g))
      g←g+1
   End While
End
```

3.1 Generation of the Initial Population and Genetic Operators

In order to avoid the generation of infeasible individuals (a machine in different cells) we use a randomized generation without replacement to generate the initial population, the partially matched crossover [29], the remainder stochastic sampling without replacement selection operator [21], and the uniform mutation [20].

3.2 Encoding the solutions

Each individual is a string of characters from a determined alphabet. In our case, we use an integer alphabet, $\{1,...,m\}$ where m is the number of machines. An individual g is a vector of k_{max} integer variables $(g_1,g_2,...,g_{kmax})$, where k_{max} is the number of expected families of parts and machine cells. The element g_k for an individual corresponds to the a machine assigned to cell k. Furthermore, each machine i, i=1,..,m, is assigned to a cell k according to Equation 2, where c_i is the cell of machine i, g_k is the gene k of chromosome g, and $\delta(v, w)$ corresponds to the Hamming distance between machines v and w as specified in Equation 3. Cells C^t_k, k=1,...,k_{max}, of the iteration t, with t=0, are determined according to Equation 4.

$$c_i = \begin{cases} k & if \quad i = g_k \\ \arg\min_{k \le k_{\max}} \{\delta(i, g_k\} & otherwise \end{cases} \tag{2}$$

$$\delta(v,w) = \sum_{s=1}^{n} |a_{vs} - a_{ws}| \tag{3}$$

$$C^t_k = \bigcup_{i=1|k=c_i}^{m} i \tag{4}$$

3.3 Formation of a Family of Parts on the Basis of Machine Cells

STEP 1. For each part j, j=1,...,n, calculate the number of operations in the cell C_k^t, k=1,..., k_{max}, using Equation 5.

$$\alpha_{kj} = \sum_{i \in C_k^t} a_{ij} \tag{5}$$

STEP 2. For each part j, j=1,...,n, calculate the number of un-required operations in cell C_k^t, k=1,..., k_{max}, using Equation 6.

$$\beta_{kj} = \left|C_k^t\right| - \alpha_{kj} \tag{6}$$

STEP 3. For each part j, j=1,...,n, calculate the number of required operations for machines not present in the cell C_k^t, k=1,..., k_{max}, using Equation 7,

$$\omega_{kj} = \sum_{q=1|q \neq k}^{k_{\max}} \alpha_{qj} \tag{7}$$

STEP 4. Associate part j with cell C_k^t according to Equation 8, where f_j is the family of part j

$$f_j = \arg\max_{k \leq k_{\max}} \{\frac{\alpha_{kj}}{\beta_{kj} + \omega_{kj}}\} \tag{8}$$

STEP 5. Each family of parts F^t_k, k=1....,k_{max}, is formed according to Equation 9.

$$F_k^t = \bigcup_{j=1|k=f_j}^{n} j \tag{9}$$

3.4 Formation of Machine Cells on the Basis of Part Families

STEP 1. For each machine i, i=1,...,m, calculate the number of operations in the family F_k^{t-1}, k=1,..., k_{max}, using Equation 10.

$$\eta_{ki} = \sum_{j \in F_k^{t-1}} a_{ij} \tag{10}$$

STEP 2. For each machine i, i=1,...,m, calculate the number of un-required operations in the family F_k^{t-1}, k=1,..., k_{max}, using Equation 11.

$$\mu_{ki} = \left|F_k^{t-1}\right| - \eta_{ki} \tag{11}$$

STEP 3. For each machine i, i=1,...,m, calculate the number of operations required by parts not present in the family F_k^{t-1}, k=1,..., k_{max}, using Equation 12.

$$\xi_{ki} = \sum_{q=1|q \neq k}^{k_{\max}} \eta_{qi} \tag{12}$$

STEP 4. Associate machine i with family F_k^{t-1} according to Equation 13.

$$c_i = \arg\max_{k \leq k_{\max}} \{\frac{\eta_{ki}}{\mu_{ki} + \xi_{ki}}\} \tag{13}$$

STEP 5. Form cell C_k^t, k=1,...,k_{max}, according to Equation 4.

3.5 Fitness

The fitness function is designed to determine a number which is supposed to be proportional to the ability of the individual to survive [21]. For the Cell Formation problem, we used a recursive fitness function *f*(*g*) based on the Baldwin effects [20] allowing to increase the individual survival opportunity without changing its genetic structured. The pseudo-code for *f(g)* can be summarized as follows. *f*(*g*) denotes the fitness of individual *g*, and Γ^t denotes the grouping efficacy at time *t*.

```
Begin
   t←0, iteration t
   f(g)←Γ⁰ //grouping efficacy
   Do
      t←t+1
      Form groups of machines Cᵗ
      Form groups of part families Fᵗ
      Calculate Γᵗ
   While (Γᵗ⁻¹ <Γᵗ)
   f(g)←Γᵗ ⁻¹
End
```

Accordingly, *f*(*g*), the fitness of an individual (genotype), is the largest value in the grouping efficacy evaluated by Equation 14 in the phenotype space. Also, it is possible that some individuals include fewer cells and families that expected, and thus their fitness is penalized according to Equation 15, where $|C_k|$ and $|F_k|$ it is the cardinality of the cell and family *k*, respectively.

$$\Gamma^t = \varphi \frac{\sum_{j=1}^{n} \alpha_{f_j j}}{\sum_{i=1}^{m} \sum_{j=1}^{n} a_{ij} + \sum_{j=1}^{n} \beta_{f_j j}} \tag{14}$$

$$\varphi = \frac{1}{1 + \sum_{k=1}^{k_{\max}} 4 - \min\{|C_k|, 2\} - \min\{|F_k|, 2\}} \tag{15}$$

4 Analysis of Results

The performance of our proposed method is analyzed by solving, a set of problems available in the literature. Each problem was solved twenty times and the following results are reported: the mean value of the best solutions, mean solution time (*t*) measured in seconds, and the mean generation (*g*) where the best value is obtained. The genetic algorithm was written in C++, and executed on a personal computer, Pentium IV, 1.6 GHz and 256 MB of RAM.

The genetic parameters used were: population size is equal to 3*m, probability of mutation is equal to 0.03, probability of crossover is equal 1, maximum number of generation is equal to 150, and length of chromosome is equal to k_{max}. The values selected allow an acceptable mean performance in all of the solved problems. For the test problems, the value of k_{max} corresponds to the best known cell value found in the literature.

Table 1 summarizes the results for the 25 problems in [14] obtained with the following algorithms: TSP-GA [13], ZODIAC [9], GA [14], and pGA, our proposed algorithm. When using Wilcoxon's test to study performance differences between algorithm pGA and the first three, one sees statistical evidence of our Genetic Algorithm superiority. Our proposed algorithm reaches better solutions in 19 problems with respect to ZODIAC, in 17 problems with respect to TSP-GA and in17 problems respect to GA. The pGA algorithm get similar results in 6 problems with respect to ZODIAC, in 7 problems with respect to TSP-GA, and in 6 problems with respect to GA. Finally, pGA obtains worst results in only 1 problem with respect TSP-GA, and 2 problems with respect to GA.

Table 1 Comparative Results

N°	Reference	*mxn*	k_{max}	Grouping Efficacy (%)				g	$t[s]$
				ZODIAC	TSP-GA	GA	*p*GA		
1	[30]	5x7	2	56.62	68.00	62.50	62.50	1	0.02
2	[31]	5x18	2	77.36	77.36	77.36	79.59	0	0.05
3	[32]	6x8	2	76.92	76.92	76.92	76.92	0	0.03
4	[33]	7x11	3	39.14	46.88	50.00	53.13	0	0.07
5	[26]	7x11	3	70.37	70.37	70.37	70.37	0	0.07
6	[8]	8x20	2	58.33	58.33	55.91	58.72	0	0.12
7	[34]	8x20	3	85.24	85.24	85.24	85.25	0	0.13
8	[35]	10x10	3	70.59	70.59	72.79	70.59	0	0.11
9	[36]	14x24	5	65.55	67.44	63.48	69.33	0	0.53
10	[2]	10x15	3	92.00	92.00	92.00	92.00	0	0.15
11	[1]	16x43	5	53.76	53.89	86.25	54.86	0	1.43
12	[35]	20x20	5	21.63	37.12	34.16	42.96	37	1.01
13	[5]	20x35	4	75.14	75.28	66.30	76.14	0	1.33
14	[37]	20x35	5	51.13	55.14	44.44	58.06	1	1.33
15	[38]	20x23	5	38.66	46.62	39.02	49.65	0	1.30
16	[39]	24x40	7	100.00	100.00	100.00	100.00	0	0.01
17	[39]	24x40	7	85.11	85.11	85.11	85.11	0	3.18
18	[39]	24x40	7	73.03	73.03	73.03	73.51	0	3.86
19	[39]	24x40	9	20.42	49.37	37.62	51.88	3	5.44
20	[39]	24x40	9	18.23	44.67	34.76	46.95	37	5.91
21	[39]	24x40	9	17.61	42.50	34.06	44.85	45	5.59
22	[39]	30x41	11	33.46	53.80	40.96	58.33	30	5.81
23	[39]	30x50	12	46.06	56.51	48.28	59.66	8	9.79
24	[36]	30x50	11	21.11	45.93	37.55	50.51	3	12.24
25	[8]	40x100	10	83.92	84.03	83.90	84.03	0	22.38

Table 1 indicates the average number of generations (g) necessary to achieve the best value, and it is extremely low. A value of zero indicates that the solution is achieved in the initial populations. In the worst case 45 generations are required. With respect to the average time required to reach the best solution, pGA requires less than two seconds for small problems, and less than 23 seconds for large problems such as problem nº 25.

Figure 2 show the average fitness function for generation over twenty different runs. The incidence matrix has 20 machines, 20 parts, and k_{max} is equal to 5. The maximum fitness is equal to 42.96% and the minimum fitness is 42.00%.

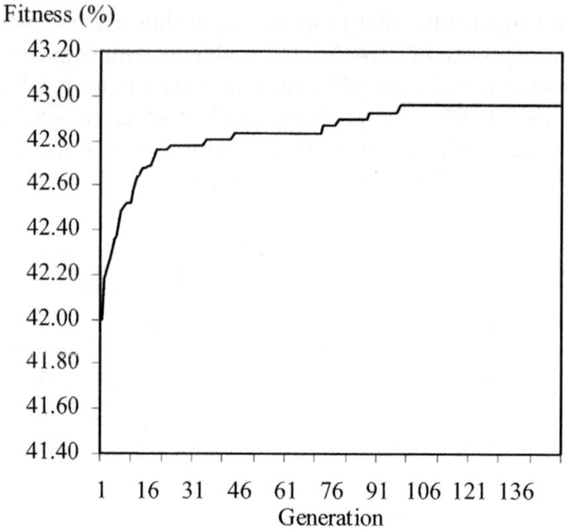

Figure 2. A Convergence of genetic search process.

Finally for some of the problems analyzed, two solutions having a similar fitness can differ substantially in terms of the components in each cell (see Figure 3). This makes it extremely difficult to escape local optimums.

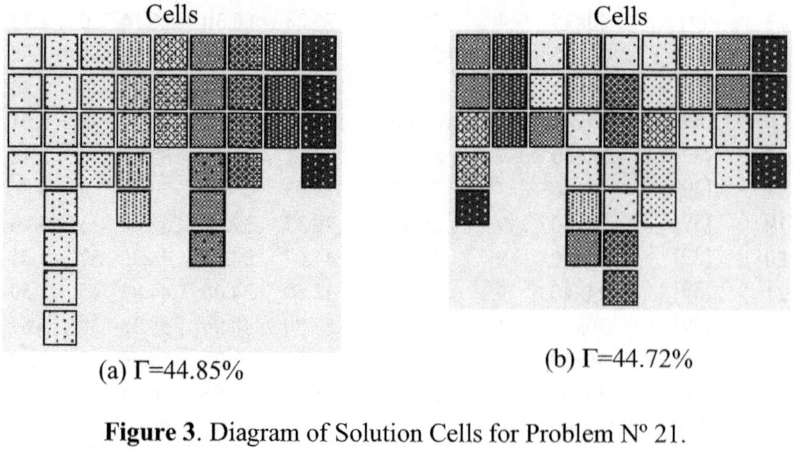

Figure 3. Diagram of Solution Cells for Problem Nº 21.

5 Conclusions

In this paper we introduce a genetic algorithm designed to maximize the efficiency of manufacturing cell grouping. The performance of the proposed algorithm was measured against 25 problems taken from the literature, and compared against the performance of the algorithms of Cheng et al. [13], ZODIAC [9], and Onwubolu and Mutingi [14]. In 88% of the examples the GA algorithm either equals or improves on the known solutions, (independent of the algorithm), and only in 12% of the cases does it achieve an inferior performance.

From the point of view of the design of a genetic algorithm, the inclusion of the Hamming distance reduces the long of the individuals and the solution time for highly structured incidence matrices. The space size of our algorithm is significantly smaller than that of the genetic algorithms presented in the Section 1. Furthermore, the Baldwin effect allows to maintain the population's diversity, thus reducing the probability that the algorithm stops in a low quality local optimum (see Figure 3).

Finally, since the recursive function fitness does not change the genetic structure of the individuals, our algorithm it is not able to escape from low quality local optimum without the collaboration of the crossover and the mutation. Clearly, our approach changes the test problems in problems of easy solution, being its main advantage its scaling capacity and its ability to solve large problems - generating a reduced number of quality solutions.

Acknowledgment

The authors thank the reviewer's comments that allowed to clarify and to improve the content of this work. This work is financed partly by the doctorate program in computer science of the University of Santiago of Chile.

References

1 King JR, (1980) Machine-Component Grouping In Production Flow Analysis: An Approach Using a Rank Order Clustering Algorithm. International Journal Of Production Research 18; 213- 232.

2 Chan HM, Milner DA, (1982) Direct Clustering Algorithm for Group Formation in Cellular Manufacture. Journal of Manufacturing Systems 1; 65-75.

3 Chu CH, Tsai M, (1990) A Comparison of Three Array-Based Clustering Techniques for Manufacturing Cell Formation. International Journal of Production Research 28 (8); 1417-1433.

4 McAuley J, (1972) Machine Grouping for Efficient Production. Production Engineer 51; 53-57.

5 Carrie AS, (1973) Numerical Taxonomy Applied to Group Technology and Plant Layout. International Journal of Production Research 4; 399-416.

6 Seifoddini H, Wolfe PM, (1987) Selection of a Threshold Value Based On Material Handling Cost in Machine-Component Grouping. IIE Transactions 19; 266-270.
7 Yasuda K, Yin Y, (2001) A Dissimilarity Measure for Solving the Cell Formation Problem in Cellular Manufacturing. Computers and Industrial Engineering 39; 1-17.
8 Chandrasekharan MP, Rajagopalan R, (1986a) An Ideal-Seed Non-Hierarchical Clustering Algorithm for Cellular Manufacturing. International Journal of Production Research, 24; 451- 464.
9 Chandrasekharan MP, Rajagopalan R, (1987) Zodiac: An Algorithm for Concurrent Formation of Part Families and Machine-Cells. International Journal of Production Research 25; 835- 850.
10 Srinivasan G, Narendran, TT, (1991) Grafics: A Non-Hierarchical Clustering Algorithm for Group Technology. International Journal of Production Research 29; 463-478.
11 Srinivasan G, (1994) A Clustering Algorithm for Machine Cell Formation in Group Technology Using Minimum Spanning Trees. International Journal of Production Research 32; 2149-2158.
12 Joines JA, Culbreth CT, King RE, (1996) Manufacturing Cell Design: An Integer Programming Model Employing Genetic, IIE Transactions 28; 69-85.
13 Cheng CH, Gupta YP, Lee WH, Wong KF, (1998) A Tsp-Based Heuristic for Forming Machine Groups and Part Families. International Journal of Production Research 36; 1325-1337.
14 Onwubolu GC, Mutingi M, (2001) A Genetic Algorithm Approach to Cellular Manufacturing Systems, Computers & Industrial Engineering 39; 125-144.
15 Dimopoulos C, Mort N, (2001) A Hierarchical Clustering Methodology Based On Genetic Programming for the Solution of Simple Cell-Formation Problems. International Journal of Production Research 39; 1-19.
16 Mark KL, Wong YS, Wang XX, (2000) An Adaptive Genetic Algorithm for Manufacturing Cell Formation. The International Journal of Advanced Manufacturing Technology 16; 491-497.
17 Ribeiro G, Lorena L, (2000) A Constructive Evolutionary Approach to the Machine-Part Cell Formation Problem in Buildings Competencies for International Manufacturing - Perspectives for Developing Countries, A. Fleury, H. Yoshizaki, L. B. M. Guimaraes And J. L. D. Ribeiro (Eds.) Ufrgs/Feeng, Porto Alegre Brazil, 340-348.
18 Gonçalves JF, Resende MG, (2002) A Hybrid Genetic Algorithm for Manufacturing Cell Formation, Submitted to Computers and Industrial Engineering.
19 Whitley D, (1989) The Genitor Algorithm and Selection Pressure: Why Rank-Based Allocation of Reproductive Trials is Best- Proceedings of The Third International Conference of Genetic Algorithms, Don Mateo, California, Los Altos, Ca: Morgan Kaufmann, 116-121.
20 Joines JA, King RE, Culbreth CT, (1997) Cell Formation Using Genetic Algorithms. In Group Technology And Cellular Manufacturing, Ed. N. C. Suresh And J. M. Kay, Boston: Kluwer, Inc.
21 Goldberg D, (1989) Genetic Algorithms in Search, Optimization, and Machine Learning, Addison Wesley Publishing Company, Massachusetts.

22 Michalewicz Z, (1996) Genetic Algorithms + Data Structures=Evolution Programs. New York, (Springer-Verlag).
23 Moon C, Kim J, (1999) Genetic Algorithm for Maximizing the Parts Flow Within Cells in Manufacturing Cell Design, Computers & Industrial Engineering 36; 379-389.
24 Moon C, Gen M, (1999) A Genetic Algorithm-Based Approach for Design of Independent Manufacturing Cells, International Journal of Production Economics 60-61; 421-426.
25 Gravel M, Nsakanda A, Price W, (1998) Efficient Solutions to the Cell-Formation Problem with Multiple Routing via Double-Loop Genetic Algorithm. European Journal of Operational Research 109; 286-298.
26 Boctor FF, (1991) A Linear Formulation of the Machine-Part Cell Formation Problem. International Journal of Production Research 29; 343-356.
27 Sarker B, Khan M, (2001) A Comparison of Existing Grouping Efficiency Measures and a New Weighted Grouping Efficiency Measure. IIE Transactions 33; 11-27.
28 Kumar CS, Chandrasekharan MP, (1990) Grouping Efficacy: A Quantitative Criterion for Goodness of Block Diagonal Forms of Binary Matrices In Group Technology. International Journal of Production Research 28; 233-243.
29 Goldberg, D, Lingle R, (1985) Alleles, Loci, and the Traveling Salesman Problem. Proceedings of an International Conference on Genetic Algorithms and Their Applications; 155-159.
30 Waghodekar PH, Sahu S, (1984) Machine-Component Cell Formation in Group Technology Mace. International Journal of Production Research 22; 937-948.
31 Seifoddini H, Wolfe PM, (1986) Application of the Similarity Coefficient Method in Group Technology. IIE Transactions 18; 271-277.
32 Kusiak A, Cho M, (1992) Similarity Coefficient Algorithm for Solving the Group Technology Problem. International Journal Of Production Research 30; 2633-2646.
33 Kusiak A, Chow WS, (1987) Efficient Solving of The Group Technology Problem. Journal of Manufacturing Systems 6; 117-124.
34 Chandrasekharan MP, Rajagopalan R, (1986) Modroc: An Extension of Rank Order Clustering for Group Technology. International Journal of Production Research 24; 1221- 1233.
35 Mosier L, Taube C, (1985) Weighted Similarity Measure Heuristics for the Group Technology Clustering Problem. Omega 13; 577-579.
36 Stanfel LE, (1985) Machine Clustering for Economic Production. Engineering Costs and Production Economics 9; 73-81.
37 Boe WJ, Cheng CH, (1991) A Close Neighbour Algorithm for Designing a Cellular Manufacturing System. International Journal of Production Research, 29; 2097-2116.
38 Kumar KR, Kusiak A, Vannelli A, (1986) Grouping Parts and Components in Flexible Manufacturing System. European Journal of Operational Research , 24; 387-397.
39 Chandrasekharan MP, Rajagopalan R, (1989) Groupability: An Analysis of the Properties of Binary Data Matrices for Group Technology. International Journal of Production Research 27; 1035-1052.

Evolutionary Simulated Annealing Algorithms for Uncapacitated Facility Location Problems

Vecihi Yigit [1], M. Emin Aydin [2], and Orhan Turkbey [3]

[1]Ataturk University, Faculty of Engineering, Dept. of Industrial Engineering, Erzurum, Turkey
vyigit@atauni.edu.tr

[2] University of the West of England, Advanced Computation for Design and Decision making Group, Frenchey Campus, Bristol, UK.
mehmet.aydin@uwe.ac.uk

[3] Gazi University, Dept. of Industrial Engineering, Ankara, Turkey
turkbey@gazi.edu.tr

Abstract

Simulated annealing (SA) is one of the potentially powerful probabilistic metaheuristics to solve large-scale combinatorial optimisation problems. The main drawback with this metaheuristic is its time consuming nature, although it gives more confidence to reach the global optimum. The aim of this paper is to examine an evolutionary approach to simulated annealing for Uncapacitated Facility Location (UFL) problems with some useful comparisons with the latest genetic algorithm approach by Jaramillo et al [17]. The approach presented in this paper seeks to combine the power of both SA and the evolutionary approach to get a desirable quality of solution within a shorter time. For this purpose, SA is incorporated with evolutionary approach in order to cut down the processing time needed.

1 Introduction

Facility Location Problems that have NP-Hard nature play an important role in operation research and the manufacturing context [1]. The uncapacitated facility location (UFL) problem, also known as simple plant location problem, is basically a member of the family of location problems. In a general form, the problem is to determine the optimal number of facility echelons, the number and location of facilities in each facility echelon, the assignment of distribution activities/commodities to facilities, and the allocation of customers' demands among the facilities. UFL problems have several applications, such as the bank account location problem, the clustering problem, economic lot sizing, vehicle routing, network design, distributed data and communication networks.

UFL problems have been studied for many years providing a very rich literature in operations research (OR). Since they are NP-Hard problems [1,2], the larger the size of the problem, the harder to find the optimal solution and furthermore, the longer to reach reasonable results. During the past three decades, these UFL problems have been triggered and examined extensively by various attempts and approaches [1,2]. There are mainly two categories of approaches to this type of problems: classical OR, (branch and bound, primal and dual ascent methods, linear programming and Lagrangean relaxation algorithms) [3, 4, 5, 6, 7, 8, 9, 10, 11, 12] and meta-heuristic based methods [13, 14, 15, 16, 17, 18].

The first meta-heuristic based method is Alves and Almeida's simulated annealing algorithm [13]. In this work, the results of four versions of their simulated annealing algorithm were compared with those of some well-known heuristic methods. The quality of the solutions is good, but the computing time is significantly longer in comparison to the others. TANN (Taboo Neural Network) by Vaithyanathan [14] integrated an analogue version of the short term memory component of taboo search with neural networks to generate a massively parallel, analogue global search strategy, which was applied on UFLP instances of small dimensions (10x10, 20x20 and 30x15). The obtained solutions have errors of more than 20% from optimal ones. Kratica et al. [16] applied genetic algorithms to UFL problems to solve 1000x1000-sized customer-facility pairs. They considered many benchmarks from the literature to be solved by their algorithm in addition to those large sized problems they generated. They compared their results with Erlenkotter's [3] dual-based algorithm (DUALOC) showing that their algorithm outperforms DUALOC for the problems larger than 100x1000. Although DUALOC performs well for some benchmarks, it is very time consuming. Jaramillo et al. [17] applied a genetic algorithm that is mainly based on the operators that Beasley and Chu [15] applied in their genetic algorithms for covering problems. Ghosh [18] applied neighbourhood search heuristics based on taboo search and complete local search with memory to solve large instances of the UFL problem. His computational experiments show that on medium sized problem instances, both heuristics return solutions with costs within 0.075% of the optimum with execution times that are often several orders of magnitude less than exact algorithms.

The aim of this paper is to examine an evolutionary approach to the two different simulated annealing algorithms for solving UFL problems. As mentioned previously, simulated annealing approaches may be very successful in terms of the quality of solutions but not so impressive with respect to CPU times. The approach presented within this paper seeks to use the power of the evolutionary approach in getting quality of solutions within shorter times. For this purpose, simulated annealing algorithms are incorporated with an evolutionary approach to minimise the time needed.

The rest of this paper is organised as follows. The tackled problem is briefly introduced in Section 2. The foundations of the evolutionary simulated annealing are discussed in Section 3. Section 4 provides the experimental results with analysis and discussions some for improved SA implementation and finally Section 5 presents the conclusions.

2 Problem Description

The combinatorial problem tackled here is a very well-known in operations research; Uncapacitated Facility Location (UFL) problem. The mathematical formulation of this problem type as mixed integer programming models has proven very fruitful in the derivation of solution methods [1, 2]. To formalise the UFL problems, consider a set of candidate sites for facility location, $I=\{1,..,m)$, and a set of customers, $J=\{1,...,n\}$. Each facility $i \in I$ has a fixed cost, f_i. Every customer $j \in J$ has a demand, b_j, and c_{ij} is the unit transportation cost from facility i to customer j. Without a loss of generality, the customer demands can be normalised with $b_j = 1$. Thus, the problem is formulated in the following way:

$$Z = \min \sum_{i=1}^{m} \sum_{j=1}^{n} c_{ij} x_{ij} + \sum_{i=1}^{m} f_i y_i \tag{1}$$

subject to:

$$\sum_{i=1}^{m} x_{ij} = 1, \qquad \text{for } \forall j \in J; \tag{2}$$

$$0 \le x_{ij} \le y_i \quad \text{or} \quad x_{ij}, y_i \in \{0; 1\}; \qquad \text{for } \forall i \in I \text{ and } \forall j \in J; \tag{3}$$

where x_{ij} represents the quantity supplied from facility i to customer j, y_i indicates whether facility i is established ($y_i = 1$) or not ($y_i = 0$). In this case, x_{ij} is kept integer as the capacities and the size of demand ($b_j = 1$) are ignored. Thus, once the demand of one client allocated to a particular site, it is assumed that all demands of the client is assigned to that site. In this sense, the constraint (2) makes sure that all demands have been met by a particular open site, and the constraint (3) is to keep variables integer. By this model, both (i) the number of facility sites to be established and (ii) the quantities to be supplied from facility i to customer j must be determined, such that the total cost (including fixed and variable costs) is minimised.

3 Evolutionary Simulated Annealing (ESA)

Simulated annealing can be viewed as a probabilistic decision making process in which a control parameter called temperature is employed to evaluate the probability of accepting an uphill move (in the case of minimisation). As pointed out in the literature [24]. SA is able to converge to the optimum value, but it could be very expensive to get a desired solution in terms of computational time. For that reason, SA has been studied by hybridising with other methods such as the genetic algorithms [21] or by parallelising the algorithm [23]. The evolutionary simulated annealing algorithm was developed to overcome such problems. It offers an evolutionary process in which a shorter SA algorithm substitutes the genetic operators to evolve a population of solutions. The SA algorithm is so compacted that one can easily use it in any evolutionary process, where SA can manipulate the solutions selected from the population. This makes ESA easily implementable in various environments and together with different methods.

Begin
- ⇨ *Initialise* the **population (*P*)**,
- ⇨ *Set* the **number of evaluations (*N*)**

Repeat:
- *Select* an **individual (p_n)**
- *Operate* by the **SA** and *get* the **new individual (p_n')**
- *Evaluate* the **new individual** for **replacement**

Until ***n>=N***

End.

Figure 1; a pseudo code for ESA algorithm

ESA adopts a SA as an evolutionary operator together with other evolutionary operators. We embedded a shortened SA into an evolutionary algorithm, which does not contain any reproduction operators (crossover, mutation). The algorithm is sketched in Fig. 1. After initialisation and parameter setting, the algorithm repeats the following steps: (i) selects one individual subject to the running selection rule, (ii) operates it by the SA algorithm, and (iii) evaluates whether to replace it in the population or not according to a particular replacement rule. The details of SA operator are given in Fig. 2, where a shortened SA process is working. The usual elements of a typical simulated annealing algorithm are contained. Depending on the presence of inner repetitions, which are implemented during the acceptance stage before decreasing the level of temperature, the neighbourhood function works once or several times to cool the temperature per iteration. Therefore, the total number of moves per SA operation could be 200, as the highest temperature of 100 is decayed by 0.955 per iteration until it reaches 0.01 with no inner repetitions. The implementation will change if one change the set up parameters outlined above.

Obviously, the only running operator together with selection is the SA. Since the SA operator re-operates on particular solutions several times, the whole method works as if it is reheated after a number of iterations. That is set up by specifying *f(t)* function. If we suppose that there is a single solution operated by this SA, it will become a multi-start (not multi-run) algorithm that reruns repeatedly. Thus, the novelty of ESA can be viewed from two points of view: firstly its multi-start property, and secondly its evolutionary approach. The multi-start property provides ESA with a more uniform distribution of random moves along the whole procedure that helps to diversify the solutions. In fact, typical SA works in such a way that the search space is explored by exponentially distributed random moves, where each random move starts a new hill climbing process to reach the global minimum. However, the random moves are condensed in the earlier stages of the cooling process and the probability of having random moves goes down exponentially. Since it almost behaves like a hill climber in the later stages of the process, it becomes harder to escape from local minima, especially, when it is applied to very difficult combinatorial optimisation problems. The idea is to distribute the random moves more uniformly than exponentially across the whole process.

```
Begin:
  • pick one feasible solution (p_n),
  • set the highest temperature (t=100),
  • set i=0 and p_i = p_n
  Repeat:-
       repeat
          ▪ conduct a move by neighbourhood function to get a new solution (p_i')
          ▪ if (p_i'-p_i)<0 then p_i+1 = p_i'
          ▪ else
                 – generate a random number (r)
                 – if exp(-(p_i'-p_i)/t)>r then p_i+1 = p_i'
                 – else p_i+1 = p_i
                 – endif
          ▪ endif
       until pre-defined number of inner iterations
          ▪ t=f(t) and i=i+1
  Until t<0.01
  • p_n' = p_i
End.
```

Figure 2; a pseudo code for SA operator

The other property of ESA is to tackle a population of solutions rather than an individual. This reduces the effects of the initial solutions on the optimisation process. There are many works to solve combinatorial optimisation problems by heuristics has focused on the effects of the initial solutions. When an initial solution has been chosen, there arise limited possible paths to proceed since the optimisation process behaves as a Markov chain and each chain offers limited paths to the destination, as widely shown in the literature [21]. Looking at the initial conditions, one can estimate the probability of getting an optimal or useful near optimal solution with a particular initial solution. In fact, under all given initial conditions, it is quite difficult to avoid the local optima in reasonable time. Therefore, it can be said that a diverse population of initial solutions gives higher probability than a single initial solution to catch the optimum or a useful near optimum within a reasonable time. Moreover, if useful selection and replacement strategies can be utilised, it will certainly help the process to improve the quality of solutions. For this reason, the ESA algorithm is run on a population of solutions rather than an individual using various selection and replacement strategies.

3.1 Representation and neighbourhood function

UFL problems are usually represented in a binary way; the open facilities are denoted by 1 and closed ones by 0. In order to move from one solution to another, the digits are simply converted to each other. Depending on the heuristics employed, this conversion is done either by a neighbourhood function, as happens here, or a genetic operation. We represent the state of solutions by a list of integers, say F, denoting the enumerated open facility, (f_i). For instance, $F=\{1,4,12\}$ means the open facilities are Number *1*, *4* and *12*, the rest are considered as closed. As we denote the particular solution state at time t with s_t, one state of solution will be $s_t = \{F\} = \{f_i | 0 < i \leq m\}$ where m is the total number facilities.

The neighbourhood function employed allows us to modify the states by three different operations: (1) change one integer on the list with another possible one; (2) add up a new integer to the list; (3) remove one integer from the list, meaning exchange the state of one facility with another, or open a new facility or close one open facility, respectively. It allows only one of these operations to be applied at each iteration. One operation is randomly selected according to the following rule:-

$$operator \leftarrow \begin{cases} \text{Exchange()} & (\lambda(L)=1 \cap 0 \leq \rho < 0.7) \cup (\lambda(L)>1 \cap 0 \leq \rho \leq 0.5) \\ \text{Add()} & (\lambda(L)=1 \cap 0.7 \leq \rho < 1) \cup (\lambda(L)>1 \cap 0.5 \leq \rho \leq 0.7) \\ \mathit{Remove()} & (\lambda(L)=m) \cup (\lambda(L)<m \cap 0.7 \leq \rho \leq 1) \end{cases}$$

where $\lambda(F)$ is the length of the list, ρ is a uniformly generated random number, and m is the maximum number of facilities as usual. By applying this function, we move to a neighbouring state. This is a preventive neighbourhood function that keeps the solution feasible by letting the operators manipulate on convenience. The convenience of one situation is determined by both the length of the list and the random number ($\lambda(F)$, ρ). *Exchange()* and *Add()* are allowed to work when $\lambda(F)=1$, on the other hand, *Remove()* is allowed to work when $\lambda(F)=m$. For instance, if $\lambda(F)>1$ then any of the operations can be selected according to ρ, on the other hand, if $\lambda(F)=m$ then only *Remove()* operator is allowed to operate. The idea here is to prevent the solution from moving to an infeasible state, i.e. close all sites.

4 Experimental Results

This experimental study was performed using an AMD Athlon 1.7 GHz processor computer, running Windows XP. All the software is developed using Sun Java JDK1.3.1. In the following sections, we present an analytic comparison for particular SA implementations, and then compare with a very recently published GA application for UFL problems.

The problems tackled are very well known benchmarks that are accessible on the OR Library [20]. The name of benchmarks are indicated in the first column of each table. The first four benchmarks are 16 x 50, the second four are 25 x 50, the third four are 50 x 50 and finally, and the last three are 100 x 1000 sized problems as given in the second column of Table 1.

4.1 Comparative study

In this study, a number of ESA implementations have been developed to solve UFL problems. The implementations differ from each other by the population size, the number of generations and the numbers of inner iterations. In order to have a fair comparison with the latest work, a genetic algorithm that was reported by Jaramillo et al [17] was re-encoded to run on the same resources and using the same language (Java). In fact, the software was originally written in FORTRAN 90 and executed on PC equipped with a Intel Pentium running at 200MHz. Regarding the big difference between our experimental conditions and the experimental conditions they had, it would not be fair, if we just consider their results as standing in the paper. We encoded the algorithm in Java without missing any details given in their paper.

On the other hand, we have implemented two different simulated annealing algorithms based on ESA. The main difference between two implementations is the presence of inner repetitions. In first algorithm we set the population size to 5 and number of generations to 10. Initial temperature is *100*, the stopping level of temperature is *0*.01 and the cooling function is;

$$f(t) = \begin{cases} 0.955t & if \;\; m < 25 \\ 0.995t & otherwise \end{cases}$$

where m is the total number of facilities and t is the current level of temperature. In this case the first implementation (SA-I) does not have any inner repetitions, while the second implementation, SA-II, has 1, and 3 for the benchmarks have less than 25 candidate facilities and more than 25, respectively. Another difference of second algorithm is the number of generations. In second implementation we set the number of generations is to 5. Experiments are done 100 times for each benchmark. The CPU time consumed is measured as the time of the last best result found.

Table 1 presents the results gained by the genetic algorithm specified by Jaramillo et al [17], SA-I and SA-II. The columns show the information in the following order: the name of the benchmark, the size of the benchmark, optimum value, the results found with respect to both the percentage of deviation from the optimum (quality of the solution) and CPU time spent. The percentage of deviation has been calculated as the difference between the optimum and the result found divided by the optimum. The quality of results we found with GA are slightly worse than those produced by [17], as 7 out of 15 are slightly deviated from the optimum. This may be because of the random numbers used in the execution. The most important aspect is the time consumed in finding these solutions, which are much longer than the results that given in [17]. The reason behind this is the time consuming nature of Java over the other languages. The other results gained by SA-I and SA-II are all optimum with respect to the quality of solution and far better than GA in terms of CPU time. The SA-II is slightly better than SA-I in terms of CPU time. That means the inner repetitions are slightly beneficial. Depending on the hardness of the problems, it may help significantly.

Table 1 Experimental results obtained by various algorithms

Problem	Size	Optimum	Diff %			CPU		
			GA	SA-I	SA-II	GA	SA-I	SA-II
Cap 71	16 x 50	932615.75	0.00	0.00	0.00	0.287	0.0740	0.0121
Cap 72	16 x 50	977799.40	0.00	0.00	0.00	0.322	0.0617	0.0076
Cap 73	16 x 50	1010641.45	0.00033	0.00	0.00	0.773	0.0506	0.0084
Cap 74	16 x 50	1034976.98	0.00	0.00	0.00	0.200	0.0309	0.0058
Cap 101	25 x 50	796648.44	0.00020	0.00	0.00	0.801	0.1198	0.0534
Cap 102	25 x 50	854704.20	0.00	0.00	0.00	0.896	0.0718	0.0248
Cap 103	25 x 50	893782.11	0.00015	0.00	0.00	1.371	0.1062	0.0287
Cap 104	25 x 50	928941.75	0.00	0.00	0.00	0.514	0.0317	0.0104
Cap 131	50 x 50	793439.56	0.00065	0.00	0.00	6.663	0.3278	0.4254
Cap 132	50 x 50	851495.33	0.00	0.00	0.00	5.274	0.0895	0.2164
Cap 133	50 x 50	893076.71	0.00037	0.00	0.00	7.189	0.1458	0.3085
Cap 134	50 x 50	928941.75	0.00	0.00	0.00	2.573	0.0336	0.0926
Cap A	100 x 1000	17156454.5	0.00	0.00	0.00	184.422	1.0971	1.8138
Cap B	100 x 1000	12979071.6	0.00172	0.00	0.00	510.445	9.6294	7.2766
Cap C	100 x 1000	11505594.3	0.00131	0.00	0.00	591.516	26.6694	22.5028

5 Conclusion

Facility location problems have been studied for many years and thus there is a very rich literature in operations research for these kinds of problems. Since UFL problems are NP-Hard, the larger the size of the problem, the harder to find the optimal solution and furthermore, the longer to reach a reasonable results. This paper discussed the examination of an evolutionary approach to simulated annealing for UFL problems with some useful comparisons with the latest genetic algorithm approach [17]. As is mentioned in many works, the simulated annealing approaches may be very successful in terms of quality of the solutions but not so impressive with respect to the CPU times. The approach presented in this paper seeks to use the power of evolutionary computation in getting quality of solutions within a shorter time. The ESA implementation with evolutionary approach outperformed the GA of Jaramillo et al. [17], which is the latest method published as a successful UFL problem solver.

References

1. Krarup J. and Pruzan, P.M., The Simple Plant Location Problem: Survey and Synthesis, European J. Oper. Res. 12, 36-81,1983.
2. Cornuejols, G., Nemhauser, G.L., Wolsey, L.A., The uncapacitated facility location problem, in: Mirchandani P.B. and Francis R.L.(Eds), Discrete Location Theory, Wiley & Sons, 1990, pp. 119-171.
3. Erlenkotter, D. A Dual-Based Procedure for Uncapacitated Facility Location. *Oper. Res.***26** (1978) 992-1009.
4. Christofides, N, Beasley JE, Extensions to a Lagrangean relaxation approach for the capacitated warehouse location problem , European Journal of Operational Research, vol.12, (1983), pp19-28
5. Guignard, M. A Lagrangean Dual Ascent Algorithm for Simple Plant Location Problems, *European J. Oper. Res.* **35** (1988) 193-200.
6. Simao H.P. and Thizy, J.M., A Dual Simplex Algorithm for the Canonical Representation of the Uncapacitated Facility Location Problem. *Oper. Res. Lett.* **8** (1989) 279-286
7. Koerkel, M. On the Exact Solution of Large-Scale Simple Plant Location Problems. *European J. Oper. Res.* **39** (1989) 157-173.
8. Beasley, J.E. Lagrangean Heuristic for Location Problems. European J. Oper. Res., 65, 383-399, 1993.
9. Gao, L.L. Robinson E. and Powell, Jr. Uncapacitated Facility Location: General Solution Procedure and Computational Experience, European J. Oper. Res., 76, 410-427, 1994.
10. Holmberg, K., and Jörnsten, K., "Dual Search Procedures for The Exact Formulation of The Simple Plant Location Problem with Spatial Interaction", *Location Science* 4 (1996) 83 – 100
11. Goldengorin, B., Tussen, G.A., Ghosh, D., and Sierksma, G., "Solving The Simple Plant Location Problem Using a Data Correcting Approach", J. Global Optimization, 25 (2003) 377-406
12. Goldengorin, B., Ghosh, Diptesh and Sierksma, G., "Branch and Peg Algorithms for The Simple Plant Location Problem", *Computers & Operations Research*30 (2003), 967-981.
13. Alves M.L. and Almeida, M.T. Simulated Annealing Algorithm for the Simple Plant Location Problem: A Computational Study. *Rev. Invest.* **12** (1992).
14. Vaithyanathan, S., Burke, L. and Magent, M.A., Massively Parallel Analog Tabu Search Using Neural Networks Applied to Simple Plant Location Problem, *European J. Oper. Res.* 93 (1996) 317-330.
15. Chu, P.C., Beasley, J.E., "A genetic algorithm for the set covering problem", *European J. Oper. Res.* **94** (1996) 392-404.
16. Kratica J., Tošiæ D., Filipoviæ V., Ljubiæ I., "Solving the Simple Plant Location Problem by Genetic Algorithms", RAIRO - Operations Research, 35, No. 1, (2001), 127-142.
17. Jaramillo, J.H., Bhadury, J., Batta, R., "On the use of genetic algorithms to solve location problems", *Computers & Operations Research*, 29 (2002), 761-779.
18. Ghosh, D. "Neighborhood Search Heuristics for The Uncapacitated

Facility Location Problem", ", European Journal of Operational Research, 113, (1999), 123-136.

19. Reeves, Colin, "Simulated Annealing in: Modern Heuristic Techniques for Combinatorial Problems", John Wiley, 1993, pp.20-69.
20. Beasley, J.E. Obtaining Test Problems *via* Internet. *J. Global Optim.* **8** (1996) 429-433, http://mscmga.ms.ic.ac.uk/info.html
21. Kolonko M., "Some new results on simulated annealing applied to job shop scheduling problem", European Journal of Operational Research, 113, (1999), 123-136.
22. Christofides, N, Beasley JE, A tree search algorithm for the p-median problem, European Journal of Operational Research, vol.10, (1982), pp196-204
23. Aydin, M.E.,Fogarty, T.C., " A modular simulated annealing algorithm for multi-agent systems: A job shop scheduling application", In: Proceedings of ICRM 2002 *(2nd International Conference of Responsive Manufacturing),* 26-18 June 2002, University of Gaziantep, Gaziantep, Turkey.
24. Wong, S.Y.W., "Hybrid simulated annealing/genetic algorithm approach to short term hydro-thermal scheduling with multiple thermal plants", Electrical Power & Energy Systems, 23 2001, 565-575.

Chapter 4

Interactive Evolutionary Design and Manufacture

A Technique for Evaluation of Interactive Evolutionary Systems

M. Shackelford[1], D. W. Corne[2]

[1]School of Systems Engineering,
University of Reading,
Reading RG6 6AY
Mark.Shackelford@inkfish.co.uk

[2]School of Engineering, Computer Science and Mathematics
University of Exeter
Exeter EX4 4QF
D.W.Corne@exeter.ac.uk

Abstract

Very large scale scheduling and planning tasks cannot be effectively addressed by fully automated schedule optimisation systems, since many key factors which govern 'fitness' in such cases are unformalisable. This raises the question of an interactive (or collaborative) approach, where fitness is assigned by the expert user. Though well-researched in the domains of interactively evolved art and music, this method is as yet rarely used in logistics. This paper concerns a difficulty shared by all interactive evolutionary systems (IESs), but especially those used for logistics or design problems. The difficulty is that *objective evaluation* of IESs is severely hampered by the need for expert humans in the loop. This makes it effectively impossible to, for example, determine with statistical confidence any ranking among a decent number of configurations for the parameters and strategy choices. We make headway into this difficulty with an *Automated Tester* (AT) for such systems. The AT replaces the human in experiments, and has parameters controlling its decision-making accuracy (modelling human error) and a built-in notion of a target solution which may typically be at odds with the solution which is optimal in terms of formalisable fitness. Using the AT, plausible evaluations of alternative designs for the IES can be done, allowing for (and examining the effects of) different levels of user error. We describe such an AT for evaluating an IES for very large scale planning.

1 Introduction

Large-scale optimisation problems are often distinct from smaller scale problems in that the various factors pertinent to the quality of a candidate solution include several which cannot easily (or at all) be expressed formally. As logistics problems get larger (hence involving greater amounts of resource and larger timescales), planners need to think increasingly in terms of political, environmental, and social factors, which are typically indeterminate. This has long been recognised; going back to one of the seminal works in algorithmic project planning [1]. E.g. a five-year plan for a large consultancy may be affected by intervening national budgets, expected EU labour laws, the expansion plans of a nearby university, and half-expected mid-plan water shortages in the Midlands. Planners must make their best guess about how these (and many other) other factors will affect their plans, and will typically use these guesses to shape their organisation's programmes in salient ways. E.g. the planner may wish to ensure that their overall manpower demand is low during times of potential flooding, and high in September and October, but will not worry about an unduly high demand in September if he or she knows from experience that placement student availability is typically strong and likely to improve in the plan period. These and similar factors generally defy any practically useful attempts to formalise them. Even if we could imagine formalising each and every possible one of these considerations, the desirability and practicality of doing so is zero – large scale project planners simply will not be persuaded to wade through screens of difficult questions before being allowed to start building a plan.

The quantity and importance of such non-formalisable factors tasks means that such large scale problems cannot be effectively addressed by fully automated schedule optimisation systems. Indeed, such systems are very rarely used in commerce and industry. Instead, tools are used which do limited `linear' scheduling with no attempt to optimise, but allow the user to see the consequences of any particular priority ordering among projects, and provide an interface which allows the planner to craft the overall plan by hand. With respect to the quality of a large-scale plan, what such systems can (and do) usefully do is arrange a given list of priority ordered projects in time according to given dependencies between projects, and in line with a matrix of resource availabilities (usually manpower) over time. This inevitably involves necessary shifts in the preferred start dates and durations of the various projects being scheduled, and the planner is typically able to see the extent of these 'slippages'. Beyond these easily formalised factors, however, it is up to the planner to try to arrange the projects in such a way that the slippages are acceptable (although they may be unacceptable in varying degrees to the project managers in question – these are among the unformalisable factors in the planner's expertise), *and* such that the many external issues are taken into account.

As part of a project involving a number of organisations who face such large-scale planning problems, we have developed an Interactive Evolutionary System (IES), for use by an organisation's highest level planner. Candidate plans, in the form of Gantt charts augmented by resource usage profiles, are shown on the screen together with indication of their formalisable fitness (slippage in preferred start dates and durations). The planner then supplies a score (or leaves the default score) for each plan, via a simple drop-down menu, which is designed to be an evaluation of the plan in terms of its overall quality, taking the formalisable and unformalisable factors into account. The IES uses the user-supplied scores in its assignment of selective fitness to each plan, and then produces a new population of plans for the user to assess, and the process repeats. In previous work [2], we have described the basic approach and reported on the quality of the underlying evolutionary algorithm in solving a number of large scale tasks using only internal fitness. This work also went into detail about the commercial desirability of such an IES for large scale logistics, following the results of a questionnaire survey of multi-project planners in large organisations. In summary, the basic EA we use in the IES (which uses an indirect priority based encoding via real-number random keys [3] has been shown to outperform both Hillclimbing and Simulated Annealing on problems of realistic size, and the raw ability of the IES to achieve results *acceptable to real users* has also been proven [2]. That is, planners have been able to verify, in limited testing, that using the IES over acceptable time can lead to plans which meet their unformalised criteria with acceptable levels of slippage.

When following up that work, with the task of optimising many aspects of the IES, we were faced with an immediate concern: that of objectively evaluating versions of an optimisation system which involves intensive human collaboration, without the need to shackle experts to their PCs for months at a time. Our approach is the concept of an *Automated Tester* (AT). The AT replaces the human in the loop during experimental evaluations, and has parameters controlling its decision-making accuracy (modelling human error), and a built-in notion of a target solution which may typically be at odds with the solution which is optimal in terms of internal, formalisable fitness. The idea is that, using the AT, we can do sufficient experiments for statistical evaluation, and plausible rankings of alternative designs for the IES can be achieved. We can also allow for (and examine the effects of) different levels of human error in such studies, with the potential to design IESs which are as robust as possible to expected levels of human error. The tests so far done, and reported herein, are limited to evaluate (via the AT) the IES in two respects; whether it can be expected to work effectively on realistic problems for different kinds of target 'shape' which the user may drive the plan towards, and how robust the basic IES is to different levels of user error.

The remainder is set out as follows. Section 2 describes certain details of the large-scale scheduling problem for which our IES was developed, and sets out some key aspects of the problem and the approach that support understanding later sections. In section 3, we detail the flow of control in our IES, and then introduce the AT. Section 4 details experiments using the AT to determine the ability of a user to drive the plan towards specific shapes, and to evaluate robustness to different levels of human error. Section 5 is a concluding discussion.

2 Multi-Project Programme Scheduling

The application area we address here concerns the problems encountered when scheduling a large number of independent resource-constrained projects, making up an organisation's overall programme, involving perhaps thousands of individual tasks. This is a common real-world problem, faced by many different industries (such as Manufacturing, Construction, Consultancy, R&D.), in which each project requires resources from the same finite resource pool, but needs to be scheduled as timely and efficiently as possible, based on estimated durations and preferred start and end dates specified by individual project planners. As illustrated in figure 1, an organisation's overall programme thus consists of a number of projects, each with its own desired plan of subtasks, arranged by the manager of that project. Each project is viewed as a single 'summary task' which inherits the preferred start time and duration, and overall resource requirements, of the tasks within it taken together. (lower half of the figure 1.). A single department in the organisation may have several such projects, and the Programme Manager has the job of arranging an overall plan for the entire organisation, which means organising together the summary tasks for each project in each department (upper half of figure 1.)

The nature of an organisation's overall programme varies between industries. For example, an aerospace company would have a few very large programmes, each lasting several years, made up from complex projects, each with hundreds or thousands of individual tasks. At the other end of the scale, a support department would have hundreds of individual, small projects each consisting a few tasks, with a duration of a few days or weeks at most. The individual projects are developed independently by project planners who use their knowledge and experience in defining the dates and durations as well as the layout of the project plan. The project plan represents the preferences of the project planner, who wants to minimise the potential disruption to the dates or layout of his plan, caused by the programme scheduling process. Resources tend to be scarce, and/or are required by several projects at once. These typically include people (usually defined by skills such as Lecturer, Programmer, Designer), Equipment (such as cranes, factory machines or vehicles) and Facilities (e.g. laboratories, workshops or rooms).

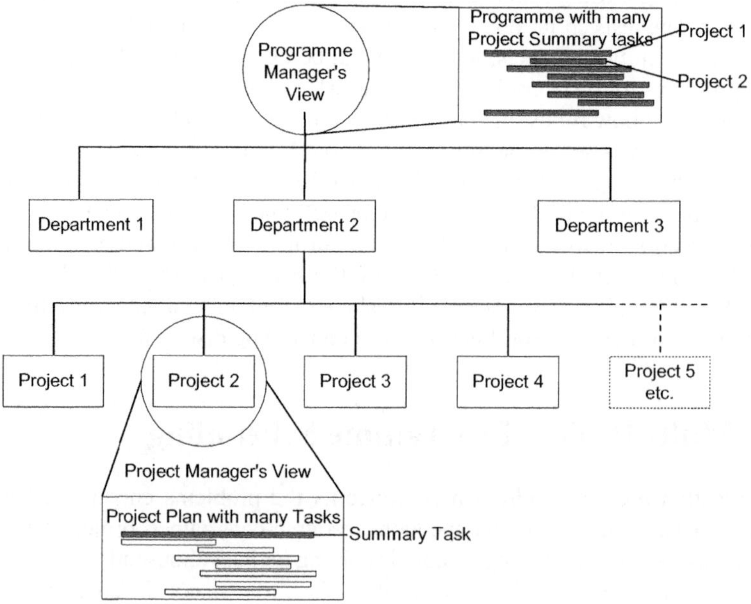

Figure 1. The structure of an organisation's overall programme.

Although evolutionary computation and other methods are well-known to work effectively on a variety of scheduling problems (e.g. [4–10]), the programme manager's concerns in the type of project in question (as explained in section 1) go well beyond what can be formalised within a fitness function, or even a multiobjective fitness function. In this context, fully automated scheduling techniques are seldom fully acceptable to the end user, since they do not take into account this wide range of experience and intuition (or 'gut-feel') which the programme manager uses to judge the overall fitness of the solutions, and consequently the user often finds serious fault with the scheduler's solutions, and does not 'buy-into' the process or trust the output.

However, several of the intuitive and unformalised issues which affect the plan can be cast in terms of a desired overall 'target shape' for the resource (usually mainly manpower) usage profile. This was hinted at in our example at the start of section 1, and is indeed a common consideration. Whether or not we can use this fact to argue for an extra formalised element to the fitness function is very arguable. That is, we could possibly construct a fully automated scheduling system which allowed the user to enter a target resource usage profile, thus compiling together many of the unformalised factors. However, this provides (undesirably) additional requirements for input, and ignores the fact that internal weighting of fit to the target profile against slippage must be determined somehow, and this, as well as the shape of the ideal profile itself, may be dependent on other aspects of the plan. We proceed in the view that formal specification of a profile would *limit* the applicability of our IES, although this remains a research issue.

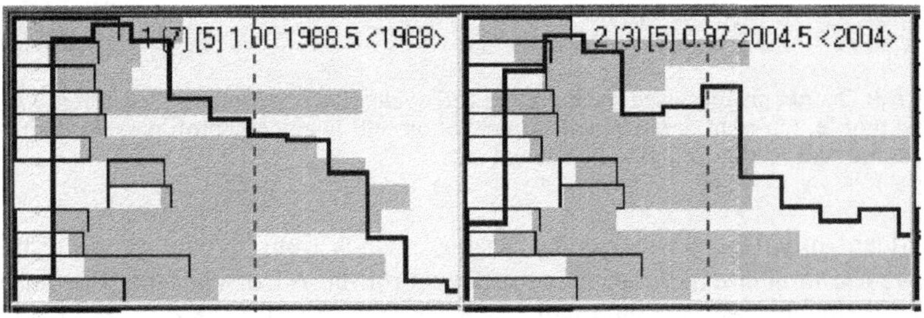

Figure 2. Two example project plans as they are shown to the user (a screen would normally contain 9 or 16 such plans in rank order).

The notion of a 'target shape' is illustrated in figure 2, which shows two example plans as they are viewed in our IES. Each box to the left of a row envelopes a summary task, in terms of its desired start time and duration. The grey boxes represent the placement and duration of this summary task in time in the plan itself. These are (quite typically and expectedly) far longer in duration than the project manager's plans. The stepped line across each plan represents the resource usage profile over time (if the summary tasks all started at their planned times, this profile would need to hit the roof early on in the plan and stay there). Resource availability is the cause of plans generally stretching much further in time than planned. The series of numbers at the top of each plan summarises some key aspects, following the format **N [S] c.cc FFFF.F,** where: **N** is a rank (from 1 to size of screened population), calculated on the basis of internal fitness combined with user supplied fitness; **[S]** is the Score assigned by the user (1 [Bad] - 5 [Excellent]), **c.cc** is the scaled fitness value (0.000 – 1.000) used for selection purposes, and **FFFF.F** is the internal Fitness value, in terms of days of slippage. For example, if a single project was desired to start on day 30 and last 70 days, but in a particular plan it starts on day 35 and lasts 100 days, its slippage is (35–30)+(100–70) = 35 days. A plan's slippage is simply the summed slippage of its tasks. In real cases, it almost never happens that any aspect of a project's slippage is negative.

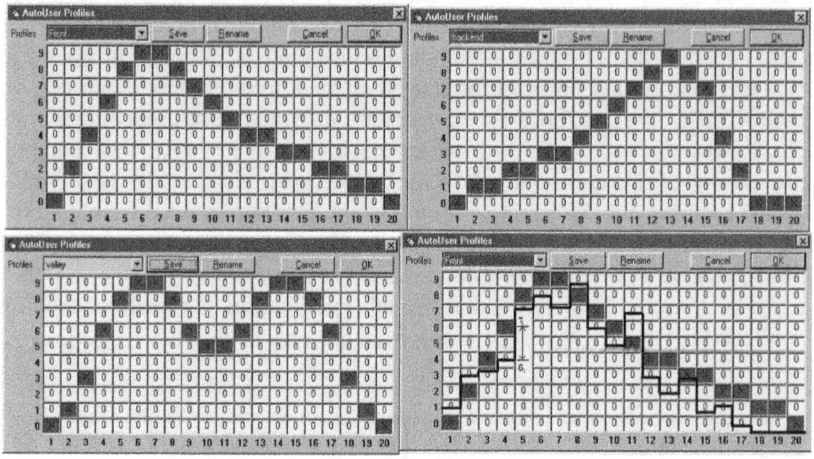

Figure 3: Target profiles used by the AT in IES evaluation experiments. Top left is a *front-loaded* profile, top-right is *back-loaded*, and bottom left is a *valley* profile. At bottom right is a desired profile aligned against an actual one.

Meanwhile, figure 3 shows the 'shapes', which from hereon we call profiles, that we use in later experiments as the targets for the AT driving the IES. Each of *front-loaded, back-loaded*, and *valley* is a typical style of desired overall resource usage profile. Another which doesn't need illustration is a *flat* profile. At bottom right of the figure a desired profile is aligned against an actual one; the absolute differences between the two in each column are summed to yield a measure of a plan's fit to a desired profile. In the experiments described later, the AT will typically assign Scores (from 1 to 5) to the 16 plans it sees in each generation according to the fit of each to its target profile.

3 The Interactive Evolutionary Scheduler and the Automated Tester

Figure 4 shows the flow of control in the IES. There is a growing collection of applications for evolutionary systems in which a human user essentially supplies the fitness assessment throughout the process, and our pseudocode is entirely typical of such Interactive Evolutionary Systems (e.g. see [11–19] for just a small selection). Meanwhile, an equally well-developing area includes systems where the human interaction is sporadic, providing occasional (and necessary) guidance to the direction of an otherwise mainly automated search (e.g. [20,21,22]). This wide range of interactive EA research is testament to the hypothesis that a human user can successfully collaborate with an EA using an interactive interface, and is able to derive solutions that would have been very unlikely (i.e. very difficult and time-consuming) to have been derived by either element (human or algorithm) alone. This in turn adds further weight to the question and difficulty of objectively evaluating alternative IES designs.

> **IES :** Initialise the Population with random chromosomes
> **While** Not Finished
>   **IES :** Determine Internal Fitness Values of Population
>   [*uses simple algorithm based on slippage*]
>   **IES** : Display Population for Human Evaluation
>   [*graphical display of Gantt charts aligned against resource profiles*]
>   **USER**: Inspect Individuals to derive 'personal' evaluation
>   [*using intuition, experience and 'gut-feel'*]
>   **USER**: Enter User Fitness Values
>   [*new values entered directly on screen by mouse*]
> **IES** : Determine Selective Fitness Value
>   [*from the collaborative fitness function, combining internal and user values*]
> **IES** : Breed next Generation using Selective Fitness values
>   [*standard EA mechanisms of selection, recombination, mutation and replacement*]
> Repeat

Figure 4. The Interactive Evolutionary Scheduler algorithm.

> **AUTOMATED TESTER**: Inspect Individuals to calculate fit between target profile and actual profile
>   [*the target profile is predefined for each experiment; the fit is mapped to 1, 2, 3, 4 or 5*]
> **AUTOMATED TESTER**: Enter Possibly Erroneous Scores
>   [*according to error range and concentration parameters, a perturbed score may be entered*]

Figure 5: The IES flow of control when the AT is in use has these lines replacing the "USER" lines in Figure 4.

When the AT is in use, the two "USER" lines in Figure 4 are replaced by those in Figure 5. The AT has parameters as follows. It's *Inaccuracy* varies between 0% and 20% representing the proportion of erroneous Scores it will assign. Its *Error range*, either 1 or 2, represents how wrong erroneous scores might be. E.g. if an error is to be made and the error range is 1, it will randomly assign a score either 1 better or 1 worse than the correct score it should assign on the basis of the target profile. It also has a *Concentration* parameter designed to make the error range and accuracy change over time in different ways; we have not yet experimented with this (awaiting research which identifies what would be plausible variations of concentration), and this is hence set to *Constant* in experiments reported here.

4 Experiments and Results

Trials of the AT looked first at the case of the 'perfect' user, who never makes a mistake in their relative ranking of the candidate plans on show (and also assigns those rankings against a maintained consistent target plan shape throughout the process). These tests also served to assess whether it was possible at all for reasonable target plans to be achieved (which is not guaranteed in the light of potential mismatch between the AT's target profile and the natural shape of profiles which score well against the internal fitness measure). In all of the experiments, data for a real programme scheduling problem with 269 tasks was used, with a total Resource Usage requirement of 7252 man days. The IES had a population of 16 (as found to be suitable in preliminary experiments with human experts). In all cases, the target profile was scaled and aligned so as to define an area consistent with the total resource usage requirement of the programme being scheduled. The first set of trials were run using the 'perfect' user parameters in the AT; these were 0% errors and Constant Concentration. Four different target profiles were used, and 10 different runs of the AT were performed for each target. The summarised results are shown in Table 1.

Table 1. Applying the IES with the AT to a 269-task/7252 man-day programme, with varying AT internal target profiles.

Target Profile	Best Diff.	Best Percent	Average Diff.	Average Percent	Worst Diff.	Worst Percent	Std. Dev.
Front	633	8.7%	871	12.0%	1062	14.6%	117
Back	749	10.3%	903	12.4%	1074	14.8%	123
Flat	408	5.6%	605	8.3%	808	11.1%	111
Valley	858	11.8%	1033	14.2%	1345	18.5%	140

The 'difference' values are in units of "Man Days". The percentage values give the difference as a percentage of the total resource usage requirement. In each case the resulting profile of the best schedule in the final population shows a surprisingly accurate fit with the target. The second set of tests was to determine how effective the algorithm could be when run with an inconsistent user, one who makes mistakes in assigning fitness to the generated solutions. These tests were done with the two more difficult profiles (those least in tune with internal fitness), which were the "Back End" and "Valley" profiles. The first tests were against the "Back End" profile with error range of 1. 10 experiments were carried out for each error level and ran for 100 iterations. The summarised results are shown in Table 2.

Table 2: Applying the IES with the AT to a 269-task/7252 man-day programme targeting a back-loaded profile, with varying AT inaccuracies and an error range of 1. In the Average column, the results for error range 2 are also given (to the right of the hyphen).

Error Rate	Best Diff.	Best %	Average Diff.	Average %	Worst Diff.	Worst %	Std. Dev.
0	749	10.3%	903	12.4%	1074	14.8%	123
1%	648	8.9%	863—883	11.9%	1160	16.0%	166
2%	565	7.8%	865—865	11.9%	1193	16.5%	186
3%	573	7.9%	843—946	11.6%	1137	15.7%	177
4%	536	7.4%	889—1076	12.3%	1208	16.7%	201
5%	699	9.6%	902—966	12.4%	1284	17.7%	176
6%	837	11.5%	970—908	13.4%	1176	16.2%	118
7%	890	12.3%	1058—1032	14.6%	1249	17.2%	127
8%	687	9.5%	1031—997	14.2%	1214	16.7%	171
10%	825	11.4%	1060—1013	14.6%	1362	18.8%	159
15%	883	12.2%	1142—1130	15.7%	1374	18.9%	173
20%	1068	14.7%	1259—1315	17.4%	1603	22.1%	174

Table 3: Applying the IES with the AT to a 269-task/7252 man-day programme targeting a valley profile, with varying AT inaccuracies.

Error Rate	Best Diff.	Best %	Average Diff.	Average %	Worst Diff.	Worst %	Std. Dev.
0	858	11.8%	1033	14.2%	1345	18.5%	123
1%	812	11.2%	1047—1070	14.4%	1294	17.8%	164
2%	783	10.8%	1100—1137	15.2%	1770	24.4%	277
3%	708	9.8%	1163—1188	16.0%	1470	20.3%	235
4%	1073	14.8%	1301—1211	17.9%	1596	22.0%	179
5%	936	12.9%	1267—1170	17.5%	1646	22.7%	280
6%	1010	13.9%	1313—1234	18.1%	1617	22.3%	197
7%	1141	15.7%	1450—1295	20.0%	1993	27.5%	279
8%	1121	15.5%	1474—1417	20.3%	1918	26.4%	227
10%	1185	16.3%	1478—1410	20.4%	1816	25.0%	234
15%	1070	14.8%	1641—1775	22.6%	2072	28.6%	346
20%	1041	14.4%	1391—1706	19.2%	1700	23.4%	254

Table 2 (and table 3) also shows the average difference in experiments with an error range of 2. For space and readability we don't give the error range 2 figures for every column, Of great interest is that the scores for the best and average differences improve compared to the perfect user for low levels of error, suggesting that low levels of user error introduce an appropriate amount of diversity which aids the search process. The second set of experiments was carried out on the same programme, with the same parameters, but this time aiming for the "Valley" profile. Table 3 summarises the results. The tests on the Valley profile also show improvements in initial results over the prefect (0% error) user for low error rates. However in this case the average difference is clearly best in the case of a perfect user, but the best differences recorded for each of the error ranges from 1% to 5% were all better than the best recorded for the perfect user. In both cases, the IES appears surprisingly robust to an error range of 2.

5 Concluding Discussion

By replacing the human user with an AT, extensive trials can be easily done. However, before the results of such trials can be analysed, we need a way of objectively evaluating the result of a single trial. This is problematic, because the very fact that an IES is being used suggests that the quality of a solution is not readily formalisable. There are two responses to this issue: (1) We could use human expert evaluation to categorise the solutions from each trial; this is costly in terms of human expert time, however it remains a very significant saving in time compared with not using the AT at all. This is because, instead of human categorisation of a candidate solution being needed perhaps a hundred times (maybe many more) per trial, it is now needed only once per trial, to categorise the end result: (2) As we have done here, we can find some way to 'fake' the informal criteria by furnishing the AT with a target against which it can simply evaluate candidates, but where the conflict between that target and the formalisable criteria clearly echoes the conflict we expect between a human expert's informal criteria and the formalisable criteria. That is, we implement a simple model of the scenario in which the user is attempting to drive the solution towards regions of the search space using criteria which are unknown to the rest of the system, and the results of a variety of experiments using the Auto-User suggested that the IGA scheduler can generally converge on a pre-defined Target profile to within about 10% of the desired 'target' profile, measured in man days. This has to take into account that the target profile is not necessarily feasible, and so may not be able to be exactly matched by any calculated solution.

By using the AT, we have been able to conduct some extensive experiments into the usefulness of the IES and have a number of interesting and favourable findings. First, at least given the parameters and design of the basic IES in question, low levels of human error seem to aid the process (presumably supplying extra diversity), and so such a system can be seen as robust to human error, and at (we believe in this case) levels which may be reasonable to expect from an expert user. This may not be true of a different IES configuration which perhaps has a higher mutation rate and lower pressure selection, but then again such a system may find itself *not* robust to human error, since the extra diversity will be too disruptive. Hence this finding points to a possible design principle for IESs. Similar can be said concerning the error range. The other main finding is that the IES can clearly drive a large-scale plan towards a given target profile, even a notoriously difficult one. With the IES/AT experiments this process represents a model of a user driving the plan towards acceptability to intuitive unformalised factors, and as such (and also given that details of duration and slippage were generally acceptable – details omitted for space reasons) this provides an experimental validation of the general capability and fit-for-purpose of a basic IES system for large scale programme planning.

References

1. Woodgate, H.S. (1964) *Planning by Network*, Business Publications Ltd.
2. Shackelford, M., Corne, D. (2001) Collaborative Evolutionary Multi-Project Resource Scheduling, in *Proceedings of the 2001 Congress on Evolutionary Computation*, IEEE Press, pp. 1131—1138.
3. Hartmann, S. (1999) *Project Scheduling under Limited Resources*, Springer-Verlag,
4. Tsang, E., Voudouris, C. (1995) Fast Local Search and Guided Local Search and their Application to British Telecom's Workforce Scheduling Problem, *Operations Research Letters*, **20**, pp119—127.
5. Louis, S., Xu, Z. (1996) Genetic Algorithms for Open-Shop Scheduling and Re-Scheduling, In M. E. Cohen and D. L. Hudson, editors, *Proc. of the ISCA 11th Int'l Conf. on Computers and their Applications*, pages 99–102. ISCA.
6. Bruns, R. (1997) Evolutionary Computation Applications: Scheduling, Ch 1.5, *Handbook of Evolutionary Computation*, OUP.
7. Corne, D.W., Ross, P. (1997) Practical Issues and Recent Advances in Job- and Open-Shop *Scheduling,* in D.Dasgupta and Z.Michalewicz (eds), *Evolutionary Algorithms in Engineering Applications*, Springer-Verlag, pp. 531—546.
8. Montana, D. (1998) Introduction to the Special Issue: Evolutionary Algorithms for Scheduling, *Evolutionary Computation*, **6**(1)
9. Bierwirth, C., Mattfield, D.C. (1999) Production Scheduling and Rescheduling with Genetic Algorithms. *Evolutionary Computation*, **7**(1), pp.1—17.
10. Husbands, P. (1999) Genetic Algorithms for Scheduling, *AISB Quarterly*, **89**
11. Dawkins, R. (1988) *The Blind Watchmaker*, Penguin.
12. Sims, K. (1991) Artificial Evolution for Computer Graphics, *Computer Graphics,* **25**(4), pp. 319—328.
13. Smith, J.R. (1991) Designing Biomorphs with an Interactive Genetic Algorithm, *Proc. 4th ICGA*, Morgan Kaufmann.
14. Graf, J., Banzhaf, W. (1995) Interactive Evolution of Images, *Proc. of Int'l Conf on Evolutionary Programming*, San Diego.
15. Aoki, K., Takagi, H. (1997) 3-D CG Lighting with an Interactive GA, *1st Int'l Conf. On Conventional and Knowledge-based Intelligent Elec. Sys.* (KES '97).
16. Banzhaf, W. (1997) Interactive Evolution, *Handbook of Evolutionary Computing*, C2.10, OUP.
17. Johnston, V., Caldwell, C. (1997) *Tracking a Criminal Suspect through Face Space with a Genetic Algorithm*, Ch 8.3, Handbook of Evolutionary Computation, OUP.
18. Louis, S., Tang, R. (1999) Interactive Genetic Algorithms for the Travelling Salesman Problem, In *Proc. of the Genetic and Evolutionary Computation Conference, GECCO'99*, Morgan Kaufmann.

19. Oliver, A., Regragui, O., Monmarche, N., Venturini, G. (2002) Genetic and Interactive Optimization of Web Sites, In *Proc. Of 11th Int'l WWW Conf*, Hawaii.
20. Inoue T., Furuhashi T., Fujii M. et al., 1999, Development of Nurse Scheduling Support System using Interactive Evolutionary Algorithms. *Proceedings IEEE International Conference on Systems, Man and Cybernetics (SMC'99)*; pp 533-537
21. Parmee IC, Cvetkovic D, Watson AH, Bonham C, (2000) Multi-objective Satisfaction within an Interactive Evolutionary Design Environment. *Evolutionary Computation*, **8** (2); 197-222.
22. Parmee IC, Cvetkovic D, Bonham C, Packham I, (2001) Introducing Prototype Interactive Evolutionary Systems for Ill-defined Design Environments. *Journal of Advances in Engineering Software*, Elsevier, **32** (6);pp 429-441.

Interactive Evolutionary Strategy Based Discovery of Image Segmentation Parameters

Praminda Caleb-Solly, Jim Smith

Faculty of Computing, Engineering and Mathematical Sciences,
University of the West of England
Bristol, BS 16 1QY
Praminda.Caleb-Solly@uwe.ac.uk

Abstract

The symbiosis of human expertise, in terms of creativity and pattern recognition, with evolutionary algorithms for user controlled and directed search is now a rapidly emerging model.

One of the main issues that need to be addressed is the development of techniques to ensure that the power of the evolutionary search is exploited without compromising its efficiency by introducing too much noise in the form of human assessment. Human assessment is likely to have a high component of subjectivity and non-linearity of focus. This implies that in the first instance it is necessary to analyse the nature of the variability of the human assessment. Another important issue that needs to be addressed is ensuring that the evolutionary progress is rapid without compromising the granularity of the search. Rapid convergence is important to the practical applicability of the system and also prevents the process from becoming tedious for the human participant, resulting in loss of concentration.

This paper explores appropriate strategies for the interactive evolution of parameter sets for image segmentation and examines issues relating to reliability of user scores for selection of parents. The nature of user scoring is analysed both in terms of the evolutionary strategy adopted and the temporal progression of the runs. The correlations between number and type of images seen at each generation, the time taken to achieve satisfactory results and the quality of the resulting solutions are analysed in terms of their ability to generalise.

1 Introduction

This work stems from the development of an automatic surface inspection system for classifying defects in sheet steel. In the first instance this involves identifying regions of interest (ROI) on the surface of the steel. Delineating regions of interest and being able to identify them as defects, involves being able to separate them from the background accurately. This segmentation process is crucial as the next step involves extracting appropriate descriptors, or features, for these regions which are then used for the classification. The main problem associated with segmentation arises from the noisy nature of the surfaces, in terms of amount of debris and an inherently irregular texture. Due to the nature of the defects found most commonly on hot rolled steel surfaces, there is often a compromise between segmenting the whole defect accurately and falsely segmenting areas of the noisy background. In these cases, the decision on the location of the segmentation boundaries is a compromise to give an acceptable level of ROI extraction over a whole set of images, using a single set of image processing (IP) parameters. This problem may be further complicated by the fact that the end user specifications for the segmentation might vary over time in non-quantitative terms. In such situations, it is not possible for the system to be delivered with a universal set of parameters and therefore the system needs to have the ability to be re-tuned easily and effectively by the user.

If such automated systems are to be of use in an industrial setting, it is clearly unreasonable to expect the end-users to have sufficient IP expertise to specify a new set of parameters. One possible solution being explored is the use Evolutionary Algorithms (EAs) to search the space of image processing parameters and to evolve an optimum set based on the user's visual evaluation and grading of the resulting segmentation. Thus the question being asked of the user changes from *"specify your requirements"* to "*how well do the regions outlined in this set of images correspond to the ones that you want to detect and identify?*"

In literature regarding other applications using "interactive evolution" user intervention can either be "subjective selection", where the user is simply selecting the offspring that will be the parents in the next generation as in the selection of cherry brandy colouring [1], the evolution of "biomorphs" [2] and evolving line drawings [3] or the user intervention can be "subjective evaluation", where the user assigns fitness values based on their subjective ranking of the offspring. Applications using this second approach range from generating Jazz solos [4] to personalising hearing aid characteristics [5]. The latter approach is adopted in this study.

2 Background

The method that was found to perform consistent and robust region detection for the steel images was developed using texture-based segmentation. Texture describes the spatial distribution of sub-patterns of grey levels in an image, these sub-patterns are also called texels. Visually, texture can be described as fine, coarse, smooth, speckled etc. Mathematically, one of the ways in which textural information can be extracted is by computing the statistical relationships of the spatial distribution of the texels. This method, developed by Haralick [6], involves the calculation of a set of grey level spatial dependence probability distribution matrices or grey level co-occurrence matrices (GLCMs). From these are calculated textural features: statistical measures relating to specific textural characteristics such as homogeneity, contrast, correlation, variance and entropy.

In our system the image is simultaneously processed by two texture derivation modules. Both of the resultant images are normalised, low-pass filtered, and thresholded to extract the ROI, before being recombined using a logical OR function. The variables that need to be specified for calculating each texture image are summarised as follows:

- Texture Option: the texture measure that is to be calculated.
- Kernel Dimensions: Size of the sliding window in which the texture parameter is calculated.
- Line and Sample Increment: the number of pixels in the y and x direction that the sliding window moves each time.
- Orientation angle: the direction in which the pixels for the GLCM are evaluated within the window. Four angles are used, from 0 to 135 degrees measured counter-clockwise.
- Sample Distance: the distance between co-occurrence samples.
- Threshold: the cut-off value used to define ROIs.

The main aims of this study were to analyse the consistency of user scoring, explore appropriate evolutionary strategy (ES) selection schemes and consider the effect of reducing user interaction time by reducing the number of images scored in each evaluation. In order to objectively assess these attributes it was necessary to have a standardised and quantified set of images to conduct the experiments. A set of 10 images were created each with an embedded defect of a different irregular shape and size. The edges of the defect were blended with the background texture using a non-linear gradient around the perimeter of the defects. The objective was to create a set of images with a high level of technical complexity and uncertainty/ambiguity with regards to visual clarity of the boundaries of the defects. All the experiments described here were conducted using the synthetic set of defect images.

3 Evolutionary Algorithms

The key issues that need to be considered for this application when selecting an appropriate evolutionary algorithm are:

- Monitor size and resolution: coupled with the visual clarity required for segmentation, these place a limit on the number of images that can be simultaneously displayed
- Time constraints on the users availability and, more importantly, on their mental concentration

Given the extent of interaction between the (real-valued or integer) parameters, the search space is suspected to be highly complex and is known to contain several local minima, which prompted the use of (μ, λ) or $(\mu + \lambda)$ evolutionary strategies for generating optimised parameter sets. In the former, the parent dies off and only the children undergo selection, whereas in the latter the parental generation is also taken into account in the selection [7, 8]. We examined the use of both strategies to select one parent, and used eight offspring, i.e. (1, 8) and (1+8).

In this study each individual is comprised of 2 sets of variables, one for each of the 2 texture modules. Each set of variables include a texture option, the height and width of the sliding window and a threshold value, which makes a total of 8 variables in each individual. The increment, orientation and sample distance were fixed at 2, 0 and 1 respectively.

Each set of variables within the offspring is created by mutation from the corresponding ones in the parent as follows:

- With probability 0.15, a new random texture option is chosen, a new threshold randomly chosen uniformly from (0,256) and the height and width of the sliding window chosen at random uniformly from (0,28).
- Otherwise the texture option from the parent is retained and the threshold mutated by the addition of a random deviate from a $N(0, \sigma_1)$ distribution. Similarly the window height and width is mutated by the addition of random deviates from $N(0,\sigma_2)$ distributions. Both σ_1 and σ_2 are 15% of the range of the respective variable.

The scheme of using two mutation operators as described above was devised in order to help maintain a certain amount of linkage that exists between the variables within a set, while still allowing for reasonable levels of perturbation. The values for the probability and mutation step sizes were experimentally verified as resulting in the shortest convergence times.

4 Design of Experiments

For this work the ten images were split into two groups, the first of which (1,...5) were used for evaluation, and the second of which (6,...,10) were used as a test set. We considered three different training sets: Set A comprising the fixed group of images (1, 2, 3), Set B, comprising images (3, 4, 5) and Set C, where a random group of 3 images was selected at each generation from the group (1, 2, 3, 4, 5). For each set five runs were made.

During the user scoring phase, (i.e. the interactive assignment of fitness values to the parameter sets), the user is shown a screen displaying 8 segmented images arising from the application of the parameter sets of the current generation to one of the images. The user assigns to each a quality rating, between 0 and 10. After all three images have been segmented and the results scored in this way, the system calculates a score for each offspring that is an average of the user allotted scores over the three images for the corresponding parameter sets. The parameter set with the highest score is the parent for the next generation. The process continues until a set of parameters are generated that result in a set of segmented images attaining a high target score.

In order to examine the issue of the reliability of user-scoring, after each generation we also computed a static quality value for the selected parameter set. This was done both on the set of all five training images and on the (unseen) test images. The accuracy metric used to calculate the score is as follows: $Accuracy = (N_c / N_s) * (N_c / N_d)$ where N_c is the number of correctly segmented pixels, N_s is the total number of segmented pixels, and N_d is the number of defect pixels. The terms in this metric penalise over and under segmenting of the image respectively, so that a perfectly segmented image will achieve a score of 1.0.

Typical results of segmentations at various stages of a run and corresponding user and system accuracy scores are shown in Figure 1 (note that the width of the segmenting line has been increased for visual clarity in this paper).

5 Results

For this particular application of interactive evolution, the feasibility of the approach will be impacted by the amount of user involvement that is required to attain a desired solution.

Image I

Image II

Image III

Figure 1. Segmented images at various stages of a run. Image I (System Accuracy = 0.117, User Score = 0.33), Image II (System Accuracy = 0.242, User Score = 3.33), Image III (System Accuracy = 0.663, User Score = 10)

A comparison of maximum number of generations per run between Comma (1, 8) and Plus (1+8) Evolutionary Strategies was conducted using these results. As can be seen from Figure 2, the box plots show that even though there is high level of variability which can be attributed to the composition of the sets A, B and C, comparing the mean values of the number of generations between each set shows that the Plus strategy results in a lower mean.

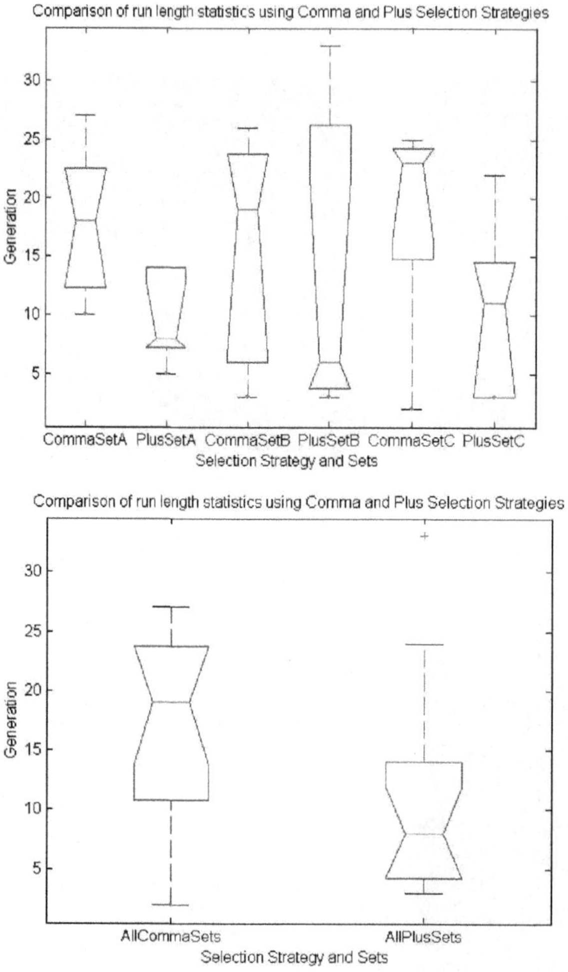

Figure 2. Comparison of run length statistics between Comma and Plus Evolutionary Strategy

This is mainly due to the retention of a good solution in an elitist approach. The heights of the upper whiskers of the box plots of the Plus strategy runs demonstrate one of the negative aspects of this approach, which is an early settling in a sub-optimal solution which can be difficult to escape from.

Setting the fixed mutation step size, to a larger value was found to have a negative effect on time taken to converge.

Another approach to overcoming early sub-optimal convergence would be setting a limit on the number of generations the same parent is selected and then increasing the mutation step size for the subsequent generation. Pooling the results from all three sets, and using the non-parametric Wilcoxon test, we can say that with 92% confidence the plus strategy leads to shorter mean run times.

We analysed the consistency of user scoring behaviour by splitting the number of generations in each run into thirds to define a beginning, middle and end stage for each run. Figures 3 and 4 show the quality of the best set from every generation in every run as scatter plots of user score vs. system score for the two different selection strategies.

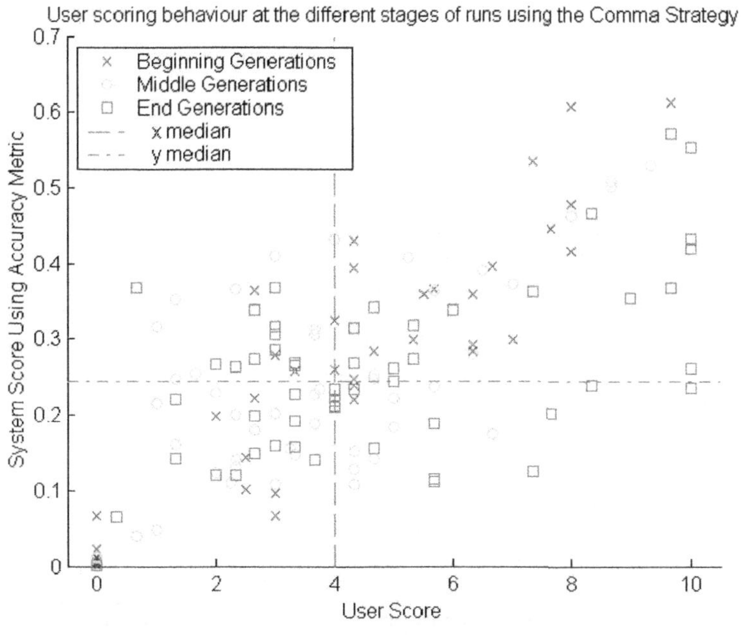

Figure 3. User Scoring at the different stages of runs using the Comma Strategy

As can be seen from Figure 4, there is a clear trend within the plus strategy for the user to assign higher scores at the end of the runs than at the middle and the beginning, whereas as shown in Figure 3 the temporal relationship is far less clear for the commas strategy, partly as a result of the periodic loss of the current best parameter sets.

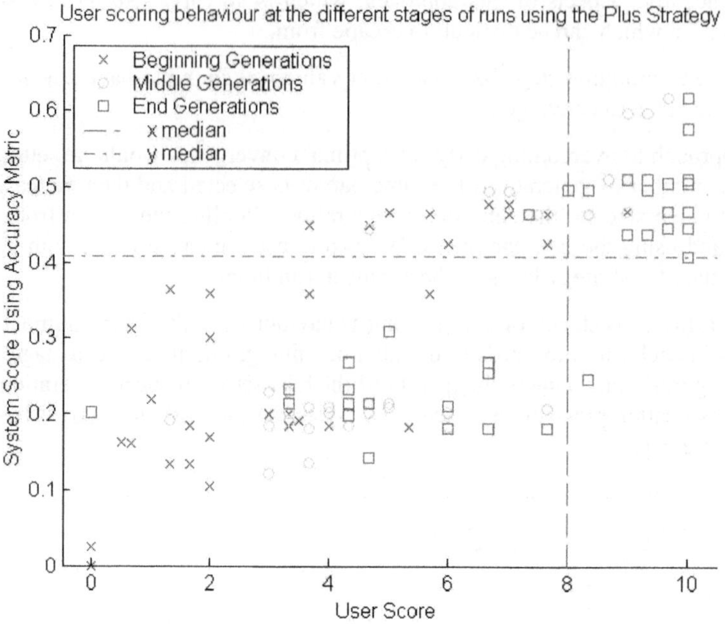

Figure 4. User Scoring at the different stages of runs using the Plus Strategy

Table 1 shows a comparison of median values of user score for each of the 3 stages using both the comma and plus strategies. Median values of user scores for runs conducted using the plus strategy show a more marked increment between stages and are higher than those using the comma strategy indicating a possible loss of good solutions. In addition, correlation coefficients were calculated to assess the correspondence of user scores on the training image Sets A and B with system accuracy on the same images. These can be seen in Table 1 as well.

Table 1. Comparison of user scoring behaviour between comma and plus strategy runs. (Generations abbrv. to Gens.)

	Comma Strategy			Plus Strategy		
	Beginning Gens.	**Middle Gens.**	**End Gens.**	**Beginning Gens.**	**Middle Gens.**	**End Gens.**
Median US	2.667	3	4	3	5	8
Std Dev US	2.956	2.49	2.837	2.594	2.832	2.79
Correlation Coefficient:	0.943	0.766	0.580	0.828	0.855	0.788

When we examine the correlation between user scores and the fixed metric, we see that in both cases this is quite high in the beginning, suggesting that the metric reasonably accurately reflected the user's subjective requirements. For the plus strategy, this correlation actually improves in the middle phase, but then tails off in the end-game, possibly reflecting increased user-expectations from the system.

Although this effect probably occurs in the comma strategy as well, it is swamped by a much faster and more dramatic fall of in correlation. Recalling the observed difference in mean runs times leads to the reason for this – the user's concentration is wavering, and it is quite simply harder to make consistent subjective judgements over a longer period of time.

As stated earlier, one of the potential disadvantages of interactive evolution is the relatively high commitment and reliance on user time for scoring the images. It is therefore important to determine what level of accuracy and generalisation can be achieved using a fewer number of training images at each generation.

Table 2 shows results for the plus strategy obtained using 3 images per generation and one image per generation (where the single image was held constant, or was randomly selected from (1,...5) each generation). Maximum test set system accuracy results and the mean number of generations for achieving the highest user allotted scores on the training set and the system test set accuracy over all the runs are shown. Using just one image per generation does not result in as high test set accuracies as using three images per generation, although the results using the random methods (set C or the last column) are similar. Visually, the difference between a system test set accuracy of 0.663 and 0.58 is minimal. It is also important to note is that this level of performance is achieved more rapidly as well.

Table 2. Comparison of statistics between runs scored using 3 images/generation and 1 image/generation

	3 images / generation			**1 image/generation**					
	Set A	**Set B**	**Set C**	**Img 1**	**Img 2**	**Img 3**	**Img 4**	**Img 5**	**Rand**
Max Test Set System Accuracy	0.66	0.61	0.59	0.58	0.58	0.49	0.58	0.47	0.54
Mean Generations to Max User Score	9.4	12.8	9.6	10.6	8.4	5.4	6.2	6.4	10.2
Mean Generations to Max System Test Accuracy	5.4	12.2	8.4	6.4	5.6	4	4.6	4	5.6

6 Conclusions

Using the (μ,λ) evolution strategy is usually recommended for multi-modal landscapes as it enables escape from local minima. In an eight variable search space conducted within a limited number of generations however, this can also result in a good solution being lost and not being recoverable within a reasonable time thus necessitating an increase in the time the user is required to interact with the system.

The results demonstrate that this can be overcome using a $(\mu + \lambda)$ strategy, but the compromise related to using an elitist strategy is a possibility of the system settling in local solutions. An alternative approach might be to increase the granularity of the search by incorporating system assessment of solutions between showing the user only results of those parameter sets which attain a test accuracy that is over a certain system acceptance threshold. This approach however assumes a high correlation between the user criteria for segmentation and the system metric for calculating accuracy of the segmentation. This has been shown to exist for a single objective problem domain, but might be difficult to implement for multi-objective task or new user criteria. As one of the main advantages of this interactive evolutionary approach lies in the fact that the quality of segmentation in terms of the defects or regions that have to be isolated does not have to be quantitatively defined, but is instead naturally elicited from the users bias towards particular segmented sets of images. Therefore using a system accuracy metric to enable a finer grained exploration of the search space might be limiting.

User scoring of images in terms of consistency was found to vary at different stages of runs, particularly when using the comma strategy. The low median user scores at the end of runs indicate loss of good solutions, and perhaps the users' altered scoring strategy is an attempt at a recovery. The reduced correlation seen towards the end of runs could be the result of loss of concentration over longer periods of time, which no doubt will affect the consistency of subjective scoring. This is also confounded by raised user expectations when a solution seems imminent, particularly in the case of the plus strategy. Having identified these sources of noise in the system has been an important step, however these issues need to be further investigated.

Current work is underway to explore the efficacy of user controlled mutation by linking the user allotted scores with the selection of the mutation step-size. This is hoped to result in rapid convergence with a reduced risk of settling in a local minima. The variability of user scoring will be an important factor to take into account when considering user control of mutation step-size.

References

1 Herdy, M. (1996) Evolution Strategies with Subjective Selection. Lecture Notes in Computer Science 1142, Intl. Conf. On Evolutionary Comp, Parallel Problem Solving from Nature, pp 22-31.

2 Dawkins, R., (1987) The Blind Watchmaker. WW Norton and Company.

3 Baker, E., Seltzer, M., (1994): "Evolving Line Drawings" Graphics Interface 94 Procs, Morgan Kaufmann. http://citeseer.nj.nec.com/baker94evolving.html

4 Biles, J. A. (1994). GenJam: A genetic algorithm for generating jazz solos. In ICMC Proceedings 1994. The Computer Music Association.

5 Takagi, H., Ohsaki M. (1999): "IEC- based Hearing Aid Fitting". Proceedings of Int'l Conf. On System, Man and Cybernetics (SMC'99), Vol 3, 657-662 IEEE.

6 Haralick, R.M., Shanmugam, K., Dinstein, I. (1973) Texture Feature for Image Classification. IEEE Transactions on Systems, Man and Cybernetics, Vol. SMC-3, No 6, November, pp 610-620.

7 Thomas Bäck, Frank Hoffmeister, and Hans-Paul Schwefel.(1991) A survey of evolution strategies . In Lashon B. Belew, Richard K.; Booker, editor, Proceedings of the 4th International Conference on Genetic Algorithms, pages 2--9, San Diego, CA, July 1991. Morgan Kaufmann..

8 Schwefel HP (1997). Evolutionary computation - A study on collective learning. Proc. World Multiconference on Systemics, Cybernetics and Informatics (SCI'97), vol 2, pp 198-205

Extraction of Emerging Multi-Objective Design Information from COGA Data

J.A.R Abraham, I.C. Parmee
Advanced Computation in Design and Decision-making (ACDDM)
http://www.ad-comtech.co.uk/ACDDM_Group.htm
University of the West of England, Bristol
johnson.abraham@uwe.ac.uk ian.parmee@uwe.ac.uk

Abstract

The paper describes further developments of the interactive evolutionary design concept relating to the emergence of mutually inclusive regions of high performance solutions relating to differing objectives from cluster-oriented genetic algorithm (COGAs) output. These common regions are further defined by the application of clustering algorithms and relevant variable analysis. The multi-objective output of the COGA is then compared to output from a strength Pareto evolutionary algorithm (SPEA-II).

1 Introduction

This research contributes to the interactive evolutionary design (IED) concept, a human centric approach [1,2] supporting the generation and extraction of high-quality design information during conceptual design. The aim of IED is the provision and succinct presentation of information appertaining to complex relationships between the variables, objectives and constraints that define a developing design space. The IED system (IEDS) comprises several modules which are described in [1, 2, 3, 4].

Cluster-oriented Genetic Algorithms, developed by Parmee [5,6,7] and further improved by Bonham [8,9], identify high-performance (HP) regions of complex design space as opposed to converging upon single optimal solutions. Briefly, the identification of HP regions is achieved by passing the evolving COGA populations through an Adaptive Filter (AF). By first normally distributing population fitness and introducing a fitness threshold the AF can copy HP solutions from the population of each generation to a Final Clustering Set (FCS). The solutions within the FCS at the end of the COGA run then define the HP region.

Core evolutionary processes of the IEDS can comprise several concurrent COGA elements each searching design spaces relating to a different objective. HP solution sets relating to various objectives can thus be generated and these solution sets define the HP region for each included objective. The BAE Systems miniCAPs model for preliminary military airframe design [10, 3, 4] has been used for experimentation and the development of the IEDS modules. Preliminary results are presented here which is contribute to the further development of an agent-based information-gathering IEDS component again utilising miniCAPS. MiniCAPs has nine continuous input variables and twelve output parameters (objectives). The model comprises subroutines which calculate properties relating to performance, wing geometry, propulsion, fuel capacity, structural integrity etc. Three objectives, Ferry Range (FR), Subsonic Attained Turn Rate (ATR1) and Subsonic Specific Excess Power (SEP1) have been chosen from the twelve objectives of the miniCAPS model for experimentation.

Table 1: MiniCAPS Input Variables

CLMN - Climb Mach Number	**GWP** - Gross Wing Plan Area	**WLES** - Wing Lead Edge Sweep
CH - Cruise Height	**WAR** - Wing Aspect Ratio	**WTCR** -Wing T/C Ratio
CMN - Cruise Mach Number	**WTR** - Wing Taper Ratio	**BPR** - ByPass Ratio

It has been proposed [6] that, by varying AF threshold settings, COGA can be used in an exploratory manner to generate information relating to multi objective interaction. Figure 1a shows the HP regions relating to the three objectives generated from COGA runs utilizing all nine miniCAPS variables. The regions are projected onto a two variable hyperplane (Gross Wing Plan Area and Wing Aspect Ratio) of the nine variable miniCAPS design space.

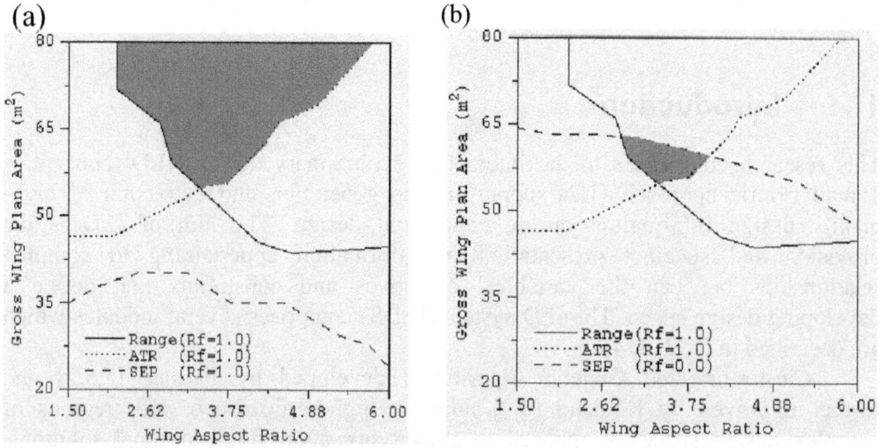

Figure 1: Utilising the AF threshold to establish a common region between the three chosen objectives.

The initial adaptive filter threshold settings for each objective are high and the spatial distribution of the resulting HP regions illustrates a high degree of conflict between objectives FR and SEP1 and between ATR1 and SEP1. A common region of high performance solutions exists between FR and ATR1 (i.e. high performance solutions satisfying both objectives) as shown in figure 1a. In order to attain a common region between all three objectives the filter threshold of the SEP1 objective is relaxed allowing less fit solutions into the FCS and thereby expanding the bounds of the final clustering set. By doing this one can achieve a mutually inclusive region where solutions satisfy all objectives, as shown in figure 1b. Lowering the AF threshold is equivalent to introducing a lower preference rating to that objective as the average solution performance is compromised. Further search in the common region generates solutions with best compromise fitness relative to all objectives.

This approach provides a highly visual indication of the degree of objective conflict. Experimentation with AF thresholds and the resulting shifting of HP region bounds supports the setting of preference for each objective. Following sections describe common region identification processes and feature selection techniques and also illustrate the relationship between the COGA output and non-dominated Pareto fronts.

2 Common Region Identification

Each COGA run generates a final clustering set FCS_i where *i=1 to M*, (M=number of objectives) containing high performance solutions for that particular objective. The solutions in each FCS_i form a cluster with particular characteristics defining the objective C_i. Identifying the common region between all objectives' FCSs is a classification problem where classified sets of solutions share the neighbourhood with solutions satisfying other objectives.

2.1 The Classifier

The working principle of the K-Nearest Neighbour, KNN, classifier is adopted but, in this instance, the KNN classifier is used to classify solutions from all FCSs into a cluster where each solution relating to a particular objective shares a common neighborhood with solutions belonging to other objectives' FCSs. The neighborhood relationship between solutions is determined by calculating the squared Euclidean distance metric $D_{i,j} = \sqrt{\sum_{k=1}^{n} |x_{ik} - x_{jk}|^2}$ between solutions where k is the number of dimensions used and n is the number of solutions. To reduce the scaling effect between the variables, data is normalized to have zero mean and unit standard deviation.

Initially the FCS solutions of two separate COGA runs relating to objectives ATR1 and FR provided input to the classifier which then identified those solutions that have near neighbours belonging to the other objective's FCS. All the nine variables were utilized in the near-neighbour calculations with K=3. Figure 2 shows that although KNN successfully identifies the common region, outliers are evident. It was assumed that these outliers are caused by the noise introduced into the distance metric by variables that are relatively uncorrelated to the corresponding objective.

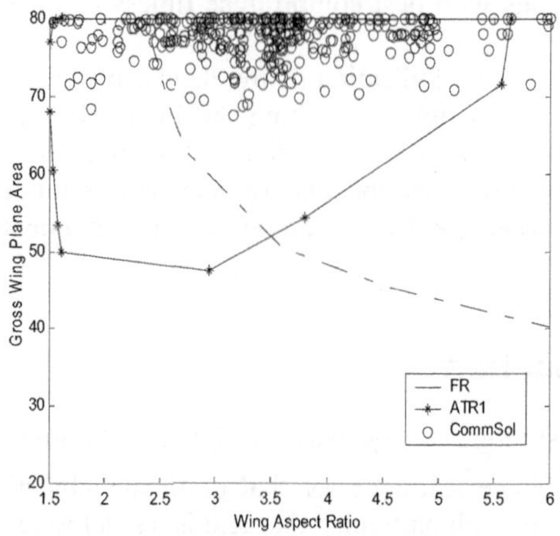

Figure 2: ATR1/FR common region identified using all the solutions and all the variables for K=3

This assumption is based upon previous work where it is obvious that although succinct HP regions exist in some variable hyperplanes, a relatively uniform distribution of the same HP solutions can be evident in others (see figure 3b). In order to improve clustering efficiency feature selection is introduced to identify highly correlated variables.

2.2 Feature Selection

Attribute Relevance Analysis [12,13] is a method originating from machine learning, statistics, and fuzzy / rough set theories to identify most relevant variables. Attribute Relevance Analysis quantifies the relevance of an attribute (i.e. variable) with respect to a given class or concept by

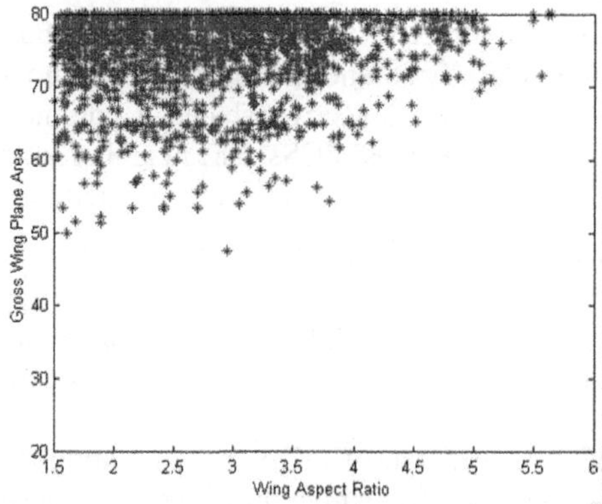

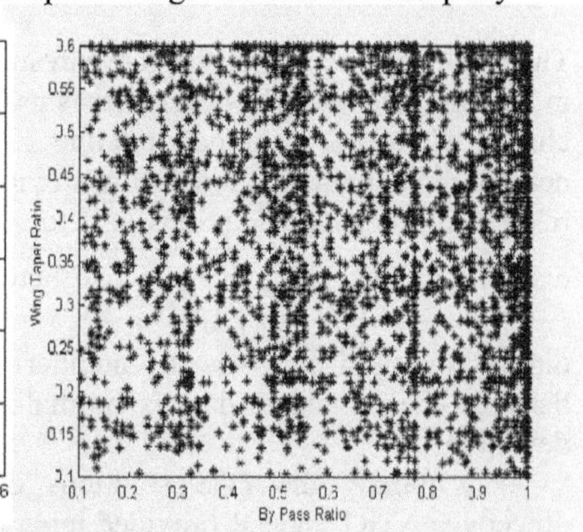

Figure 3a: Distribution of ATR1 solutions in GWP / WAR variable hyperplane

Figure 3b: Distribution of same solutions in W TR / BR variable hyperplane

measures such as information gain and correlation co-efficient. In this work an information gain analysis technique with a dimension-based data analysis method is used. Let *S* be a set of solutions belonging to all *M* objectives under consideration. Let *S* contain subset s_i solutions of each objective c_i for $i = 1...M$. An arbitrary sample of solution belongs to objective c_i with the probability s_i/s where $s = |S|$ the total number of solution in the set S. The expected information needed to classify a given sample is given by:

$$I(s_1, s_2, ... s_M) = -\sum_{i=1}^{M} \frac{s_i}{s} \log_2 \frac{s_i}{s}$$

An attribute with values $\{a_1, a_2, ... a_n\}$ can be used to partition S into subsets $S = \{S_1, S_2 ... S_v\}$ where S_j contain those solutions in S that have value a_j of attribute A. If the attributes are continuous variables then the values $S = \{S_1, S_2 ... S_v\}$ can be obtained by discretizing the attribute A. Let s_j contain s_{ij} solutions of the objective c_i. The expected information based on the partitioning by attribute A is known as the entropy of A. It is the weighed average given by:

$$E(A) = \sum_{j=1}^{v} \frac{s_{1j} + s_{2j} + ... + s_{Mj}}{s} I(s_{1j}, s_{2j}, ... s_{Mj})$$

The information gain obtained by this partitioning on A is defined as:

$$Gain(A) = I(s_1, s_2, ... s_M) - E(A).$$

Using the above procedure the information gain of each attribute is calculated. Based on the information gain attributes are ranked and the ranking is used to select the relevant variables. The attribute with the highest information gain is considered the most discriminating attribute of a given set. Information gain calculated for each variable for i) ATR1, FR; ii) ATR1, FR and SEP1 are tabulated in table 2. The ranking of the variables changes depending upon the considered objectives. Figure 3a shows the distribution of solutions in a high information gain variable hyper plane whereas figure 3b shows solution distribution in a low information gain variable hyperplane for the objective ATR1.

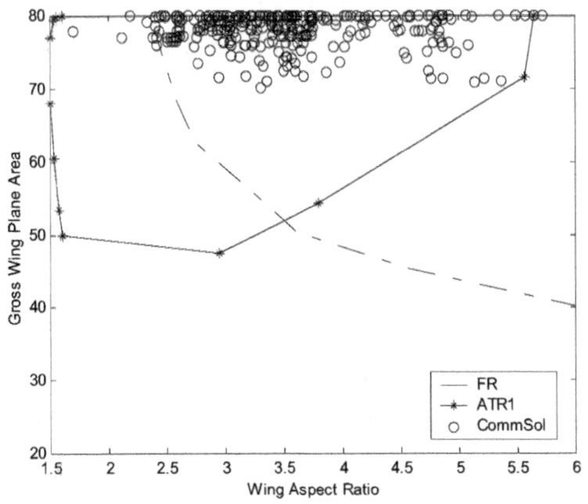

Figure 4: Common Region identified u sing high information gain variables 4,5,7 and 8 and all the solutions for K=3

Input variables with an information gain greater than the mean (5,7 and 8) were then used to identify the common region for K=3. As illustrated in figure 4 this resulted in a reduction in the occurrence of outliers whilst also significantly reducing computational effort. However, as the K-value is increased to improve solution cover in the common region so the number of outliers also increases. This problem can be addressed through the introduction of a data reduction process.

2.3 Data Reduction

By discarding solutions that are not potential common region solutions computational effort can be further reduced and outliers eliminated even at high K values thereby allowing a higher degree of solution set cover within the identified common region. Data reduction is a rule-based method used to remove solutions that are not likely to lie in the common region. The minimum bound $Min_i = Max[Min(A_i)_{j=1}^{m}]_{i=1}^{n}$ of the rule for each variable A_i $i=1$ *to* n (where n is the

Table 2: Details about skewness, correlation coefficient and information gain measurement of all the variables for i) objectives FR and ATR1 ii) FR, ATR1 and SEP1.

Input Var	Skewness			Correlation Coefficient			Inf Gain 1	RANK 1	Inf Gain 2	RANK 2
	ATR1	FR	SEP1	ATR1	FR	SEP1	ATR 1 &FR		ATR1, FR& SEP1	
WLES	-1.351	1.098	0.315	0.478	-0.349	-0.071	0.595	1	0.265	3
WTCR	1.059	-0.922	1.073	-0.55	0.249	-0.521	0.594	2	0.419	2
WAR	0.501	-0.404	0.761	-0.257	0.251	-0.207	0.593	3	0.255	4
CMN	-0.475	-1.123	-0.151	0.051	0.181	0.049	0.104	4	0.118	5
CH	-0.566	-0.193	-0.430	0.059	0.307	0.043	0.079	5	0.068	6
CLMN	-0.481	-0.888	0.013	0.095	0.136	-0.086	0.022	6	0.026	7
BPR	-0.460	-0.757	-0.127	0.141	0.119	0.019	0.013	7	0.014	8
WTR	-0.230	0.172	-0.008	0.013	0.001	-0.018	0.010	8	0.013	9
GWPA	-1.653	-1.758	1.280	0.170	0.463	-0.566	0.007	9	0.953	1
Mean of Information Gain							0.224		0.237	

number of variables) is found by calculating the maximum of the minimum values of each variable A_i in a particular cluster C_j where $j=1$ *to* M, M is the number of clusters.

The maximum bound $Max_i = Min[Max(A_i)_{j=1}^{m}]_{i=1}^{n}$ of the rule for each variable A_i is calculated by calculating the minimum of the maximum values of each variable A_i in a particular cluster C_j. Figure 5 shows the standard normal distribution of solutions in the FCSs of objectives FR and ATR1 with respect to

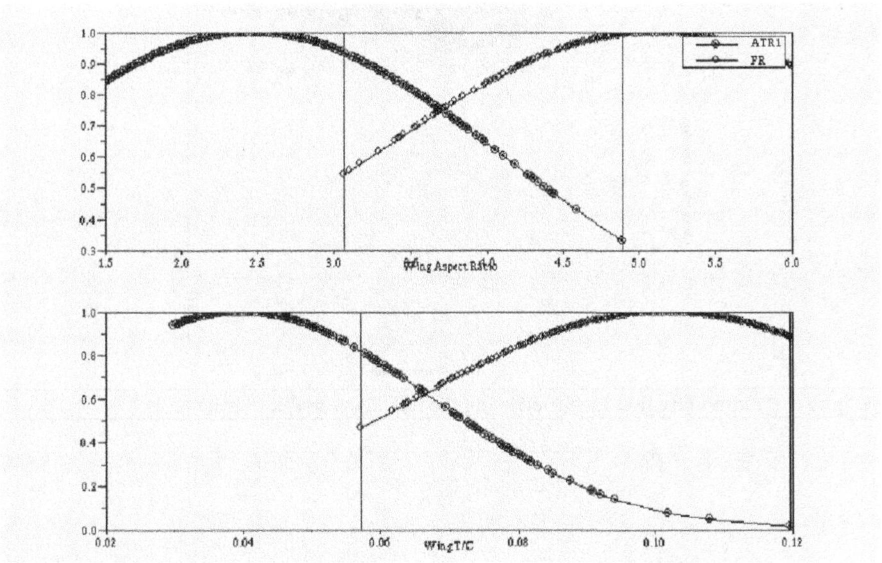

Figure 5: The upper and lower bounds of the variables of Wing aspect ratio and Wing T/C

WAR, and WTCR. The vertical lines indicate the maximum and minimum bounds of the variables common to all the objectives under consideration.

Figure 6 illustrates the outcome of the classifier algorithm subsequent to both variable relevance analysis and data reduction. The outlier problem has been eliminated since all the common region solutions identified lie within the line defining the convex hull [16] of the objectives ATR1 and FR. The above experimentation was repeated for the three objectives FR, ATR1 and SEP1. The classifier performed best at K=15. Although the classifier identified near neighbour solutions, these lie outside identified convex hull intersections of the common region for the different objectives in a high information variable hyperplane as illustrated in figure 7. This is due to the sparsity of solutions in the approximated common region defined by the rule based data reduction technique. This sparsity is evident in the outer areas of the individual objective HP regions as shown in figure 3a. Hence, when attempting to identify common regions that comprise these outer areas sparsity affects classifier efficiency. Various solutions to this sparsity problem are possible such as perturbing the solutions in the approximated common region to increase set cover. Experimentation relating to the Adaptive Filter settings during the early stages of a COGA run are currently underway to also determine whether improved set cover in the outer areas can be achieved 'on-line' during a COGA run. The sparsity problem is not considered to be serious and is being further addressed in research relaing to various scaleability issues.

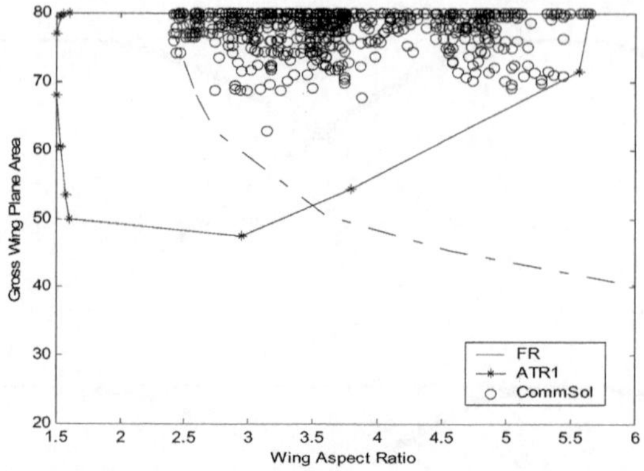

Figure 6: Common Region Identified using the relevant variables for k=15.

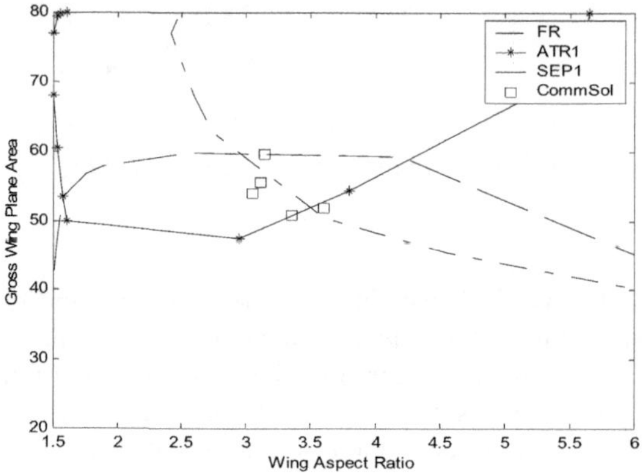

Figure 7: The common region solutions identified using three objective FR, ATR1 and SEP1

2.4 Common Region Identification – Computational Cost

The common region identification procedure is summarized and illustrated in figure 8. All the various computations were carried out using an Intel Pentium 4 PC with 2.53Ghz processing speed and 260 MB RAM. The computational cost of the KNN classifier before and after data reduction is shown in Table 3. The cost increases with the number of solutions to be classified. The data reduction greatly reduces the number of potential common region solutions thereby significantly reducing the computational cost as shown. The introduction of a third objective (SEP1) greatly reduces the size of the common region and, upon the introduction of data reduction, the number of solutions to be classified and, hence, the computational cost.

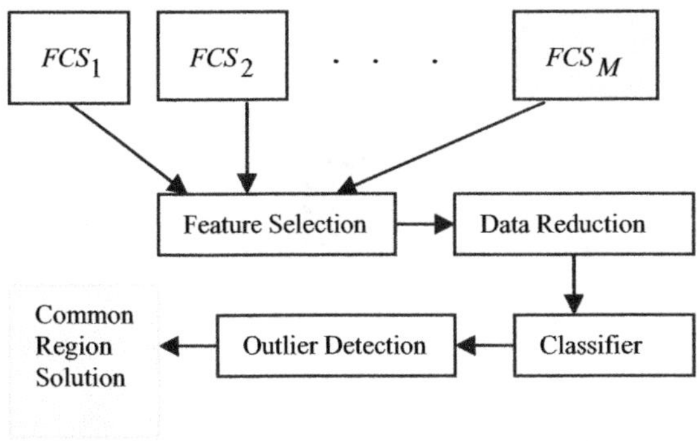

Figure 8: Block diagram of procedure involved in identification of common region between different objective in the variable space.

Table 3: Computational cost vs number of dataset solutions

Data Reduction	Number of Objective	Number of solutions	Computation time (Sec)
no	2	7051	291.8
	3	17732	2475.1
yes	2	3836	75.5
	3	120	0.5

3 Exploring the relationship Between COGA and MOGA Output.

If we take the FCS solutions and the identified common region solutions for ATR1 and FR and plot them in objective space the distributions shown in Figure 9 emerge. We have always assumed a relationship between the solutions in the FCSs and a Pareto frontier and the outer edge of the plot would seem to support this assumption [17].

The working principle of COGA for a multi-objective problem is different to that of standard evolutionary multi- objective algorithms [18] which tend to uses a Pareto dominance based approach. The principle of COGA is to generate as much information as possible concerning high performance regions relating to various objectives within a problem space. Using a standard multi-objective GA (MOGAs) it is possible to obtain solutions lying along the Pareto front but difficult to explore the relationship between variable and the objective space and to discover what is occurring close to the frontier. During the early stages of design it is quite possible that the designer is also interested in such solutions and solutions that lie around particular sections of the Pareto front.

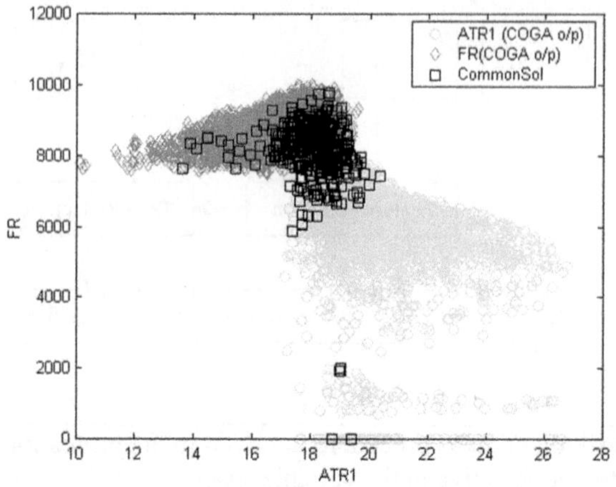

Figure 9: The distribution of common region solution in the objective space bettween the objective FR and ATR1

(a)

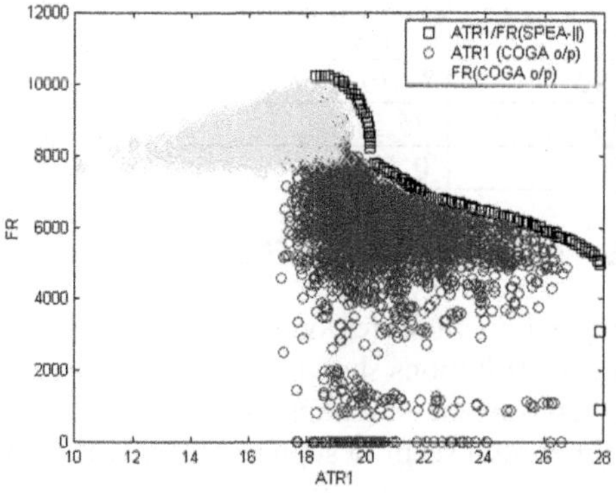

(b)

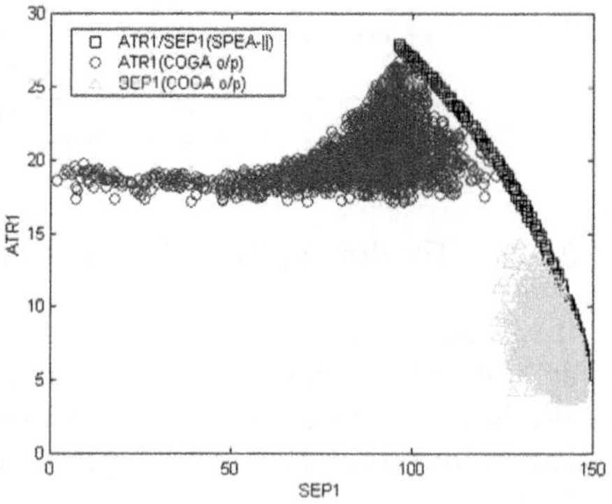

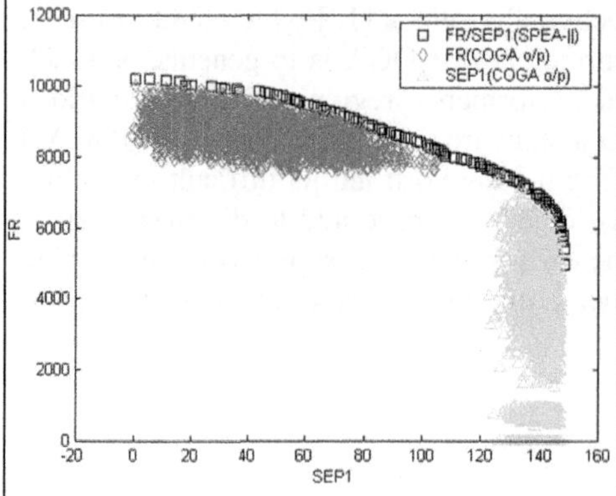

Figure 10a: Distribution of solutions for objective ATR1 and FR against SPEA-II Pareto front

Figure 10b: The distribution of solutions for objective ATR1 and SEP1 against SPEA-II Pareto front.

Figure 10c: The distribution of solutions for objective ATR1 and SEP1 against SPEA-II Pareto front.

NB: Colour versions of these plots can be viewed at:
http:/www.ac-comtech.co.uk/cogaplots.htm

The multi-objective COGA approach provides a good visual indication of the degree of conflict between objectives; an opportunity to explore varying objective preferences and view their effect upon HP region bounds and to generate an approximated Pareto front relating to the objectives under investigation plus solutions around the Pareto front. The following work confirms this whilst also providing a useful benchmark for our work as it illustrates the relationship between COGA output and standard MOGA output.

Most multi-objective genetic algorithms use the concept of Pareto dominance [18]. Deb and Ziztler [19] compared a set of these algorithms and results have shown that the Strength Pareto Evolutionary Algorithm (SPEA) [19] performs comparatively well. SPEA's strength lies in its use of elitism (the concept of storing and using the good solutions in earlier generations for future search). The SPEA-II algorithm has therefore been utilised to generate Pareto fronts for the objectives SEP1, ATR1 and FR for comparative purposes. Figure 10a, 10b 10c & 10d illustrate the distribution of COGA output and the SPEA-II output in the objective space.

Figure 10b &10c shows the conflicting relation between the objectives ATR1 and SEP1 and between objectives FR and SEP1. Figure 10a shows complete COGA cover of the SPEA Pareto front for objectives FR and ATR1 further indicating less conflict between them.

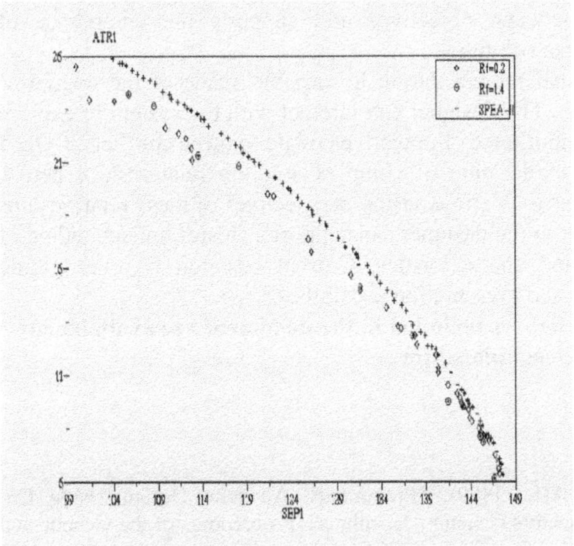

Figure 11: Comparing Pareto front of SPEA-II with that of COGA for lower and high AF threshold

Figure 11 shows that COGA can provide a good approximation to the non-dominated front identified by SPEA-II. This figure also shows how conflict between the objectives can be reduced by lowering the filter threshold. The COGA solutions in figure 11 have been obtained by identifying the non-dominated solutions in the ATR1 and SEP1 FCSs. The darker non-dominated solutions are from the FCSs generated with a higher filter threshold (Rf=1.4) whereas the lighter non-dominated solutions have been generated using a lower filter threshold (Rf = 1.2). It is clear from the figure that with a low filter threshold is possible to obtain a continuous Pareto front and the front only breaks down with an increase in filter threshold indicating the conflict between the objectives in a high information gain variable space i.e. GWPA and WA

4. Conclusion

Initial results from the mining and processing of data generated from multi-objective COGAs has been presented. It has been shown that the relatively novel usage of established data-mining techniques such as KNN clustering can result in the identification of a mutually inclusive common region in terms of the solution set that describes it. Variable Attribute Analysis reduces computational overhead by identifying prime variables and eliminating the need for the KNN algorithm to process the full variable set. Data reduction assists this process, significantly lowering computational cost by reducing the number of solutions to be classified. Further work in this area is investigating other methods of on-line identification of common region bounds during COGA runs as opposed to waiting for and processing the identified Final Clustering Sets of each objective. Scaleability in terms of the inclusion of more objectives is also an area receiving significant attention.

The ability to generate approximate Pareto frontiers from COGA output provides a useful benchmarking aspect and provides additional information re available trade-offs. The visualization of the distribution of the objectives' FCSs in relation to the Pareto front (figure 9 &10) provides a further perspective in terms of projecting solution sets onto objective space. The 'closing of the gap' in the SEP1 / ATR1 Pareto front in figure 11 confirms the assumption that filter thresholds relaxation educes conflict between objectives and supports the emergence of common regions of compromise solutions.

COGA provides a visual representation in variable space of the degree of conflict between the variables. The designer can interact with the system to explore how changes to the relative importance of objectives relate to these conflicts. COGA provides an opportunity for exploration of complex solution relationships across both variable and objective space. Various visual perspectives of these relationships presented in a succinct manner to the designer can support a clearer understanding of the overall problem space and characteristics. Current research is investigating further information extraction and presentation possibilities.

N.B. Colour versions of the figures included in the document are available at: http:/www.ac-comtech.co.uk/cogaplots.htm

References

[1] Parmee IC, Watson AH, (1999), "Preliminary Airframe Design Using Co-Evolutionary Multi-objective Genetic Algorithms", Proceedings of the Genetic and Evolutionary Computation Congress (GECCO).

[2] Cvetkovic D, Parmee IC, (1999), "Preferences and their application in Evolutionary Multi-Objective Optimization", IEEE Transactions on Evolutionary Computation 6(1), 2001, pp 442-47.

[3] Parmee IC, (2002), "Improving Problem Definition through Interactive Evolutionary Computation", Journal of Artificial Intelligence in Engineering Design, Analysis and Manufacture-Special Issue: Human-computer Interaction in Engineering Contexts 16(3).

[4] Parmee IC, Cvetkovic D, Watson AH, Bonham CR, (2000), "Multi-Objective Satisfaction within an Interactive Evolutionary Design Enviroment", Evolutionary Computation 8 (MIT 2000), pp-197:222.

[5] Parmee IC, (1996), "The Maintenance of Search Diversity for Effective Design Space Decomposition using Cluster Oriented Genetic Algorithms (COGAs) and Multi-Agent Strategies (GAANT)", Proceedings of 2nd International Conference on Adaptive Computing in Engineering Design and Control, PEDC, University of Plymouth, pp 128-138.

[6] Parmee IC, and Bonham CR, (2000), "Towards the support of innovative conceptual design through interactive designer / evolutionary computing strategies", Journal of Artificial Intelligence in Engineering Design, Analysis and Manufacture, 14. Cambridge Press, pp 3-16.

[7] Parmee IC, (1996b), "Cluster Oriented Genetic Algorithms (COGAs) for Identification of High Performance Regions of Design Spaces", First International Conference on Evolutionary Computation and Applications, EvCA 96, Presidium of the Russian Academy of Sciences, Moscow; pp 66-75.

[8] Bonham CR, Parmee IC, (2000), "Improving the Robustness of COGA: The Dynamic Adaptive filter", In Evolutionary Design and Manufacture, Selected papers from the 4th International Conference on Adaptive Computing in Design and Manufacture, April 2000, Springer Verlag.

[9] Bonham CR, Parmee IC, (1999), "Improving the Performance of Cluster Oriented Genetic Algorithms (COGAs)", In Proceedings of IEEE Congress on Evolutionary Computation, Washnigton DC, pp 554-561.

[10] Webb E., (1997), "MINICAPS- A Simplified version of CAPS for use as a research tool", Unclassified Report BAe-WOA-RP-GEN-11313, British Aerospace.

[11] Cvetkovic D, Parmee IC, (2002), "Agent-based Support within an Interactive Evolutionary Design", Proceeding of Fifth International Conference on Adaptive Computing in Design and Manufacture, Springer Verlag.

[12] Arvin LB, Langley P, (1997), "Selection of Relevant Feature and Examples in Machine Learning", Artificial Intelligence, 10:245-247.

[13] Han J, Kamber M, (2001), "Data Mining: Concepts and Techniques", Morgan Kaufmann, San Francisco, California.

[14] Aggarwal CC, Yu PS, (2002), "Outlier Detection for High Dimensional Data", In Proceedings of the Sixth European Conference on the Principles of Data Mining and Knowledge Discovery, pages 15 -26.

[15] Knorr EM, Raymond TNg, (1998), "Algorithms for Mining Distance-Based Outliers in Large Datasets", Proceedings of the 24th VLSB Conference New York, USA.

[16] Bykat A, (1978), "Convex Hull of a finite set of points in two dimensions", Inf. Process. Lett.7, 296-298.

[17] Parmee IC, Abraham JAR, (2003), "Further Developments of the Interactive Evolutionary Design System - Towards a better understanding of variable and objective space through interactive exploration". Interactive Evolutionary Computation, Series of Workshops at the Genetic and Evolutionary Computing Conference (GECCO), http://www.ad-comtech.co.uk/Workshops.htm

[18] Deb K, (2001), "Multi Objective Optimization Using Evolutionary Algorithms", John Wiley & Sons.

[19] Zitzler E, Deb K, Thiele L, (2000), "Comparison of Multi Objective Evolutionary Algorithms: Empirical Results", Evolutionary Computation, 8(2), pp: 173-195.

[20] Ziztler E, Laumanns M, Thiele L, (2002), "SPEA-II: Improving the Strength Pareto Evolutionary Algorithm for Multi Objective Optimisation", Evolutionary Methods for Design Optimisation and Control, CIMNE, Barcelona, Spain, pg 95-100.

Building Compact Rulesets for Describing Continuous-Valued Problem Spaces Using a Learning Classifier System

D. Wyatt[1], L. Bull[2], I. C. Parmee[3]

Faculty of Computing, Engineering and Mathematical Sciences,
University of the West of England
Bristol, BS 16 1QY
[1]david2.wyatt@uwe.ac.uk
[2]larry.bull@uwe.ac.uk
[3]ian.parmee@uwe.ac.uk

Abstract

Learning Classifier Systems have previously been shown to have some application in deducing the characteristics of complex multi-modal test environments to a suitable level of accuracy. In this study, the issue of presenting human-readable rulesets to a potential user is addressed. In particular, two existing ruleset compaction algorithms originally devised for rulesets with an integer-valued representation are applied to rulesets with a continuous-valued representation. The algorithms are used to reduce the size of rulesets evolved by the XCS classifier system. Following initial testing, both algorithms are modified to take into account problems associated with the new representation. Finally, the modified algorithms are shown to outperform the originals.

1 Introduction

The Learning Classifier System (LCS) [7] and, in particular, XCS [14] has demonstrated its capabilities for data-mining through rule-induction in several published studies, that is, the induction of a set of general rules from a sample of data points using a combination of Reinforcement Learning [9] and a Genetic Algorithm [6]. Wilson [16][17] demonstrated the capabilities of an interval based encoding when used to induce rules describing the Wisconsin Breast Cancer (WBC) data-set [2] and showed that XCS could improve on best known performance for that data-set. In [16], Wilson addressed the issue of presenting human-readable rulesets and showed reductions of 97.8% in ruleset size whilst being able to maintain highly competitive performance. However, the algorithm assumes that the ruleset to be compacted has a level of performance on the training data of 100%. Fu and Davis [5] continued the investigation into ruleset compaction showing that with some modifications (to Wilson's algorithm) less well-trained rulesets could be compacted. Both studies were based on rulesets trained on the WBC data-set using XCSI [17], an extension of the binary-input XCS to integer inputs for use in integer-valued problems. The work presented in this investigation focuses on rulesets trained on a continuous-valued problem using XCSR [15] an extension of the binary-input XCS to real inputs and, in particular, on issues raised by the continuous nature of evolved rules.

Initial investigations in [3] showed that the XCSR system is able to identify high performance regions from a continuous-valued multi-variable search space using a sample of training data points. Parmee [11] introduced the concept of the identification of high performance regions of complex preliminary design spaces rather than the identification of single optimal design solutions. A region of high performance is any contiguous set of points in a given design space which are considered to be exceptional solutions to a particular set of possibly conflicting design criteria. The solution provided by XCSR is a complete set of simple classification rules that define orthogonal regions of the solution space with attached classification labels. Investigations continued using a new *Simplified Learning Scheme* with the aim of improving XCSR performance with respect to learning speed and ability to respond to changes in the underlying test environment (such as class relabelling). The new system was termed sXCSR and results showed that improvements can be made under the new learning scheme. The work presented clearly demonstrated the capability of XCSR to evolve real-valued pairs to describe interval bounds for each variable in the multi-variable problem and thereby define a set of simple classification rules for the high performance regions.

The investigation was extended in [18] by applying XCSR and sXCSR to progressively more complex multi-modal test environments each with typical search space characteristics, convex/non-convex regions of high performance and complex interplay between variables. The paper provided a clear foundation for the possible use of XCS as an exploratory tool, and in particular one that could provide information from conceptual design spaces enabling a designer to identify the best direction for further investigation [12] plus a better representation of their design problem through redefinition and reformulation of the design space. In particular, two test environments were used to investigate the effects of different degrees of feature sampling, parameter sensitivity and training set size. Both test environments were constructed using a combination of functions allowing for the simple generation of sample points. Each sample point is represented by a vector of continuous values and continuous performance measure. In fact, the two dimensional Himmeblau environment used in [18] is used in this study.

2 XCSR

In [15], Wilson presents a version of XCS for problems, which can be defined by a vector of bounded continuous real-coded variables - XCSR. In that system, each rule in the classifier system population consists of the following parameters:

<condition> : *<action>* : prediction (*p*) : prediction error (ε) : fitness (*F*) : experience (*exp*) : time-stamp (*ts*) : action set size (*as*) : numerosity (*n*). Given that XCSR is an accuracy-based classifier system, the three parameters p, ε and F are used to assess the accuracy of the rule's prediction in relation to its experiences over time, that is, how accurately the rule predicts the actual reward or payoff from its use for a given environment input. The other parameters, *exp*, *ts*, *as* and *n*, are used by the classifier system to maintain the internal dynamics of the system, such as balancing resources across environmental niches, genetic algorithm invocation and computational issues.

The following description of the process actions of the XCSR system can be found in algorithmic form in [4]. A sample point is selected at random from the database and is presented to the system as an *input vector.* The system defines a subset of the *Population*, called the *Match Set*, from those rules whose *<condition>* *matches* the *input vector*, where each rule predicts one of n actions (n = 2 for this study). If there are no matching rules, the system generates, or *covers*, a rule for each possible action using the *input vector* as a template. The *Prediction Array* is *calculated* as a sum of the fitness-weighted prediction of each rule in the *Match Set*, that is, π represents the sum of fitness-weighted prediction for all rules advocating action 1 and ϕ for those advocating action 2. The action with the highest sum represents the systems "best guess" at the classification for the given *input vector.* There are two *action selection* regimes, explore and exploit. Assuming the system exploits its knowledge, the system defines a subset of the *Match Set*, called the *Action Set*, from those rules advocating the *selected action.* The *predicted class* is compared with the *actual class* for the given *input vector* and a *reward* is received, 1000 for correct and zero for incorrect. The system *reinforces* those rules in the *Action Set* using the *reward.* If the average number of time-steps since the last invocation of the *Genetic Algorithm* component is greater than some pre-defined threshold, the *Genetic Algorithm* is permitted to act upon the members of the *Action Set.*

In [15], Wilson defines a *<condition>* as consisting of interval predicates of the form $\{\{c_1,s_1\},...,\{c_n,s_n\}\}$, where c is the interval's range "centre" and s is the "spread" from that centre - termed here as the *Centre-Spread* encoding. Each interval predicate's upper and lower bounds are calculated as follows: $[c_n - s_n, c_n + s_n]$. If an interval predicate goes outside the variable's defined bounds, it is truncated. In order for a rule to match the environmental stimulus, each input vector value must sit within the interval predicate defined for that variable. In [16], Wilson describes another version of XCS which could also be used for such multi-variable problems in which a vector of integer-coded interval predicates is used in the form $\{[l_1,u_1],...,[l_n,u_n]\}$, where l and u are the intervals' lower and upper bounds, respectively - termed here as the *Interval* encoding. Mutation is implemented via a random step (range $-0.1 \leq x \leq 0.1$) and cover produces rules centred on the input value with a "spread" of s_0. There are two different *Interval* encodings, termed *Ordered Interval* and *Unordered Interval* [13]. The first way uses a repair operator to enforce an ordering restriction on the predicates by swapping the offending values to ensure that all interval predicates in the *<condition>* remain feasible. The second lifts the ordering restriction such that an interval $[l_n,u_n]$ is equivalent to $[u_n,l_n]$.

3 Ruleset Compaction Algorithms

3.1 Wilson's *Compact Ruleset Algorithm*

In [16], Wilson defines the *Compact Ruleset Algorithm* (*CRA*) and demonstrates its performance for rulesets evolved using an XCSI classifier system trained on the WBC Problem [2]. In fact, Wilson shows a reduction of some 97.8% in ruleset size for WBC data-set, from 1155 rules to 25, whilst maintaining high levels of accuracy on test data. In the description of the *CRA*, there is an assumption that the ruleset to be compacted, R_{org}, has achieved 100% performance on training data-set, D, before compaction begins, that is, $optimal_{performance} = 100\%$. In fact, as a result of this assumption Fu and Davis [5] produced an alternative compaction algorithm designed for use with less accurate rulesets (as described in Section 3.2). However, results from applying a modified version of Wilson's algorithm to rulesets evolved using an XCSR classifier system trained on a continuous-valued problem suggest that this assumption may be incorrect given the algorithm's ability to achieve large reductions in ruleset size whilst maintaining high levels of accuracy on test data. Wilson's *Compact Ruleset Algorithm* is defined as:

Step 0 - Order all rules $r \in R_{org}$ such that $r_i < r_j$, if $property(r_i) < property(r_j)$, that is, ordered in descending order of $property(r)$. Define the ordered ruleset as R_{ord}. Wilson defines $property(r)$ as an indicator of the relative importance of each rule in R_{org} to classification performance. Numerosity, fitness and experience have all been suggested. Note: in addition to Wilson's original description, $optimal_{performance}$ is defined as the performance of the uncompacted ruleset R_{ord} on training data.

Step 1 - Given that R_{ord} is of size m and consists of a series of sequential subsets $R_{ord(i)}$, where $i \leq n$ and $0 \leq n \leq m$, find the smallest sequential subset of R_{ord} that achieves $optimal_{performance}$ when tested on D. To be clear, the ith sequential subset of R_{ord} consists of all rules r_0 to r_i. Define this subset as $R_{ord(n^*)}$. In this step, i should be incremented from 1 to m.

Step 2 - Eliminate any rule r_i from $R_{ord(n^*)}$ for $0 < i \leq n^*$, where the performance for the sequential subset $R_{ord(i)}$ is not greater than the performance of $R_{ord(i-1)}$. Define the resulting subset of $R_{ord(n^*)}$ as $R_{ord(n^{**})}$. In this step, i is incremented from 1 to n^*.

Step 3 - Find the number of instances in D matched by each rule in $R_{ord(n^{**})}$. Remove the rule with the greatest number of matches, $r_{greatest}$, from $R_{ord(n^{**})}$ and place it in the initially empty set $R_{compact}$. Remove all the matched instances from D. Repeat Step 3 until D is empty. This step generates the ordered set $R_{compact}$ of rules in descending order of their contribution with respect to D from $R_{ord(n^{**})}$.

Given that the above algorithm was originally applied to rulesets evolved by a classifier system using a discrete rather than continuous rule representation, there is an underlying assumption of non-overlapping regions as described by rules in $R_{ord(i**)}$. However, this is unlikely to be a valid assumption for rulesets evolved by XCSR where there is likely to be a significant degree of overlap. This suggests an apparent anomaly in Step 3 of the *CRA* with respect to rulesets evolved by XCSR. The original description states that all those data instances matched by $r_{greatest}$ are removed from *D*. In fact, this strategy removes data instances of both classifications and so instances in regions of overlap may be removed before a more specialised and likely more accurate rule comes to the fore. A simple revision to Step 3, which should help mitigate this issue, is to limit removal of data instances to those with the same class label as $r_{greatest}$.

Step 3a - Find the number of instances in *D* matched by each rule in $R_{ord(n**)}$ ensuring only instances matching the rule's class label are counted. Remove the rule with the greatest number of matches, $r_{greatest}$, from $R_{ord(n**)}$ and place it in the initially empty set $R_{compact}$. Remove all the matched instances from *D*. Repeat Step 3 until *D* is empty. This step is similar to Wilson's Step 3 except that only those instances from *D* that have the same class label as $r_{greatest}$ are removed from *D*.

3.2 Fu and Davis' *Compact Ruleset Algorithm*

In [5], Fu and Davis define another *Compact Ruleset Algorithm* suggesting several changes to Wilson's *CRA* that are required to overcome the limitation placed on initial performance levels. The initial version of Fu and Davis' algorithm is termed *Approach 1* and is defined as:

Step 0 - Record performance of uncompacted ruleset R_{org} on training data as $optimal_{performance}$.

Step 1 - Given that R_{org} is of size m, eliminate any rule r_i from R_{org} for $0 < i \leq m$, where i is incremented from 1 to m. If resulting performance is equal to or better than $optimal_{performance}$, continue to eliminate rules, one at a time, from R_{org}. If performance is worse, replace eliminated rule, r_i, and define resulting subset as $R_{org(i*)}$, where $i*$ is the index of replaced rule. This description clarifies an anomaly in the original description - how to deal with improvements in performance on deletion of r_i. In this version, better performance is dealt with in the same way as unchanged performance.

Step 2 - Continue to eliminate any rule r_i from $R_{org(i*)}$ for $i* < i \leq m$. If elimination of rule r_i results in worse performance, place it in the initially empty set $R_{org(i**)}$. This rule is not considered further in this step. This step continues until all rules have been eliminated from $R_{org(i*)}$.

Step 3 - Find the number of instances in *D* matched by each rule in $R_{org(i**)}$. Remove the rule with the greatest number of matches, $r_{greatest}$, from $R_{org(i**)}$ and place it in the initially empty set $R_{compact}$. Remove all the matched instances from *D*. Repeat Step 3 until *D* is empty or no member of $R_{org(i**)}$ matches any remaining instances from *D*. This step generates the ordered set $R_{compact}$ of rules in descending order of their contribution with respect to *D* from $R_{org(i**)}$.

Based on the results of *Approach 1* presented in [5], the authors suggested two modifications to the basic approach. The first modification is to Step 2 of *Approach 1*, where the rule elimination procedure considers *micro-rules* rather than *macro-rules*. A *macro-rule* consists of n identical *micro-rules* as defined by its numerosity. The new approach is termed *Approach 2* and consists of Steps 0, 1, 2a and 3. However, Fu and Davis present results [5] showing that this modification seems to degrade performance on the WBC dataset. As a result, the modification is not considered further in this study.

The second modification is to Step 3 of *Approach 2* leading to *Approach 3*, that is, Fu and Davis retain the modification made to Step 2 while modifying Step 3. Here, the authors motivate the need for change by indicating that Step 3 does not consider performance when building $R_{compact}$, instead considering number of samples covered. *Approach 3* consisting of Steps 0, 1, 2a and 3a. In fact, Fu and Davis recommend a further modification to *Approach 3* in which Step 2a is replaced with the original Step 2 from *Approach 1*. The new approach has been termed here as *Approach 3a* and consists of Steps 0, 1, 2 and 3a. It is this version of Fu and Davis' *CRA* together with *Approach 1* that is used in this study.

Step 2a - In Step 2, each rule r_i is assumed a *macro-rule*, where each *macro-rule* consists of n *micro-rules* as defined by its numerosity. This modification breaks the assumption by eliminating on a *micro-rule* rather than *macro-rule* basis. Therefore, continue to eliminate any *micro-rule* r_i from $R_{org(i^*)}$ for $i^* < i \leq m$. If elimination of *micro-rule* r_i results in worse performance, place it in the initially empty set $R_{org(i^{**})}$. This *micro-rule* is not considered further. This step continues until all *micro-rules* have been eliminated from $R_{org(i^*)}$.

Step 3a - Order all rules $r \in R_{org(i^*)}$ such that $r_i > r_j$, if $property(r_i) > property(r_j)$, that is, ordered in ascending order of $property(r)$. Fu and Davis suggest that $property(r)$ should be numerosity or experience. Define this subset as $R_{ord(i^*)}$. Eliminate r_{last} from $R_{ord(i^*)}$, where r_{last} represents the last rule in $R_{ord(i^*)}$. If performance worsens, the eliminated rule is reinserted at the head of $R_{ord(i^*)}$. Continue until every rule in $R_{ord(i^*)}$ has been eliminated once. This step generates ordered set $R_{compact}$ in ascending order of $property(r)$.

4 Experimental Details

The two-dimensional test environment used in this paper is the multi-modal modified Himmelblau function [1]. The equation for the modified Himmelblau function is defined as:

$$f(x_1, x_2) = 200 - (x_1^2 + x_2 - 11)^2 - (x_2^2 + x_1 - 7)^2$$

There are four optima of approximately equal magnitude. This function is used to define a two-class classification task. In particular, an exact threshold value of

$\psi = 184$, where $\psi \in [-1986, 200]$, is used to define *High*/*Low* class decision boundaries. Figure 1 shows a contour plot of the function, clearly indicating the four regions of high performance.

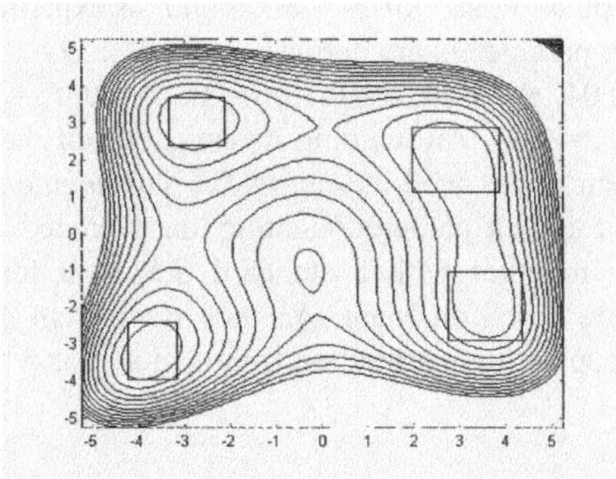

Figure 1: The modified Himmelblau function contour with four high performance regions

For all experiments presented in this study, the XCSR system is trained using a single training data-set and tested using a different test data-set generated from a uniform random distribution. All data-sets have two defined classes, *High* and *Low*. An important aspect of the experimental method identified in [18] was the class imbalance problem [8] which can be defined as a problem encountered by any inductive learning system in domains for which one class is under-represented and which assume a balanced class distribution in the training data. For a two-class problem, the class defined by the smaller set of examples is referred to as the minority class while the other class is referred to as the majority class. Initial experiments in [18] showed that without applying some form of re-balancing of the class distribution, one of the test problems presented could not be described to an acceptable level of accuracy. The approach used for the test environment presented in this study is a simple re-balancing technique. The minority class is re-sampled with replacement until there are an equal number of examples per class.

The training data-set is generated from a Halton Sequence Leaped (HSL) sequence [10], where the HSL is a quasi-random sequence that provides a set of real numbers whose degree of uniformity is high. Initially, the HSL-generated data-set is an imbalanced data-set of 2000 sample points. The data-set is re-balanced using the re-sampling technique discussed above resulting in 3798 sample points with equal points per classification (see [18] for an example).

The test data-set is manipulated in such a way as to provide an equal number of test points per classification. In particular, n points are sampled from a uniform random distribution and evaluated according to the given environment. The sample points are sorted in descending order of associated performance and the top $2m$ points are used to define the test data-set, where m equals the number of sample points of the *High* class generated. By manipulating the test data-set to include sample points from both classifications near to the classification decision boundaries, it is hoped that clear evidence of the classifier system's capability to evolve rules that define those boundaries can be gathered. The test data-set consists of 2116 sample points of which 1073 points are defined as *High* (see [18] for an example training and test dataset in graphical form).

All experiments in this study use a system based on the *Unordered Interval* encoding with a population size of N = 8000 and an experimental run of 200000 trials. XCSR's other parameters are defined as: $\beta = 0.2$, $\alpha = 0.1$, $\varepsilon_0 = 10$, $\nu = 5$, $\theta_{GA} = 12$, $\chi = 0.8$, $\mu = 0.04$, $\theta_{del} = 20$, $\delta = 0.1$, $p_I = 10$, $\varepsilon_I = 0$, $F_I = 0.01$, $\theta_{mna} = 2$, $\theta_{sub} = 20$, m = 10% and s_0 = 2%. An ongoing investigation of the *Simplified Learning Scheme* introduced in [3] as well as *Action Set Subsumption* is continued in this study. The results for each parameter setting are averaged over ten independent runs and presented together with a standard deviation for that sample. Any conclusions made are based on Mann-Whitney Rank Sum Test which makes no assumptions about population distribution from which runs where sampled.

5 Results

5.1 Wilson's *Compact Ruleset Algorithm*

Table 1 presents the results of applying Wilson's *Compact Ruleset Algorithm* together with a modified version of the algorithm (as presented in Section 3.1) to a number of different rulesets produced by four different versions of the XCSR classifier system. As discussed in Section 3.1, there is an assumption that rulesets have achieved 100% performance on training data before compaction takes place. Although, results in Table 1 shows a significant degradation in performance when using Wilson's *CRA*, this drop in performance was reversed when the modified version was used to within 3.5% of the non-compacted ruleset.

Table 1: Test accuracy and ruleset size for different parameter settings of XCSR for the modified and unmodified versions of Wilson's *CRA*

	With AS-Sub				Without AS-Sub			
	XCSR		sXCSR		XCSR		sXCSR	
	Size	Acc.	Size	Acc.	Size	Acc.	Size	Acc.
Pre-CRA	1327.2	71.4	1697.2	65.7	904.3	89.1	1073.2	85.6
Step 1	103.2	76.5	145.5	70.0	198.9	88.9	68.0	86.9
Step 2	33.7	77.3	33.9	70.4	32.2	88.1	27.1	85.9
Step 3	12.0	57.8	9.1	53.6	23.0	80.1	18.4	71.1
Step 3a	24.3	74.5	20.4	62.2	28.7	87.5	23.1	83.5

It is clear from Table 1 that performance of the non-compacted ruleset is higher in all but one case. This difference is statistically significant at a level of >99.9% for the original *CRA* and at a level of >98% for the modified *CRA* (but only when *Action Set Subsumption* was not used). Table 1 shows that the modified *CRA* outperforms the original in all four cases while the modified *CRA* has compacted rulesets which are significantly larger than those produced by the original algorithm. Both of these results are statistically significant at a level of >99%.

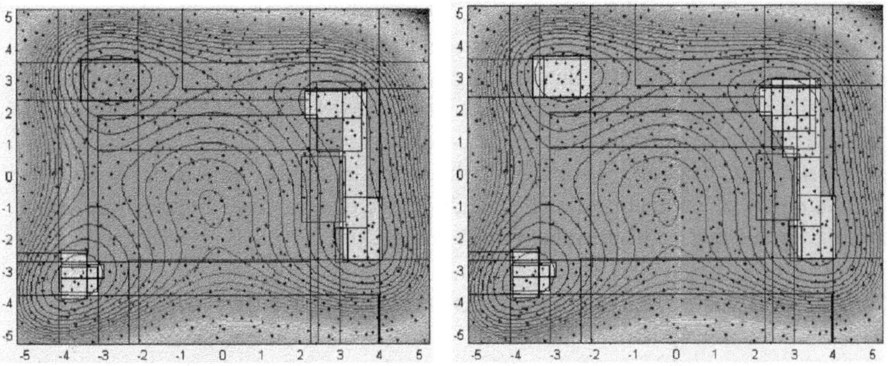

(a) (b)

Figure 2: Graphical representation of ruleset compacted with (a) unmodified and (b) modified versions of Wilson's *CRA*

There is a statistically significant difference (>99%) between versions with and without *Action Set Subsumption*. Those without *Action Set Subsumption* record much greater levels of accuracy and exhibit larger compacted ruleset sizes - although the size differential is much smaller for systems based on the modified *CRA*. Table 1 also shows that versions with *Simplified Learning Scheme* lead to worse performance and smaller compacted ruleset sizes in all four cases. This result is statistically significant at a level of >98% for the original *CRA* (but only when *Action Set Subsumption* was not used) and at a level of >99.9% for the modified *CRA*. In terms of relative ruleset sizes, the original compaction algorithm reduces the ruleset by between 97.4% and 99.5% while the modified *CRA* reductions range between 96.8% and 98.8%. These compaction rates are comparable with Wilson's results for the WBC Problem [16].

High Class Rules

	interval lower x1	interval upper x1	interval lower x2	interval upper x2
1.	(3.074970,	3.668200)	(-2.627216,	2.687206)
2.	(2.696770,	3.663909)	(0.515227,	2.975958)
3.	(3.074970,	4.056044)	(-2.627216,	-0.676490)
4.	(-3.603457,	-2.127199)	(2.387464,	3.690121)
5.	(-3.301771,	-2.127199)	(2.387464,	3.690121)
6.	(2.421430,	3.415167)	(1.349939,	2.954306)
7.	(-3.301771,	-2.389045)	(2.387464,	3.690121)
8.	(2.128220,	3.549653)	(1.823400,	2.687206)
9.	(-4.166482,	-3.178503)	(-3.622773,	-2.805600)
10.	(-4.099170,	-3.453118)	(-3.828231,	-2.320199)
11.	(-4.099170,	-3.007208)	(-3.175459,	-2.759174)
12.	(2.865044,	3.657243)	(-1.617415,	-1.603029)
13.	(-4.421262,	-3.945560)	(-3.746933,	-3.696039)

Figure 3: Textual representation of ruleset compacted with modified version of Wilson's *CRA*

Figures 2(a) and 2(b) show a graphical representation of the final compacted rulesets for both the modified and unmodified versions of a system (*Action Set Subsumption* and *Simplified Learning Scheme* were not used). In these figures, each light/dark grey box represents a region of high/low performance as defined by a single rule. Figure 2(a) shows the final compacted ruleset generated by the original *CRA*. It is clear that this ruleset has failed to adequately describe all the high performance regions of this test environment. However, Fig. 2(b) shows how the modified *CRA* is able to overcome the shortcomings of the original to describe the high performance regions with a high degree of accuracy. Figure 3 presents a textual representation of the rules shown in Fig. 2(b).

5.2 Fu and Davis' *Compact Ruleset Algorithm*

Table 2 presents the results of applying Fu and Davis' *Compact Ruleset Algorithm* (using *Approach 1* and *Approach 3a* as described in Section 3.2) to a number of different rulesets produced by four different versions of the XCSR classifier system. Results in Table 2 provide little evidence of an improvement in performance for Fu and Davis' version of the *CRA* (*Approach 1*) over Wilson's version (unmodified). It should be noted that one of the motivational factors that led to an alternate version of the *CRA*, was to overcome an assumption of 100% performance on training data before compaction, that is, Fu and Davis wanted an algorithm that did not rely on that assumption. However, any differences in performance between Fu and Davis' *CRA* (*Approach 1*) and Wilson's version (unmodified) are not statistically significant (except for the system without *Action Set Subsumption* or *Simplified Learning Scheme* where the level was >98%). A comparison of performance for the modified versions of both algorithms indicate little or no difference in performance and where there is a small difference, there is no statistical significance (except for the system without *Action Set Subsumption* and with *Simplified Learning Scheme* where the level was >95%).

Table 2: Test accuracy and ruleset size for different parameter settings of XCSR for the modified and unmodified versions of Fu and Davis' *CRA*

	With AS-Sub				Without AS-Sub			
	XCSR		sXCSR		XCSR		sXCSR	
	Size	Acc.	Size	Acc.	Size	Acc.	Size	Acc.
Pre-CRA	1327.2	71.4	1697.2	65.7	904.3	89.1	1073.2	85.6
Step 1	1120.8	77.6	1610.1	67.2	481.5	89.2	392.2	86.4
Step 2	100.3	75.9	87.7	67.5	40.0	87.3	38.3	85.8
Step 3	14.6	59.9	9.6	52.2	27.3	85.0	20.3	78.0
Step 3a	26.9	76.8	16.9	63.8	28.1	87.3	22.5	85.4

The performance of the non-compacted ruleset shown in Table 2 is higher than every case of *Approach 1* and *Approach 3a* (except for when *Action Set*

Subsumption was used and *Simplified Learning Scheme* was not). This difference is statistically significant at a level of >99.9% for *Approach 1* and at a level of >99% for *Approach 3a* (except when *Simplified Learning Scheme* was used). Table 1 also shows that *Approach 3a* outperforms the unmodified version in all four cases and that *Approach 3a* has compacted rulesets which are larger than those produced by *Approach 1*. Both of these results are statistically significant at a level of >99.9% (except when *Action Set Subsumption* and *Simplified Learning Scheme* were not used).

There is a statistically significant difference (>99%) between versions with and without *Action Set Subsumption*. As with results presented in Section 5.1, those versions without *Action Set Subsumption* record much greater levels of accuracy and also exhibit larger compacted ruleset sizes with corresponding size differential for systems based on *Approach 3a*. Table 1 shows that versions with *Simplified Learning Scheme* lead to worse performance and smaller compacted ruleset sizes in all four cases. This result is statistically significant at a level of >95% for the performance comparisons and >99% for the ruleset size comparisons. It can be shown that *Approach 1* exhibits reductions in ruleset size of between 97.0% and 99.4% while the *Approach 3a* reductions range between 96.9% and 99.0%. These compaction rates are comparable with those in Section 5.1 and published in [5].

6 Conclusion

The motivation for this work was to investigate the capabilities of both Wilson's and Fu and Davis' *Compact Ruleset Algorithms* for describing high performance regions in a continuous-valued problem domain. The emphasis was on a multi-modal test environment with typical search space characteristics, convex/non-convex regions of high performance and complex interplay between variables. The objective is to develop a classifier system that is capable of evolving a complete and accurate set of maximally general rules that identify and describe the high performance regions of real-world problem environments. The presentation of simple sets of human-readable rules to the designer/user of the system will provide some aid to decision-making, which is relevant to better representation of the design problem through the reformulation and redefinition of the design problem.

This work represents the first published study of the use of Wilson's and Fu and Davis' *Compact Ruleset Algorithms* to a continuous-valued problem space. The results presented in Section 5 clearly demonstrate the capabilities of the both algorithms on the two-dimensional test environment. In particular, results show that neither algorithm (in original form) is able to describe the test environment to an adequate level of accuracy, but with a simple modification, both can produce much greater levels of performance on test data. In fact, there is little evidence of a significant difference in performance or ruleset size between these algorithms in their original or modified form. Although neither algorithm achieved better performance than an uncompacted ruleset, ruleset size reductions were ~ 98%.

References

1 Beasley D, Bull D, Martin R (1993) A Sequential Niche Technique for Multimodal Function Optimisation. Evolutionary Computation, 1(2); 101-125

2 Blake C, Merz C (1998) UCI Repository of Machine Learning Databases

3 Bull L, Wyatt D, Parmee I (2002) Initial Modifications to XCS for use in Interactive Evolutionary Design. In: Parallel Problem Solving From Nature - PPSN VII, Springer Verlag; 568-577

4 Butz M, Wilson S (2001) An Algorithmic Description of XCS. In: Advances in Learning Classifier Systems. Third International Workshop (IWLCS-2000), Lecture Notes in AI (LNAI-1996). Springer-Verlag:Berlin; 253-272

5 Fu C, Davis L (2002) A Modified Classifier System Compaction Algorithm, Proc. of the Genetic and Evolutionary Computation Conference 2002; 920-925

6 Holland J (1975) Adaptation in Natural and Artificial Systems, MIT Press

7 Holland J (1986) Escaping Brittleness: the Possibilities of General-Purpose Learning Algorithms Applied to Parallel Rule-based Systems. In: Machine Learning, An Artificial Intelligence Approach. Morgan Kaufmann

8 Japkowicz N, Stephen S (2002) The Class Imbalance Problem: A Systematic Study. Intelligent Data Analysis, 6(5); 429-450

9 Kaelbling L, Littman M, Moore A (1996) Reinforcement Learning: A Survey. Journal of Artificial Intelligence Research 4; 237-285

10 Kocis L, Whiten W (1997) Computational Investigations in Low Discrepancy Sequences, ACM Trans. on Mathematical Software, 23(2); 266-294

11 Parmee IC (1996) The Maintenance of Search Diversity for Effective Design Space Decomposition using Cluster-Oriented Genetic Algorithms (COGAs) and Multi-Agent Strategies (GAANT). In: Proc. of 2nd International Conf. on Adaptive Computing in Engineering Design and Control; 128-138

12 Parmee IC (2002) Improving Problem Definition through Interactive Evolutionary Computation. In: Journal of Artificial Intelligence in Engineering Design, Analysis and Manufacture, 16(3)

13 Stone C, Bull L (2003) For Real! XCS with Continuous-Valued Inputs. Evolutionary Computation 11(3); 299-336

14 Wilson S (1995) Classifier Fitness Based on Accuracy. Evolutionary Computation 3(2); 149-175

15 Wilson S (2000) Get real! XCS with Continuous-valued inputs. In: Learning Classifier Systems. From Foundations to Applications Lecture Notes in Artificial Intelligence (LNAI-1813) Springer-Verlag: Berlin; 209-222

16 Wilson S (2001) Compact Rulesets for XCSI. In: Advances in Learning Classifier Systems. Fourth International Workshop (IWLCS-2001), Lecture Notes in Artificial Intelligence (LNAI-2321). Springer-Verlag:Berlin; 197-210

17 Wilson S (2001) Mining Oblique Data with XCS, In: Advances in Learning Classifier Systems. Third International Workshop (IWLCS-2000), Lecture Notes in Artificial Intelligence (LNAI-1996). Springer-Verlag:Berlin; 158-177

18 Wyatt D, Bull L (2003) Using XCS to Describe Continuous-Valued Problem Spaces. Technical Report UWELCSG03-004

Chapter 5

Evolutionary Multi-objective and Constraint Satisfaction

Metamodel Assisted Multiobjective Optimisation Strategies and their Application in Airfoil Design

Michael Emmerich, Boris Naujoks

Chair for Systems Analysis,
Dept. of Computer Science,
University of Dortmund,
D-44221 Dortmund, Germany
{emmerich, naujoks}@ls11.cs.uni-dortmund.de

Abstract

In this paper various metamodel-assisted multiobjective evolutionary algorithms (M-MOEA) for optimisation with time-consuming function evaluations are proposed and studied. Gaussian field (Kriging) models fitted by results from previous evaluations are used in order to pre-screen candidate solutions and decide whether they should be rejected or evaluated precisely. The approximations provide upper and lower bound estimations for the true function values. Three different rejection principles are proposed, discussed and integrated into recent MOEA variants (NSGA-II and ϵ-MOEA). Experimental studies on a theoretical test case and in airfoil design demonstrate the improvements in diversity of solutions and convergence to the pareto fronts that can be achieved by using metamodels for pre-screening.

1 Introduction

The last ten years have seen a substantial intensification of research in the field of multiobjective optimisation. In particular the approximation of Pareto fronts by means of optimisation software using simulator-based evaluations has become an important issue in industrial design. Unlike in single objective optimisation, the knowledge of the Pareto front allows to identify trade-offs and thus provides the decision maker to base decisions on a broader view of interesting alternatives.

A typical problem formulation in multiobjective optimisation is as follows:

$$\min f_1(\mathbf{x}), \ldots, \min f_m(\mathbf{x}),\ \mathbf{x} \in \mathcal{S} \tag{1}$$

Here $f_1 : \mathcal{S} \to \mathbb{R}$, ..., $f_m : \mathcal{S} \to \mathbb{R}$ denote objective functions which are typically evaluated by a simulator based evaluation program, which is more or less a black box from the perspective of the optimisation tool. $\mathcal{S}$ denotes the search space. In this study we consider box constrained continuous search spaces $S := [\mathbf{x}_{min}, \mathbf{x}_{max}] \subset \mathbb{R}^n$. However, it shall be noted that the metamodel-assisted approach can be easily extended to other search spaces, where a distance measure can be declared that correlates with the function value.

Multiobjective evolutionary algorithms (MOEA) are a powerful tool for approximating Pareto fronts. The population concept and the randomised search operators make them very robust for the optimisation in difficult search spaces. Moreover, they can be easily adapted to different application problems and computing environments. Recent developments like NSGA-II [1], MODES [7] and ϵ-MOEA [1] focus on achieving a good diversity of Pareto-optimal solutions as well as a close approximation of Pareto optimal fronts.

However, especially in scenarios with time-consuming evaluators EA face the problem that they often need a large number of evaluations in order to achieve a sufficient good approximation of the Pareto front. Thus, in this paper we propose to use fast approximate evaluations interpolated from a database of previous results in order to reduce the number of time-consuming precise evaluations.

In comparison to approaches that utilise artificial neural networks (ANN) [4] for approximation, here metamodels based on the Kriging method are proposed. Unlike ANN the Kriging method not only provides estimations of function values, but also confidence intervals for the predictions. Within this method, the derivation of confidence intervals is based on the assumption that the objective functions are realisations of autocorrelated Gaussian fields in a metric space.

It turns out that the interval estimates provided by the metamodel can be easily integrated in the selection procedure of recent MOEAs. The basic idea is, to reject less promising solutions, before they are evaluated precisely. Different implementations of this ideas will be discussed in this work.

In Section 2 we start with a brief description of Kriging metamodels. Then, in Section 3 recent MOEA are introduced, before in Section 4 the integration of metamodels into the MOEA by means of pre-screening procedures is discussed. In the remainder of the paper we compare various metamodel-assisted MOEA by studying their averaged behaviour on an artificial test problem (Section 5 and 6) and on an application problem - the optimisation of an airfoil at two different flight conditions (Section 7).

2 Gaussian Field Metamodels (Kriging)

In this paper we use Kriging metamodels [3, 8] in order to calculate fast approximations of function values. Kriging models interpolate the results from previous precise evaluations that are stored in a database (cf. Figure 1). Beside using information of precise evaluations from the current run, they can also utilise information from previous optimisation runs. To put things into more concrete terms we assume that a database D with $|D|$ evaluations is given. For the search points in the database $\mathbf{x}_1, \ldots, \mathbf{x}_{|D|}$ precise evaluations $\mathbf{y}_1 := (f_1(\mathbf{x}_1), \ldots, f_m(\mathbf{x}_1))^T, \ldots, \mathbf{y}_{|D|} := (f_1(\mathbf{x}_{|D|}), \ldots, f_m(\mathbf{x}_{|D|}))^T$ are given.

A Kriging approximation for a new point $\mathbf{x}' \in \mathbb{R}^n$ is specified by the mean value and the standard deviation of a Gaussian distribution, describing probability densities for different outcomes of the precise evaluation. Thus, for any new input vector $\mathbf{x}'$ the Kriging procedure calculates mean values $y_1'(\mathbf{x}'), \ldots, y_m'(\mathbf{x}')$ and corresponding standard deviations $\sigma_1', \ldots, \sigma_m'$ for the multivariate Gaussian distribution. By choosing an appropriate ω and setting $l_i(\mathbf{x}') := y_i' - \omega\sigma_i'$ and $u_i(\mathbf{x}') = y_i' + \omega\sigma_i'$ for $i = 1, \ldots, m$, an interval box $[\mathbf{l}(\mathbf{x}), \mathbf{u}(\mathbf{x})] \in \mathbb{R}^m$ can be specified, such that for a given probability p_α the model assumptions lead to $Pr((f_1, \ldots, f_m)^T(\mathbf{x}') \in [\mathbf{l}(\mathbf{x}'), \mathbf{u}(\mathbf{x}')]) \geq p_\alpha$. For two independently distributed outputs we choose $\omega = \phi^{-1}(\frac{1-p_\alpha}{4})$, where $\phi \equiv \Phi - \Phi^2$ and $\Phi(x) = \frac{1}{2}[1 + \mathrm{erf}(\frac{x}{\sqrt{2}})]$ is the standardised Gaussian distribution function. In this context, the drawing in Figure 2 illustrates the idea, how the interval approximations of the metamodel are constructed and how they can be used to screen approximate function values.

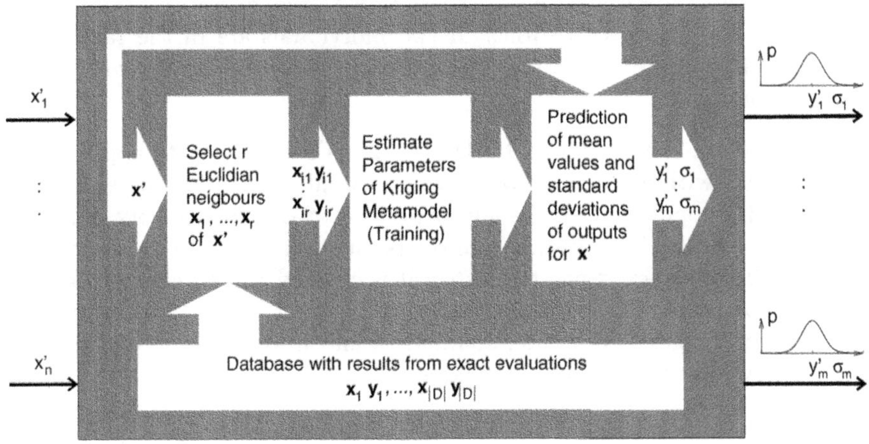

Figure 1: Data flow of the Kriging approximation. By interpolating the outputs for the new input vector $\mathbf{x}'$ from previous evaluations in its proximity a prediction $\mathbf{y}'$ of the output is calculated, accompanied with a confidence measure σ. This confidence measure is the standard deviation of the estimated Gaussian probability distribution for each output under the model assumptions.

The calculation time of Kriging grows in the order of magnitude $\mathcal{O}(r^3 + nr^2)$. Here r denotes the number of points considered for the interpolation. Thus, the Kriging procedure gets time-consuming, if the number of points in the database is large. In order to keep calculation time low, it has been suggested [3, 5, 6] to use local metamodels, i. e. to consider only a few evaluations in the proximity of the new point. In this paper the 25 nearest neighbours in the search space due to the Euclidian distance have been considered (cf. Figure 1). Moreover, ordinary Kriging based on a Gaussian correlation function [8], i. e. $\rho(|\mathbf{x}-\mathbf{x}'|_2) = \exp(-\theta \cdot (|\mathbf{x}-\mathbf{x}'|_2)^2)$, has been used. The hyperparameter θ has been estimated by minimisation of the cross validation error.

3 Evolutionary Multiobjective Optimisation

Two multiobjective optimisation algorithms have been considered. The first one is a very frequently used algorithm in design optimisation that has been developed by Deb [2]. The second is a very recent MOEA [1] that uses a steady state selection and extends the concept of the first algorithm by using an archive and the concept of ϵ-dominance.

NSGA-II [2]: This algorithm uses a $(\mu+\lambda)-$selection mechanism. In order to achieve a strict order on the offspring population, non-dominated sorting and a diversity criterion are used. This means that individuals of the k-th Pareto front are selected with higher priority than those of the $k+1$-th Pareto front. If two individuals are in the same Pareto front, a crowding criterion decides on whether the individual is selected or not.

ϵ-MOEA [1]: The ϵ-MOEA by Deb et al. is a steady state approach that works with an archive, which contains only ϵ non-dominated solutions and a population for the approximation of the Pareto front (cf. Figure 4). By combining values from the population and from the archive, new individuals are generated. Then it is checked if they can be integrated into the population or the archive. In order to enter the population, solutions need to be non-dominated, and in order to enter the archive they have to be ϵ non-dominated.

4 Pre-screening Strategies

We propose to integrate the metamodel into the $(\mu+\lambda)$-selection of the different MOEA, by rejecting less promising offspring individuals before they are precisely evaluated. Only the non-rejected individuals are considered for precise evaluation and can be integrated into the final population. Thus, in the new parent population there are only precisely evaluated solutions.

The suggested approach is contrary to most recently proposed methods for coupling metamodels with multicriteria optimisation methods. Typically the optimisation is performed only on the metamodel, which is trained a-priori [6]. Iteratively, after optimisations on the metamodel, the accuracy of the metamodel is locally improved by resampling in a reduced search space around promising points.

Three different ways of rejection have been identified:

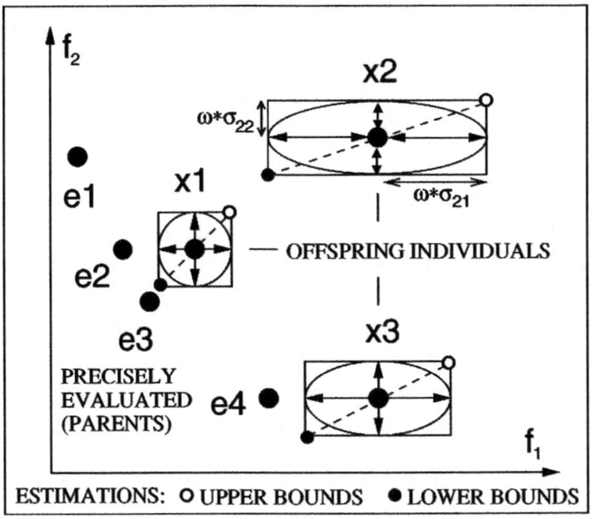

Figure 2: Rejection principles in the metamodel assisted evolution strategy. The points $\mathbf{e}_1, \mathbf{e}_2, \ldots, \mathbf{e}_4$ depict precisely evaluated parent individuals in the solution space $(f_1, f_2)(\mathbf{S})$. These individuals are to be compared to approximately evaluated offspring individuals $\mathbf{x}_1$, $\mathbf{x}_2$ and $\mathbf{x}_3$. For each approximate individual the confidence ellipsoids and their bounding boxes are depicted.

CR ***Constant ratio selection:*** A common rejection principle in single objective optimisation [3, 4] is to choose a small constant number ν ($\nu \ll \lambda$) of the most promising offspring for precise evaluation. Such a selection will be denoted as $(\mu+\nu < \lambda)$-*selection*. In MOEA the ν most promising individuals can be determined by non-dominated sorting of the approximated mean values. Recent results in single-objective optimisation [3] show that it makes the metamodel-assisted strategies even more robust if the pre-screening of values is based on an estimated lower bound and not on the estimated mean value. Since the margin between lower bound and mean value is an indicator for the degree of exploration, this pre-screening strategy can lead the population into the more unexplored regions of the search space. The CR strategy makes extensive use of the Metamodel information and thus it has the potential to improve the convergence significantly. The user has to provide an extra parameter ν that might have an significant influence on the performance of the search. This is why two further strategies with adaptive ν are suggested next. We will term them *variable ratio rejection* methods, since they work with a variable ratio of preselected individuals.

VR$^-$ ***Non-error principle:*** The first way is to reject all those individuals which have a small probability to be competitive with the parents and the other offspring individuals. In other words, we compare the new individual with the other individuals by being optimistic about the true value of the new individual and pessimistic about the true values of its competitors. This can be done by comparing the lower bound edge of the new individual with the upper bound edge of its competitors, having in mind the intervals depicted in Figure 2.

To put things into more concrete terms, given a set of (precisely and approximately) evaluated solutions $\mathbf{x}_1, \ldots, \mathbf{x}_{\mu+\lambda}$ we reject an individual $\mathbf{x}$ from this set, if at least μ individuals $\mathbf{x}_{i_1}, \ldots, \mathbf{x}_{i_\mu}$ can be found, for which $\forall j \in 1, \ldots, \mu : \mathbf{u}(\mathbf{x}_{i_j}) \prec \mathbf{l}(\mathbf{x})$ In this case the probability of a wrong rejection should be lower than $1 - p_\alpha$, if p_α is the user provided confidence level. The VR$^-$ selection is interesting, since it is the most effective rejection strategy that - assumed all prediction intervals would be precise - does not affect the convergence behaviour of the MOEA measured in terms of precise evaluations.

VR$^+$ ***High quality principle*****:** Rejecting with this principle will maximise the probability that the pre-screened solutions have a high quality and thus will be selected after precise evaluation. In order to achieve this for the new point, prediction values closer to the upper bound edge of the prediction interval box (cf. Figure 2) should be used and values close to the lower bound edge for the competitors should be chosen. This can be done by *comparing only mean values*, as it is done in this work. It can be expected that this strategy rejects much more evaluations and thus it has a higher potential of improving the convergence than the conservative VR$^-$ strategy. In contrast to the first strategy, this rejection strategy has a greater influence to change the behaviour of the MOEA, since less promising points are not considered, although they might have been integrated into subsequent populations. Especially in the first stage of the evolutionary algorithm, this strategy is expected to be superior to the VR$^-$ strategy, because in this phase we are more interested in large improvements that take the strategy near to the Pareto front than in small improvements.

The integration of these rejection criteria is depicted in Figures 3 and 4. In NSGA-II a $(\mu + \lambda)$-selection is used. The rejection with the metamodel is done before the evaluation of the λ individuals. Within this rejection step the offspring population is reduced to $\nu < \lambda$ individuals by using one of the rejection methods described above.

In the CR-NSGA-II, the offspring population is sorted by means of non-dominated sorting and only the $\nu = const$ individuals with the best lower bounds enter the next generation. If the VR$^-$ or VR$^+$ rejections are applied within these algorithms, for each individual it is decided, whether it is considered for further evaluation or rejected.

In the steady state ϵ-MOEA algorithm the rejection principles VR$^-$ and VR$^+$ can be applied. Each approximately evaluated individual is compared to the values in the population and in the archive (cf. Figure 4). For deciding,

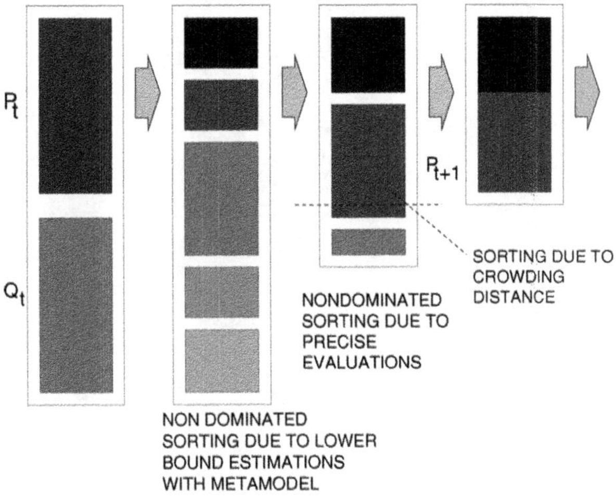

Figure 3: Schematic view of metamodel-assisted NSGA-II.

if an approximate solution is competitive with a solution in the ϵ-archive, the lower left node in the ϵ-grid next to the lower bound edge is considered when using the VR^- criterion and its upper bound edge is set to the nearest upper right edge of the ϵ-grid, when considering the VR^+ criterion.

5 Averaging Pareto fronts

The comparison of Pareto fronts received from multiple runs with the same parameterisation has been carried out with the result taken from [7]. In short, parallel lines are drawn through the corresponding region of the search space. From the Pareto front of each run, the points with the shortest distance to these lines are considered for the calculation of the averaged front. If we receive more points than half of the number of runs within a predefined maximum distance to each line, these points are averaged to become a member of the averaged Pareto front. This procedure does not benefit high density of points on the Pareto front, but could lead an engineer to filter designs with significant changes and reduce the amount of designs to choose from.

The left hand part of Figure 5 illustrates the method shortly. Here, some parallel lines can be seen next to the points from different runs, which are considered for the calculation of an averaged point. The resulting averaged points are highlighted using an other point style. In the right hand part of Figure 5 the five lines describe five Pareto fronts from different runs and the points describe the received averaged Pareto front.

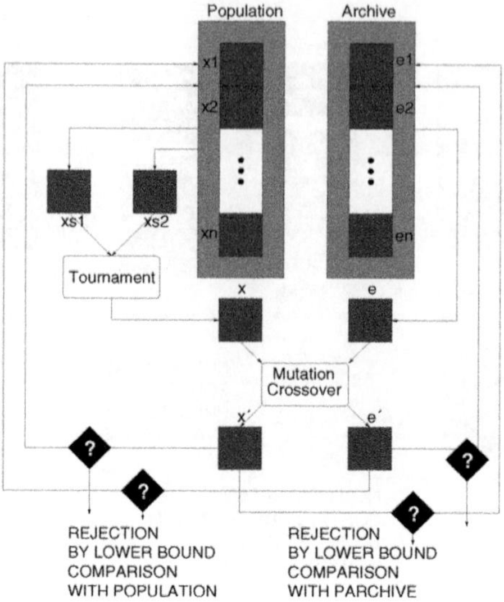

Figure 4: The ϵ-MOEA with pre-screening of solutions.

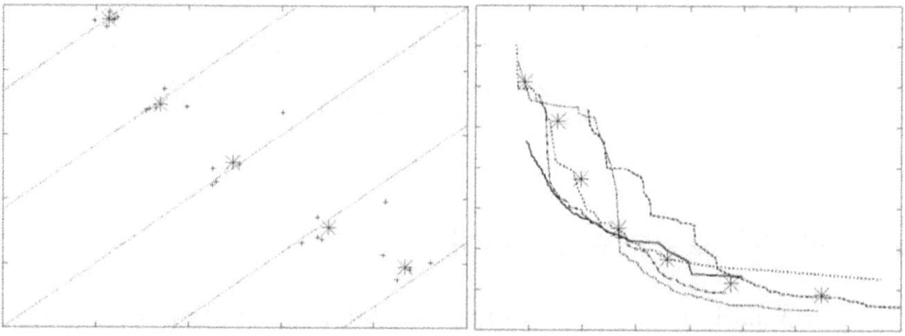

Figure 5: Illustration of the averaging Pareto fronts method. The left hand part shows the lines and corresponding points of different runs considered for the averaging as well as the resulting points. The right hand part shows the averaged points of the five results presented using lines.

6 Multiobjective Sphere Model

As a simple test case for testing the applicability of the new approach the 20 dimensional multiobjective sphere problem has been considered:

$$f_1(\mathbf{x}) := \sum_{i=1}^{20} x_i^2 \to \min,\ f_2(\mathbf{x}) := \sum_{i=1}^{20} (x_i - 2.0)^2 \to \min \tag{2}$$

This function has a convex Pareto front. Furthermore both functions are differentiable and convex. A 20 dimensional search space has been chosen in order to get a similar number of decision variables like in the subsequently presented airfoil optimisation example. The entire search space is defined by the hypercube $[0, 2.0]^{20}$. Note, that the assumption that f_1 and f_2 are orders of magnitude more time-consuming than the Kriging evaluation does not correspond to the reality. This assumption has been made, in order to have a simple test problem.

As algorithmic variants the $(20 + 4 < 20)$-NSGA-II, the $(20 + 20 < 100)$-NSGA-II and the ϵ-MOEA with a step size of 20 have been employed with $p_\alpha \approx 0.911$ corresponding to $\omega = 2.0$. For each algorithmic variant 20 runs have been conducted and averaged. Within the plot of NSGA-II only the CR selection schemes have been plotted. The VR selection schemes performed worse than the CR selection schemes and have been left out within the plots. For the ϵ-MOEA only the VR^+ and the VR^- selection have been plotted, since the algorithmic construction of the ϵ-MOEA does not allow for a straightforward application of the CR selection.

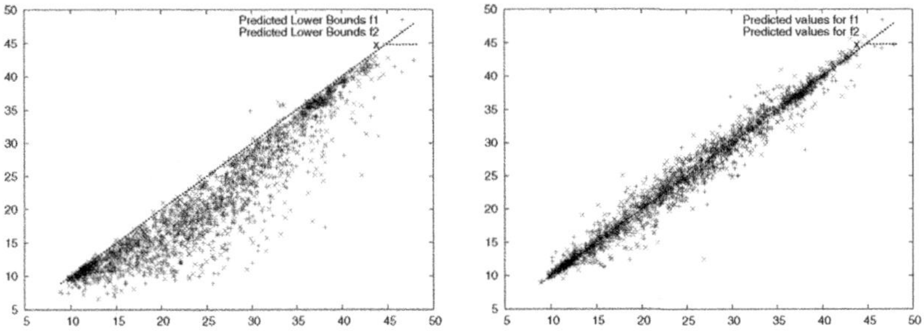

Figure 6: Original function values have been plotted against lower bounds for $p_\alpha \approx 0.911$ (left) and mean values of the predictions (right). The samples have been obtained during the optimisation with the (20+20<100)-CR-NSGA-II on the 20 dimensional multiobjective sphere model.

Before running statistical comparisons, the prediction capabilities of the metamodel during an optimisation have been tested. The results have been summarised in Figure 6, showing that the number of incorrect lower bound estimations is very low.

Statistical comparisons between optimisation strategies are summarised in Figure 7. They show first of all that the NSGA-II is much faster than the ϵ-MOEA when approaching the Pareto front. By using the metamodel it can be further accelerated. Furthermore the pre-screening with the metamodel improved the results of the ϵ-MOEA only slightly.

Within the CR-NSGA-II pre-screening it seems to be important to use the lower edge bounds as pre-selection criterion. This might be explained by the effect that the lower bound criterion drives the search into the unexplored regions of the search space. Thus a better diversity of solutions on the Pareto front can be achieved and false optima [5] can be avoided.

Table 1: Summarised design conditions

Property	Case	high lift	low drag
M_∞	[–]	0.20	0.77
Re_c	[–]	$5 \cdot 10^6$	10^7
α	[°]	10.8	1.0

7 Airfoil Design Testcase

The real-world application under investigation is an airfoil re-design testcase which has been studied several times before [7]. The two criteria stem from the task to re-design two target airfoils which themselves are nearly optimal for given flow conditions. These flow conditions can be taken from Table 1 and correspond to the take-off phase (high-lift condition: low Mach- and Reynolds-number and high angle of attack) and the cruising phase (low-drag condition: high Mach- and Reynolds-number and low angle of attack) during an aircraft flight. For both conditions the pressure distribution around the airfoil is calculated. Here, all design and simulation conditions have been fixed for the study, e.g. turbulence models, mesh generation method, etc. This results in the following 2 dimensional design problem:

$$f_{1,2}(s) = \int_0^1 (C_p(s) - C_{p,target_{1,2}}(s))^2 ds, \tag{3}$$

where s is the airfoil arch length, C_p and $C_{p,target}$ describe the pressure distributions along the airfoil. The parameterisation is performed using Bezier-splines. Upper and lower surface of the airfoil are described by 7 Bezier-points each. Next to the y-coordinates of these points, the x-coordinates of two of them are considered for design parameters of the optimisation. This results in 9 parameters for each surface and an 18-dimensional problem all together.

In Figure 7 the first results from studies with the metamodel-assisted MOEA variants are depicted. All runs have been done with 1000 precise evaluations and they all needed about 16 hours to be computed on a PENTIUM 2GhZ machine under Linux operating system. All metamodel evaluations together took less than 1% of the computation time.

8 Discussion and Outlook

For the ϵ-MOEA the results show a good convergence to the Pareto front, but their diversity is relatively small. The convergence is further improved by using the metamodel. The VR^+ criterion that uses the mean value of the

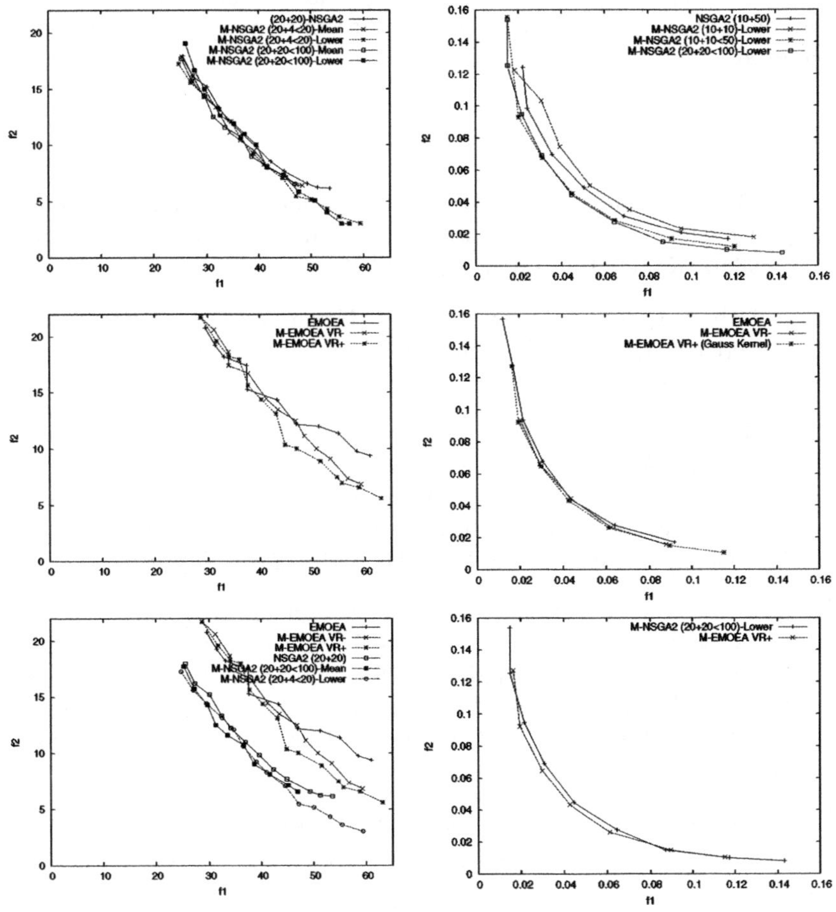

Figure 7: Averaged results for sphere model (left) and on the airfoil test case (right).

prediction leads to better results here than the conservative VR^- criterion. The latter method rejects only 10 percent of the individuals, while VR^+ rejects up to 50 percent. This is probably the reason why the effect of the metamodel is more visible.

For the NSGA variants a more significant effect of the metamodel is visible. Both, the convergence and the diversity have been increased by the use of the CR rejection principle. In the runs 20 percent of the offspring individuals are preselected for precise evaluation. The best results have been achieved with the $(20 + 20 < 100)$-CR-NSGA-II, which not only improved the convergence, but also the diversity of the results, both in comparison to ϵ-MOEA and other NSGA-II variants.

It has been shown that the use of metamodels can improve the performance of MOEA. Especially in the NSGA-II algorithm a significant improvement of the convergence and the diversity of the Pareto front approximation has been observed.

References

[1] K. Deb, M. Mohan, and S. Mishra. A fast multiobjective evolutionary algorithm for finding well-spread pareto-optimal solutions. Technical Report 2003002, KanGAL, Kanpur, India, 2003.

[2] K. Deb, A. Pratap, S. Agarwal, and T. Meyarivan. A fast and elitist multiobjective genetic algorithm nsga-ii. Technical Report 2000001, KanGAL, Kanpur, India, 2000.

[3] M. Emmerich, A. Giotis, M. Özdemir, Th. Bäck, and K. Giannakoglou. Metamodel-Assisted Evolution Strategies. In J. J. Merelo Guervós, P. Adamidis, H.-G. Beyer, J. L. Fernández-Villacañas, and H.-P. Schwefel, editors, *Parallel Problem Solving from Nature – PPSN VII, Proc. Seventh Int'l Conf., Granada*, pages 361–371, Berlin, 2002. Springer.

[4] A. P. Giotis, K. C. Giannakoglou, and J. Périaux. A Reduced-Cost Multi-Objective Optimization Method Based On The Pareto Front Technique, Neural Networks And PVM. In *Proc. European Congress on Computational Methods in Applied Sciences and Engineering (ECCOMAS'00)*, (CD-ROM), Barcelona, 2000. Center for Numerical Methods in Engineering (CIMNE).

[5] Y. Jin. A comprehensive survey of fitness approximation in evolutionary computation. *Soft Computing Journal (in press)*, 2003.

[6] P. K. Nain and K. Deb. Computationally effective search and optimization procedure using coarse to fine approximations. In *Proc. of the Congress on Evolutionary Computation CEC 2003, Canberra, Australia*, pages 2081–2088, 2003.

[7] B. Naujoks, L. Willmes, Th. Bäck, and W. Haase. Evaluating multicriteria evolutionary algorithms for airfoil optimisation. In J. J. Merelo Guervós, P. Adamidis, H.-G. Beyer, J. L. Fernández-Villacañas, and H.-P. Schwefel, editors, *Parallel Problem Solving from Nature – PPSN VII, Proc. Seventh Int'l Conf., Granada*, pages 841–850, Berlin, 2002. Springer.

[8] J. Sacks, W.J. Welch, W.J. Mitchell, and H.-P. Wynn. Design and analysis of computer experiments. *Statistical Science*, 4:409–435, 1989.

Acknowledgements: This work was supported by the Deutsche Forschungsgemeinschaft (DFG) as part of the *Collaborative Research Center* 'Computational Intelligence' (SFB 531). Also, the support from bilateral Personnel Exchange Programme between Greece and Germany (IKYDA 2000) is acknowledged.

Mechanical Component Design for Multiple Objectives Using Generalized Differential Evolution

Saku Kukkonen, Jouni Lampinen

Department of Information Technology
Lappeenranta University of Technology
P.O. Box 20
FIN-53851 Lappeenranta, Finland
Saku.Kukkonen@lut.fi

Abstract

In this paper an Evolutionary Algorithm, the Differential Evolution algorithm, and its extension for constrained multi-objective optimization are described. The described extension is tested with a set of four constrained multi-objective mechanical component design problems. Results are compared to results obtained with the elitist Non-Dominated Sorting Genetic Algorithm and the results show that the extension performs comparably to the elitist Non-Dominated Sorting Genetic Algorithm and is applicable for solving multi-objective mechanical component design problems subject to multiple constraints.

1 Introduction

Many situations in engineering and economics deal with optimization. One may want to optimize, *e.g.*, manufacturing processes, shape of products, and number of different products to be manufactured. Typical goals are minimizing costs, maximizing profits, and improving performance. Several natural aspects limits feasible solutions, *e.g.*, resources may cause limitations and the number of products cannot be a negative number.

Optimization is an intensively studied problem field in mathematics. However, functions to be optimized in traditional mathematics are relatively simple, yet functions to be optimized in practice are often far more complicated. In these cases different stochastic optimization methods have shown their effectiveness.

Most optimization research deals with single-objective optimization problems. The basic nature of many optimization problems is, however, multi-objective and these problems are usually first converted to single-objective problems. Single-objective problems are commonly considered easier to solve but conversion from multi-objective problem to single-objective problem requires some *a priori* knowledge which is not necessarily available or which is hard to determine, *e.g.*, the relative importance of each individual sub-objective. For this reason interest exists in solving multi-objective problems in multi-objective form.

Several evolutionary algorithms for multi-objective optimization have been proposed during last years. Most well known and promising methods are the Strength Pareto Evolutionary Algorithm (SPEA) [1], the Non-Dominated Sorting Genetic Algorithm (NSGA) [2, pp. 209–218], and the elitist Non-Dominated Sorting Genetic Algorithm (NSGA-II) [2, pp. 245–253].

2 Multi-Objective Optimization with Constraints

Many practical problems have multiple objectives. For example, designing a wing of an aircraft may have objectives such as maximizing strength, minimizing weight, minimizing manufacturing costs, maximizing lifting force, minimizing drag, *etc.* Multiple objectives are almost always more or less conflicting.

Several aspects cause constraints to problems. In the previous example of the wing, the thickness of the metal parts used must be a positive number, shape limitations exist, some parts are available in some predefined standard sizes, *etc.* Constraints can be divided into box or boundary constraints and constraint functions. Boundary constraints are used when the value of some optimized variable is limited to some range and constraint functions are used for more complicated constraints which are expressed as functions.

Multi-objective problems are often converted to single-objective problems by predefining weighting factors for different objectives, expressing the relative importance of each objective. However, this is impossible in many cases because a decision maker does not necessarily know beforehand how he/she wants to weight different objectives. Thus, a more convenient way is to keep multiple objectives of multi-objective problems and try to solve them in this form even though this may be harder to do in practice. Optimizing several objectives simultaneously is often called Pareto optimization [3]. An obtained solution is Pareto optimal if none of the objectives cannot be improved without impairing at least one other objective [4, p. 11–12]. A set of Pareto optimal solutions form a Pareto front. From the set of Pareto optimal solutions the decision maker may select one which has suitable values for different objectives. This can be viewed as *a posteriori* articulation of decision-makers preferences concerning the relative importance of each objective.

A mathematically constrained multi-objective optimization problem can be presented in the form [4, p. 37]

$$\begin{array}{ll} \text{minimize} & f_1(\boldsymbol{x}), f_2(\boldsymbol{x}), \ldots, f_K(\boldsymbol{x}) \\ \text{subject to } \boldsymbol{x} & S = \quad \boldsymbol{x} \quad \mathbf{R}^D \; \boldsymbol{g}(\boldsymbol{x}) = (g_1(\boldsymbol{x}), g_2(\boldsymbol{x}), \ldots, g_M(\boldsymbol{x}))^T \quad \mathbf{0} \; . \end{array} \tag{1}$$

Thus, there are K functions to be optimized and M constraint functions.

The major part of earlier mathematical research has concentrated on optimization problems where the functions are linear, differentiable, convex, or otherwise mathematically well behaving. However, in practical problems objective functions are often nonlinear, non-differentiable, discontinuous, multi-modal, *etc.* and no presumptions can be made about their behavior. Variables may also be integers or discrete instead of being continuous. Most traditional optimization methods cannot handle such complexity or do not perform in these cases in which the assumptions they are based on do not hold. For such problems stochastic optimization methods such as Simulated Annealing (SA) and Evolution Algorithms (EAs) have been demonstrated to be effective because they do not rely assumptions concerning the objective and constraint functions.

3 Differential Evolution

The Differential Evolution (DE) algorithm [5,6] [7, pp. 79–108] belongs to the family of Evolution Algorithms and was introduced by Rainer Storn and Kenneth Price in 1995. Design principles in the DE algorithm were simplicity, efficiency, and use of floating-point encoding instead of binary numbers, which is usual way of coding in Genetic Algorithms (GA).

Like in a typical EA, the idea in the DE algorithm is to have some random initial population which is then improved using selection, mutation, and crossover operations. Several ways exist to determine a stopping criteria for EAs but typical for DE is the use simply of some predefined upper limit G_{max} for the number of generations to be computed.

3.1 Initialization of Population

Values for the initial population in DE are typically drawn from uniform distribution. Formally this can be presented as [7, p. 81]:

$$\begin{array}{l} P_G = \quad \boldsymbol{x}_{1\,G}, \boldsymbol{x}_{2\,G}, \ldots, \boldsymbol{x}_{NP\,G} \quad , \quad \boldsymbol{x}_{i\,G} = x_{j\,i\,G} \\ x_{j\,i\,G=0} = x_j^{(lo)} + rand_j[0,1] \left(x_j^{(hi)} \quad x_j^{(lo)} \right) \\ i = 1, 2, \ldots, NP, \quad NP \quad 4, \quad j = 1, 2, \ldots, D \end{array} \tag{2}$$

In this representation P_G denotes a population after G generations (0 is an initial generation), $\boldsymbol{x}_{i\,G}$ denotes an object variable vector (or individual) of the population, and $rand_j[0,1]$ denotes a uniformly distributed random variable in the range $[0,1]$. Terms $x_j^{(lo)}$ and $x_j^{(hi)}$ denote lower and upper parameter bounds, respectively. The size of the population is denoted by NP and the dimension of objective vectors is denoted by D.

Other ways of initialization also exist, *e.g.*, if some knowledge exist about the position of the optimum, part of the initial population may be initialized around the possible position of the optimum using normal distribution.

3.2 Mutation and Crossover

DE goes through each objective vector $\boldsymbol{x}_{i\,G}$ of the population and creates a corresponding trial vector $\boldsymbol{u}_{i\,G+1}$ as follows [7, p. 82]:

$$
\begin{array}{l}
r_1, r_2, r_3 \quad 1, 2, \ldots, NP \;\; , \\
\text{(randomly selected, except: } r_1 = r_2 = r_3 = i) \\
j_{rand} = \text{int}\,(rand_i[0,1) \quad D) + 1 \\
\text{for}(j = 1; j \quad D; j = j + 1) \\
\\
\qquad \text{if}(rand_j[0,1) < CR \quad j = j_{rand}) \\
\qquad\qquad u_{j\,i\,G+1} = x_{j\,r_3\,G} + F \quad (x_{j\,r_1\,G} \quad x_{j\,r_2\,G}) \\
\qquad \text{else} \\
\qquad\qquad u_{j\,i\,G+1} = x_{j\,i\,G}
\end{array}
\tag{3}
$$

Indices r_1, r_2, and r_3 are mutually different and drawn from the set of the population indices. Both CR and F are user defined control parameters for the DE algorithm and they remain fixed during the whole execution of the algorithm. Parameter CR represents the probability that an element for the trial vector is chosen from a linear combination of three randomly chosen vectors instead of from the old objective vector $\boldsymbol{x}_{i\,G}$. Parameter F is a scaling factor for mutation and its value is typically $(0, 1+]$. The condition "$j = j_{rand}$" is to make sure that at least one element is different compared to elements of the old population member.

3.3 Selection

After each mutation and crossover operation the trial vector $\boldsymbol{u}_{i\,G+1}$ is compared to the old objective vector $\boldsymbol{x}_{i\,G}$. If the trial vector has equal or lower cost value, then it replaces the old vector. This can be presented as follows [7, p. 82]:

$$
\boldsymbol{x}_{i\,G+1} = \begin{cases} \boldsymbol{u}_{i\,G+1} & \text{if} \quad f(\boldsymbol{u}_{i\,G+1}) \quad f(\boldsymbol{x}_{i\,G}) \\ \boldsymbol{x}_{i\,G} & \text{otherwise} \end{cases}
\tag{4}
$$

The average cost value of the population will never increase, because the trial vector replaces the old vector only if it has equal or lower cost value.

The algorithm presented above is known as DE/*rand*/*1*/*bin* in the literature. Several variations of the basic DE algorithm exist and this is a reason for complicated notation. In the notation DE/x/y/z x indicates how the mutated vector is selected (it could be selected also to be *best* among the current population), y indicates number of vector differences used in the mutation, and z indicates the way the old vector and trial vector are recombined (an alternative exponential recombination procedure is also mentioned in the literature [7, p. 98]).

4 Extension of DE for Constrained Multi-Objective Optimization

Several extensions of DE for multi-objective optimization exists as well for constrained optimization [8–15]. The approach presented in this paper combines these by modifying the selection operation of the basic DE algorithm [16–18]. Compared to other DE extensions for multi-objective optimization, this approach makes DE suitable for constrained multi-objective optimization with minimum changes to the original algorithm. The modified selection operation for M constraint and K objective functions is presented formally in Eq. 5 [17].

$$\boldsymbol{x}_{i\,G+1} = \begin{cases} \boldsymbol{u}_{i\,G+1} \text{ if } \begin{cases} \begin{cases} j \quad 1,\ldots,M & : g_j(\boldsymbol{u}_{i\,G+1}) > 0 \\ j \quad 1,\ldots,M & : g'_j(\boldsymbol{u}_{i\,G+1}) \quad g'_j(\boldsymbol{x}_{i\,G}) \end{cases} \\ \begin{cases} j \quad 1,\ldots,M & : g_j(\boldsymbol{u}_{i\,G+1}) \quad 0 \\ j \quad 1,\ldots,M & : g_j(\boldsymbol{x}_{i\,G}) > 0 \end{cases} \\ \begin{cases} j \quad 1,\ldots,M & : g_j(\boldsymbol{u}_{i\,G+1}) \quad 0 \quad g_j(\boldsymbol{x}_{i\,G}) \quad 0 \\ k \quad 1,\ldots,K & : f_k(\boldsymbol{u}_{i\,G+1}) \quad f_k(\boldsymbol{x}_{i\,G}) \end{cases} \end{cases} \\ \boldsymbol{x}_{i\,G} \quad \text{otherwise} \end{cases}$$

where $g'_j(\boldsymbol{x}_{i\,G}) = \max\left(g_j(\boldsymbol{x}_{i\,G}), 0\right)$ and $g'_j(\boldsymbol{u}_{i\,G+1}) = \max\left(g_j(\boldsymbol{u}_{i\,G+1}), 0\right)$ are representing the constraint violations

(5)

The selection rule given in Eq. 5 selects the trial vector $\boldsymbol{u}_{i\,G+1}$ to replace the old vector $\boldsymbol{x}_{i\,G}$ in three cases:

1. Both the trial vector $\boldsymbol{u}_{i\,G+1}$ and the old vector $\boldsymbol{x}_{i\,G}$ violate at least one constraint but the trial vector does not violate any of the constraints more than the old vector.
2. The old vector $\boldsymbol{x}_{i\,G}$ violates at least one constraint whereas the trial vector $\boldsymbol{u}_{i\,G+1}$ is feasible.
3. Both vectors are feasible and the trial vector $\boldsymbol{u}_{i\,G+1}$ has less or equal cost value for each objective than the old vector $\boldsymbol{x}_{i\,G}$ has.

Otherwise the old vector $\boldsymbol{x}_{i\,G}$ is preserved.

The basic idea in the selection rule is that the old vector $\boldsymbol{x}_{i\,G}$ is not allowed to dominate the trial vector $\boldsymbol{u}_{i\,G+1}$ in constraint violation space or in objective function space. The principle is effectively rather similar to the method described in [2, pp. 131–132] even though the formulation is different. In this other method selection is based on value of a penalized objective function $F(\boldsymbol{x})$ [19]:

$$F(\boldsymbol{x}) = \begin{cases} f(\boldsymbol{x}) & \text{if } \boldsymbol{x} \text{ is feasible} \\ f_{max} + \sum_{j=1}^{M} \max\left(g_j(\boldsymbol{x}), 0\right) & \text{otherwise} \end{cases} \tag{6}$$

Here, f_{max} denotes the objective function value of the worst feasible solution in the population. The selection rules given in Eq. 5 and based on Eq. 6 perform same way when two feasible solutions are compared or when a feasible solution is compared to an infeasible solution. In the case of two infeasible solutions, the selection rule in Eq. 5 compares solutions based on dominance of the constraint violations whereas the selection method based on Eq. 6 compares solutions based on a sum of the constraint violations. The selection method based on Eq. 6 needs search of the objective function value of the worst feasible solution in the population, calculation of all constraint function values in the case of infeasible solution, and it also permits worsening of individual constraint function values because sums of the constraint violations are compared instead of individual constraint violations. The selection method based on Eq. 6 may also need normalization of different constraints in the case of different order of magnitude whereas this is not needed in the selection rule in Eq. 5.

The selection rule in Eq. 5 can be implemented in such a way that the number of function evaluations is reduced because not always all the constraints and objectives need to be evaluated, *e.g.*, inspecting constraint violations (even one constraint) is often enough to determine which vector to select for the next generation [17, 18].

One should note that the selection rule in Eq. 5 handles any number M of constraints and any number K of objectives, including cases $M = 0$ (unconstrained problem) and $K = 0$ (constraint satisfaction problem). When $M = 0$ and $K = 1$, the selection rule is identical to the selection rule of the basic DE algorithm. Because the described selection method extends DE for constrained multi-objective optimization and the basic DE algorithm is a special case, the method with the selection rule described is named *Generalized Di erential Evolution (GDE)*.

After the selected number of generations the final population presents a solution for the optimization problem. The non-dominated solutions can be separated from the final population if desired. There is no sorting of non-dominated solutions during the optimization process or explicit mechanism for maintaining diversity of solutions as in the NSGA and NSGA-II methods.

5 Mechanical Component Design Problems

GDE was implemented in C and tested with four mechanical component design problems which are often seen in the literature. Problems were taken from [20] where they are solved using the NSGA-II method which used the simulated binary crossover (SBX) [2, pp. 113–116] and polynomial mutation operators [2, pp. 124]. SBX and the mutation & crossover operator of DE have similar property of being self-adaptive to current population even though SBX is more complicated. According to [20] and [2, pp. 450–455] constraints were normalized and the sums of constraint violations multiplied with possible penalty parameters were added to objective functions.

In all tests population size of 100 was used and the number of generations was 500. Control parameter values were $CR = 0.5$ and $F = 0.3$ for all problems and these were selected based on visual evaluation of preliminary results. More suitable common control parameter values for all problems and especially individual control parameter values for each individual problem can probably be found. Results were obtained with a single run, *i.e.*, several runs were not run to obtain different solutions. Test were executed in Sun Sparc Ultra2.

5.1 Two-bar truss design

Idea of the problem is to design a truss which has to carry a certain load without elastic failure. Originally the problem was studied using the ϵ-constraint method [21]. Objectives are to minimize volume of the truss and minimize stresses. The problem has been formalized as [20]:

$$\begin{array}{ll} \text{Minimize} & f_1(\boldsymbol{x}) = x_1\sqrt{16+y^2} + x_2\sqrt{1+y^2} \\ \text{Minimize} & f_2(\boldsymbol{x}) = \max(\sigma_{AC}, \sigma_{BC}) \\ \text{Subject to} & \max(\sigma_{AC}, \sigma_{BC}) \quad 1(10^5) \\ & 1 \quad y \quad 3 \text{ and } 0 \quad x_i \quad 0.01 \end{array} \tag{7}$$

The stresses are calculated as follows:

$$\sigma_{AC} = \frac{20\sqrt{16+y^2}}{yx_1}, \qquad \sigma_{BC} = \frac{80\sqrt{1+y^2}}{yx_2}. \tag{8}$$

Obtained solution with GDE is shown in Figure 1 and Table 1 contains spread comparison between different methods. When obtained results are compared, it can be noticed that GDE provides solution which is comparable with the other methods. As can be seen from Table 1 GDE is able to find a solution which has smaller value for the volume than solutions of other methods have. When maximal stress is minimized, NSGA-II finds better solution than GDE, but both solutions are still of same order of magnitude whereas the ϵ-constraint method has found a solution order of magnitude worse.

Table 1. The spread of solutions with the -constraint method, NSGA-II, and GDE for the two-bar truss design problem

	Minimal volume		Minimal max stress	
	Volume	Max stress	Volume	Max stress
-constraint	0 004445	89983	0 004833	83268
NSGA-II	0 00407	99755	0 05304	8439
GDE	0 0031	144104	0 0450	8915

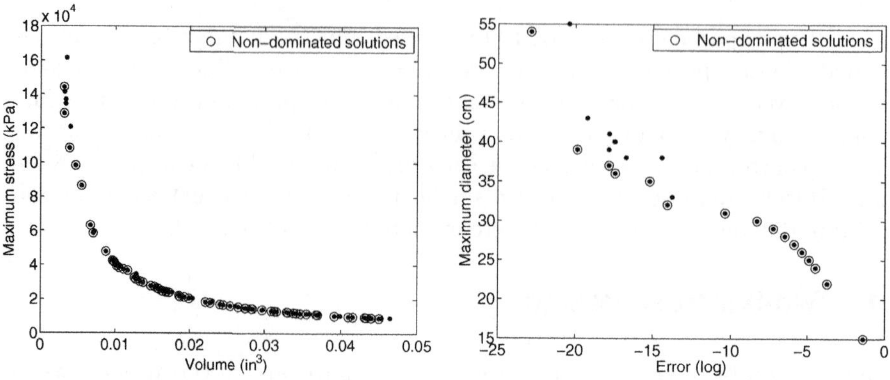

(a) (b)

Figure 1. Final population obtained with GDE for a) the two-bar truss design problem (CPU time 0.49 seconds) b) the gear train design problem (CPU time 0.55 seconds)

5.2 Gear train design

Idea of the problem is to design a gear train which has a specific gear ratio between the driver and driven shafts. The main objective is to minimize the error between the obtained gear ratio and a required gear ratio of 1/6.931. Other objective is to minimize the maximum size of any of the four gears. The problem has been formalized as [20]:

$$\begin{array}{ll} \text{Minimize} & f_1(\boldsymbol{x}) = \left[\frac{1}{6\ 931} \quad \frac{x_1 x_2}{x_3 x_4}\right]^2 \\ \text{Minimize} & f_2(\boldsymbol{x}) = \max(x_1, x_2, x_3, x_4) \\ \text{Subject to } 12 & x_1, x_2, x_3, x_4 \quad 60, \text{all } x_i\text{'s are integers.} \end{array} \qquad (9)$$

Obtained solution with GDE is shown in Figure 1 and Table 2 contains spread comparison between NSGA-II and GDE. When obtained results are compared, it can be noticed that GDE provides solution which is comparable with NSGA-II. As can be seen from Table 2 GDE is able to find a solution which has smaller error whereas NSGA-II is able to find a solution with smaller maximal diameter.

Table 2. The spread of solutions with NSGA-II and GDE for the gear train design problem

	Minimal error		Minimal max diameter	
	Error	Max diameter	Error	Max diameter
NSGA-II	$1\ 83(10^{-8})$	37	0 501	13
GDE	$1\ 166(10^{-10})$	54	0 2457	15

5.3 Spring design

The problem is to design a helical compression spring which has minimum volume and minimal stress. Problem has tree variables: the number of spring coils $N = x_1$ (integer), the wire diameter $d = x_2$ (discrete having 42 non-equispaced values [7, p. 141]), and the mean coil diameter $D = x_3$ (real). The problem has been formalized as [20]:

$$
\begin{array}{rl}
\text{Minimize} & f_1(\boldsymbol{x}) = 0.25\pi^2 {x_2}^2 x_3 (x_1 + 2) \\
\text{Minimize} & f_2(\boldsymbol{x}) = \frac{8KP_{max}x_3}{{x_2}^3} \\
\text{Subject to} & g_1(\boldsymbol{x}) = l_{max} \quad \frac{P_{max}}{k} \quad 1.05(x_1+2)x_2 \quad 0 \\
 & g_2(\boldsymbol{x}) = x_2 \quad d_{min} \quad 0 \\
 & g_3(\boldsymbol{x}) = D_{max} \quad (x_2 + x_3) \quad 0 \\
 & g_4(\boldsymbol{x}) = C \quad 3 \quad 0 \\
 & g_5(\boldsymbol{x}) = \delta_{pm} \quad \delta_p \quad 0 \\
 & g_6(\boldsymbol{x}) = \frac{P_{max} \quad P}{k} \quad \delta_w \quad 0 \\
 & g_7(\boldsymbol{x}) = S \quad \frac{8KP_{max}x_3}{{x_2}^3} \quad 0 \\
 & g_8(\boldsymbol{x}) = V_{max} \quad 0.25\pi^2 {x_2}^2 x_3 (x_1+2) \quad 0 \\
 & x_1 \text{ is integer, } x_2 \text{ in discrete, } x_3 \text{ is continuous.}
\end{array}
\tag{10}
$$

The parameters used are as follows:

$$
\begin{array}{llll}
K = \frac{4C \quad 1}{4C \quad 4} + \frac{0\ 615 x_2}{x_3}, & P = 300 \text{ lb}, & D_{max} = 3 \text{ in}, & k = \frac{G{x_2}^4}{8x_1{x_3}^3}, \\
P_{max} = 1000 \text{ lb}, & \delta_w = 1.25 \text{ in}, & \delta_p = \frac{P}{k}, & l_{max} = 14 \text{ in}, \\
\delta_{pm} = 6 \text{ in}, & S = 189 \text{ ksi}, & d_{min} = 0.12 \text{ in}, C = D/d, & \\
G = 11500000, & V_{max} = 30 \text{ in}^3. & &
\end{array}
\tag{11}
$$

Obtained solution with GDE is shown in Figure 2 and Table 3 contains spread comparison between NSGA-II and GDE. When obtained results are compared, it can be noticed that GDE provides solution similar to NSGA-II. As can be seen from Table 3 in this case NSGA-II gives more widely spread solution than GDE.

Table 3. The spread of solutions with NSGA-II and GDE for the spring design problem

	Minimal volume		Minimal stress	
	Volume	Stress	Volume	Stress
NSGA-II	2 690	187053	24 189	61949
GDE	2 8143	184209	19 1123	71080

5.4 Welded beam design

In this problem a beam needs to be welded on another beam and it must carry a certain load in such way that cost of fabrication and end deflection are minimized.

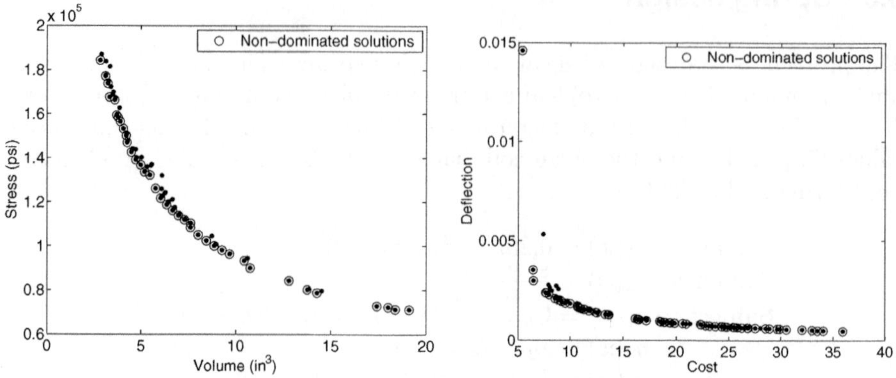

(a) (b)

Figure 2. Final population obtained with GDE for a) the spring design problem (CPU time 1.18 seconds) b) the welded beam design problem (CPU time 0.56 seconds)

The problem has been formalized as [20]:

$$\begin{array}{rl} \text{Minimize} & f_1(\boldsymbol{x}) = 1.1047h^2l + 0.04811tb(14 + l) \\ \text{Minimize} & f_2(\boldsymbol{x}) = \delta(\boldsymbol{x}) \\ \text{Subject to} & g_1(\boldsymbol{x}) = 13600 \quad \tau(\boldsymbol{x}) \quad 0 \\ & g_2(\boldsymbol{x}) = 30000 \quad \sigma(\boldsymbol{x}) \quad 0 \\ & g_3(\boldsymbol{x}) = b \quad h \quad 0 \\ & g_4(\boldsymbol{x}) = P_c(\boldsymbol{x}) \quad 6000 \quad 0 \end{array} \tag{12}$$

The end deflection, stress and buckling terms are:

$$\begin{array}{l} \delta(\boldsymbol{x}) = \frac{2\ 1952}{t^3 b}, \\ \tau(\boldsymbol{x}) = \sqrt{(\tau\ )^2 + (\tau\ \ )^2 + (l\tau\ \tau\ \ )/\sqrt{0.25(l^2 + (h+t)^2)}}, \\ \tau\ \ = \frac{6000}{\overline{2}hl}, \\ \tau\ \ = \frac{6000(14+0\ 5l)\ \ \overline{0\ 25(l^2+(h+t)^2)}}{2[0\ 707hl(l^2\ \ 12+0\ 25(h+t)^2)]}, \\ \sigma(\boldsymbol{x}) = \frac{504000}{t^2 b}, \\ P_c(\boldsymbol{x}) = 64746.022(1\quad 0.0282346t)tb^3. \end{array} \tag{13}$$

The variables are initialized in range $0.125 \quad h, b \quad 5.0$ and $0.1 \quad l, t \quad 10.0$.

Obtained solution with GDE is shown in Figure 2 and Table 4 contains spread comparison between GDE and NSGA [22, pp. 150–153]. When obtained results are compared, it can be noticed that GDE provides solution which is comparable with NSGA. As can be seen from Table 4 NSGA has found better solution according to minimal cost and GDE has found better solution according to minimal deflection. According to [20] NSGA-II spreads more widely than NSGA but total spread was not available in numerical form.

Table 4. The spread of solutions with NSGA and GDE for the welded beam design problem

	Minimal cost		Minimal deflection	
	Cost	Deflection	Cost	Deflection
NSGA	3 944	0 005	20 677	0 001
GDE	5 3720	0 0146	35 9263	0 00045

6 Conclusions and Future Research

In this paper Differential Evolution algorithm and its extension for constrained multi-objective optimization are described. Described method extends the basic DE algorithm for constrained multi-objective optimization with minor changes to the original algorithm of DE and for this reason described method is named as Generalized DE (GDE). It is simple yet effective and does not introduce any extra control parameters to the basic DE algorithm.

GDE is tested with four mechanical component design problems. Results show that the method is able to solve all test problems and performs comparable to NSGA-II in all studied Cases. Control parameter values used were coarsely selected and not carefully optimized for the problems. In particular, the same set of control parameter values was applied for solving all problems. Based on tests with different control parameter values, our preliminary recommendations for control parameter values are CR $[0.1, 0.5]$ and F $[0.1, 1+)$ for initial settings, when method is used for multi-objective optimization problems.

As future work remains more intensively study of effect of parameters for optimization process and extensive comparison of GDE with different optimization methods using numerical measurements. Also, distribution of solutions and extent of the obtained non-dominated front could be improved because now GDE does not contain any mechanism for maintaining these.

References

1. Zitzler, E., Thiele, L.: An evolutionary algorithm for multiobjective optimization: The strength Pareto approach. Technical Report 43, Computer Engineering and Networks Laboratory (TIK), Swiss Federal Institute of Technology (ETH), Zürich, Switzerland (1998)
2. Deb, K.: Multi-Objective Optimization using Evolutionary algorithms. John Wiley & Sons, Chichester, England (2001)
3. Pareto, V.: Cours D'Economie Politique. Libraire Droz, Geneve (1964 (the first edition in 1896))
4. Miettinen, K.: Nonlinear Multiobjective Optimization. Kluwer Academic Publishers, Boston (1998)
5. Storn, R., Price, K.V.: Differential evolution - a simple and efficient adaptive scheme for global optimization over continuous spaces. Technical report,

ICSI (1995) [Online] Available: ftp.icsi.berkeley.edu/pub/techreports/1995/tr-95-012.ps.gz, 15.1.2004.
6. Storn, R., Price, K.V.: Differential evolution - a simple and efficient heuristic for global optimization over continuous spaces. Journal of Global Optimization **11** (1997) 341–359
7. Corne, D., Dorigo, M., Glover, F.: New Ideas in Optimization. McGraw-Hill, London (1999)
8. Madavan, N.K.: Multiobjective optimization using a Pareto differential evolution approach. In: Proceedings of the 2002 Congress on Evolutionary Computation, CEC'02, Honolulu, Hawaii (2002) 1145–1150
9. Abbass, H.A., Sarker, R.: The Pareto differential evolution algorithm. International Journal on Artificial Intelligence Tools **11** (2002) 531–552
10. Abbass, H.A.: The self-adaptive Pareto differential evolution algorithm. In: Proceedings of the 2002 Congress on Evolutionary Computation, CEC'02, Honolulu, Hawaii (2002) 831–836
11. Chang, C., Xu, D., Quek, H.: Pareto-optimal set based multiobjective tuning of fuzzy automatic train operation for mass transit system. IEE Proceedings on Electric Power Applications **146** (1999) 577–583
12. Storn, R.: System design by constraint adaptation and differential evolution. IEEE Transactions on Evolutionary Computation **3** (1999) 22–34
13. Wang, F.S., Chiou, J.P.: Differential evolution for dynamic optimization of differential-algebraic systems. In: Proceedings of the IEEE International Conference on Evolutionary Computation, ICEC'97, Indianapolis, IN (1997) 531–536
14. Chang, T.T., Chang, H.C.: Application of differential evolution to passive shunt harmonic filter planning. In: 8th International Conference On Harmonics and Quality of Power, Athens, Greece (1998) 149–153
15. Wang, F.S., Sheu, J.W.: Multiobjective parameter estimation problems of fermentation processes using a high ethanol tolerance yeast. Chemical Engineering Science **55** (2000) 3685–3695
16. Lampinen, J.: A constraint handling approach for the differential evolution algorithm. In: Proceedings of the 2002 Congress on Evolutionary Computation, CEC'02, Honolulu, Hawaii (2002) 1468–1473
17. Lampinen, J.: DE's selection rule for multiobjective optimization. Technical report, Lappeenranta University of Technology, Department of Information Technology (2001) [Online] Available: http://www.it.lut.fi/kurssit/03-04/010778000/MODE.pdf, 15.1.2004.
18. Lampinen, J.: Multi-constrained nonlinear optimization by the differential evolution algorithm. Technical report, Lappeenranta University of Technology, Department of Information Technology (2001) [Online] Available: http://www.it.lut.fi/kurssit/03-04/010778000/DECONSTR.PDF, 15.1.2004.
19. Deb, K.: An efficient constraint handling method for genetic algorithms. Computer Methods in Applied Mechanics and Engineering **186** (2000) 311–338
20. Deb, K., Pratap, A., Moitra, S.: Mechanical component design for multiple objectives using elitist non-dominated sorting GA. In: Proceedings of the Parallel Problem Solving from Nature VI (PPSN-VI). (2000) 859–868 Also available: http://www.iitk.ac.in/kangal/papers/tech-rep2.ps, 15.1.2004.
21. Palli, N., Azram, S., McCluskey, P., Sundararajan, R.: An interactive multistage *epsilon*-inequality constraint method for multiobjective optimization. ASME Journal of Mechanical Design **120** (1999) 678–686
22. Miettinen, K., Mäkelä, M.M., Neittaanmäki, P., Périaux, J.: Evolutionary Algorithms in Engineering and Computer Science. John Wiley & Sons, Ltd (1999)

Multidisciplinary Aircraft Conceptual Design Optimisation Using a Hierarchical Asynchronous Parallel Evolutionary Algorithm (HAPEA)

L. F. González, E. J. Whitney, K. Srinivas, K. C. Wong [1], J. Périaux[2]

[1]The University of Sydney,
Sydney, NSW 2006,Australia.
gonzalez,eric,ragh,kc@aeromech.usyd.edu.au

[2] Dassault Aviation 78 Quai Marcel Dassault
99214 Saint-Cloud, France.
jacques.periaux@dassault-aviation.fr

Abstract

In this paper we present some results of continuing research into improving robustness speed and application of Hierarchical Parallel Asynchronous Evolution Algorithms (HAPEA) to multidisciplinary design optimisation (MDO) and aircraft conceptual design problems. The formulation and implementation of the HAPEA-MDO algorithm is described and can be regarded as an architecture that is applicable to either integrated or distributed system optimisation design for complex, non-linear and non-differentiable problems. In this paper the formulation for HAPEA-MDO will be described and applied to single and multi objective MDO problems. Two cases related to aircraft design are analysed. We compute the Nash and Pareto optimal configurations satisfying the specified criteria in both cases and show that the HAPEA approach provides very efficient solutions to the stated design problems.

1 Introduction

Evolutionary Algorithms (EAs) are based on Darwinian evolution, whereby populations of individuals evolve over a search space and adapt to the environment by the use of different mechanisms such as mutation, crossover and selection. EAs require no derivatives or gradients of the objective function, have the capability of finding globally optimum solutions amongst many local optima, are easily executed in parallel and can be adapted to arbitrary solver codes without major modification. Another major advantage of EAs is that they can tackle multi-objective problems directly.

EAs have been successfully applied to different aeronautical design problems [1] and there have been different efforts to explore the capabilities of EAs for MDO and aircraft design. Obayashi has applied EAs to the MDO of wing planforms [2]. Parmee and Watson [3] proposed a preliminary airframe design using co-evolutionary Multi-Objective Genetic Algorithms; this algorithm was able to find local objective optimal solutions after only a few generations. Their research also suggests that quicker less detailed runs can easily be achieved using smaller population sizes. Raymer [4] applied different MDO methods and compared them to orthogonal steepest descent search. One conclusion of Raymer's work indicates that aircraft conceptual design can be improved by proper application of MDO methods.

The paper presents a short discussion on multi-objective optimisation and the Hierarchical Asynchronous Parallel Evolution Algorithms (HAPEA). These concepts are then illustrated by two problems related to aircraft design, namely a one objective subsonic transport design and a two objective Unmanned Combat Air Vehicle (UCAV) comparison using Nash and Pareto concepts. The straightforward coupling of the optimiser and the aircraft design analysis software are given. The results are examined for practical applicability and future possibilities for the approach are outlined.

2 Multi-objective Optimisation

The purpose of a multi-objective optimiser is to provide answers to problems, which cannot or should not be posed in single objective form. To this end, a number of approaches have been developed for use in evolutionary algorithms, including traditional aggregating functions, Pareto and Nash approaches and others [5].

Pareto fronts: Originally proposed by Vilfredo Pareto [6], a solution to a given multi-objective problem is the Pareto optimal set is found using a cooperative game which computes the set of *non-dominated* solutions. This spans the complete range of compromise designs between each objective.

Nash equilibria: Originally proposed by John F. Nash [7], a Nash equilibrium point is the position in a competitive (or non-cooperative) game whereby no player can improve his position at the expense of the other, with each player optimising one fitness function over a subset of the objective variables. The Nash equilibrium is determined by *n* players competing symmetrically for *n* criteria, where each player optimises a unique set of optimisation variables, and all other variables are determined by the other players. For example, for player *i* the vector of problem variables is

$$X = (x_1, \bar{x}_2, x_3, \bar{x}_4, \bar{x}_5, \ldots, x_N) \quad (1)$$

Where all variables x_i are 'locked' by the other players. Player *i* is interested only in the objective, namely $f_i = f_i(X)$ where $F(X) = (f_1(X), f_2(X), \ldots f_n(X))$ is the entire multi-objective problem. An excellent summary with applications can be found in Sefrioui [8] The merit of the following approach is in applicability; deterministic optimisers are in general not able to compute concave Pareto fronts at all, and methods utilising Pareto front gradients are still incapable of computing mixed-type and discontinuous fronts [5]. In this research we explore the use of a Pareto approach and Nash equilibria using a hierarchical multi-objective evolutionary algorithm.

Formally, the Pareto optimal set (Figure 1) can be defined as the set of solutions that are non-dominated with respect to all other points in the search space, or that they dominate every other solution in the search space except fellow members of the Pareto optimal set. For two solutions x and y (in minimisation form):

$$\begin{aligned} rel(x,y) &= x \text{ dominates } y \text{ } if: f_i(x) \le f_i(y) : \forall_i \in 1 \ldots M \\ rel(x,y) &= x \text{ nondominated w.r.t. } y \text{ } if: f_i(x) \le l \ge f_i(y) : \forall_i \in 1 \ldots M \\ rel(x,y) &= y \text{ dominates x } if: f_i(x) \ge f_i(y) : \forall_i \in 1 \ldots M \end{aligned} \quad (2)$$

For a problem in M objectives, this is called the 'relationship' operator. In practice we compute an approximation to the continuous set, by assembling $ParetoSet = \left[x_1^*, x_2^*, x_3^* \ldots\ldots x_\mu^* \right]$.

We implement the Nash equilibrium approach using one EA in single objective mode for each player, as can be seen in Figure 2, whereby information is exchanged between the EAs after a migration period has occurred. There are two migrations present when using HAPEA for Nash optimisation. Firstly there is a migratory circulation of solutions up and down (hierarchical optimisation, section 3.3), and then a Nash migration where information between players is exchanged after an epoch and the new variables for each player are updated on each node on each hierarchical tree.

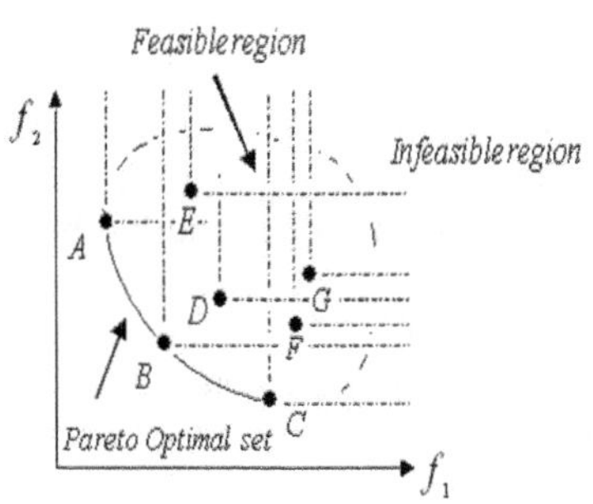

Figure 1: Pareto optimality.

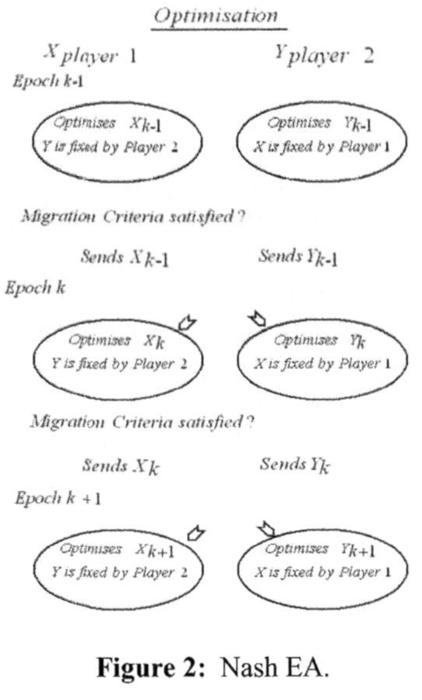

Figure 2: Nash EA.

3 The Hierarchical Asynchronous Parallel Evolutionary Algorithm (HAPEA)

Increasing complexity in engineering systems has generated a growing interest in the areas of Multi-Objective (MO) and Multidisciplinary Design Optimisation (MDO) [9,10]. As complexity in these systems arises, robust and alternative numerical tools are required, as these problems are generally multi-modal, involve approximation, are non-differentiable and involve multiple objectives.

When considering the solution to MO and MDO problems several problems arise, as many methods of solution used in engineering today may take variable times to complete their operation. The classic example of this is the modern CFD solver. In a typical industrial code used for external aerodynamic analysis of aeroplanes, the time for the residual of the solution to converge to a specified level (either machine zero or an arbitrarily selected higher value) can vary over a significant range. For example, any analysis method based upon the iterative solution of sets of non-linear partial differential equations (PDEs), it must be assumed that the time taken for each solution will vary with geometry.

The previous generation of evolutionary algorithms have mostly used a generation-based approach. The traditional genetic algorithm and evolution strategy both use this approach. A problem with generational models is that they create an unnecessary *bottleneck* when used on parallel computers. If the population size is approximately equal to the number of processors, and most candidate offspring sent for solution can be successfully evaluated, then some processors will complete their task quickly with the remainder taking more time. With a generational approach, those processors that have already completed their solutions will remain idle until all processors have completed their work.

3.1 Asynchronous Evaluation

The approach used in this research, is to ignore any concept of generation based solution. This approach is similar to work done by Wakunda and Zell [11] and other non-generational approaches, however the selection operator is quite different, as it couples one-by-one (steady-state) function evaluation with a direct multi-objective fitness criterion.

Whilst a parent population exists, offspring are not sent as a complete 'block' to the parallel slaves for solution. Instead one candidate is generated at a time, and is sent to any idle processor where it is evaluated at its own speed. When candidates have been evaluated, they are returned to the optimiser and either accepted by insertion into the main population or rejected. This requires a new selection operator because the offspring cannot now be compared one against the other, or even against the main population due to the variable-time evaluation. We compare recently evaluated offspring against a previously established rolling-benchmark and if successful, we replace (according to some rule) a pre-existing individual in the population.

We implement this benchmarking via a separate evaluation buffer, which provides a statistical 'background check' on the comparative fitness of the solution. The length of the buffer should represent a reasonable statistical sample size, but need not be too large; approximately twice the population size is more than ample. When an individual has had a fitness assigned, it is then compared to past individuals (both accepted and rejected) to determine whether or not it should be inserted into the main population. If it is to be accepted, then some replacement strategy is invoked and it replaces a member of the main population. We exclusively use the replace-worst-always method in this paper.

3.2 Pareto Tournament Selection

We implement the on-the-fly selection operator by means of a Pareto tournament selection operator. To implement an optimisation algorithm that is equally applicable to both single and multi-objective problems, a suitable selection operator capable of handling either situation must be developed. We propose the Pareto tournament selection operator, which is an extension of the standard tournament operator popular in many approaches [12].

Most evolutionary algorithms configured for multi-objective optimisation currently use the non-dominated sorting approach; This is a straightforward way to adapt an algorithm that is designed as a single objective optimiser into a multi-objective optimiser, and is used by many researchers [5,14]. The problem with sorting approaches is that the method is not a fully integrated one. Briefly, a sorting method works by computing the set of non-dominated solutions amongst a large statistical sampling (either a large population or previous data), and assigning these solutions rank one. Then ignoring these points, the process is repeated until a 'second' Pareto front is found, and this is assigned rank two. This process continues until all points are ranked, and then the value of the rank is assigned to the individual as a now single objective fitness.

A problem arises when considering whether it is fair to assign individuals in the second rank numerically half the fitness of the first, and whether the third rank deserves a third the fitness of the first. This poses a dilemma about the level of equality present amongst the solutions, as often solutions with excellent information may lie adjacent to, but not in, rank one. To solve this 'artificial scaling' problem, it is possible to introduce scaling, sharing and niching schemes, however all of these require problem-specific parameters or knowledge, even in adaptive approaches. It is of course always desirable to compose an algorithm that does not introduce such unnecessary parameters.

The current operator is a novel approach in that it requires no additional 'tuning' parameters, works seamlessly with the asynchronous selection buffer, and is very easy to encode. Simply, to determine whether a new individual x is to be accepted into the main population, we compare it with the selection buffer by assembling a small subset of the buffer called the tournament $Q=\left[q_1, q_2, q_3 .....q_n\right]$. We assemble Q by selecting individuals from the buffer, exclusively at random, until it is full. We then simply ensure that the new individual is not dominated by any in the tournament:

$$rel\left(x',q_j\right) \neq \left(q_j \text{ dominates } x'\right) : \forall_j \in Q \quad (3)$$

If this is the case, then it is immediately accepted, and is inserted according to the replacement rules. The only parameter that needs to be determined in advance is the tournament size, a parameter that would exist in a single objective optimisation anyway. Selection of this parameter requires a small amount of problem specific knowledge, and should vary between $Q = \frac{1}{2}B$ (strong selective pressure) and $Q = \frac{1}{6}B$ (weak selective pressure). The optimiser is not overly sensitive to this value, provided the user errs on the side of weak selective pressure (smaller tournaments) in the absence of better information. The egalitarian approach to the tournament (by selecting individuals at random) ensures good diversity amongst the selected individuals, and no niching or forced separation of individuals has been found necessary. It can also be seen that in the event that the fitness vectors have only one element (a single objective optimisation), this operator simplifies to the standard tournament selection operator [12].

3.3 Hierarchical Population Topology and Covariance Matrix Adaptation

A hierarchical population topology, when integrated into an evolution algorithm, means that a number of separate populations are established in a hierarchical layout to solve the given problem, rather than a single 'cure-all' type single population layout. This method was first proposed by Sefrioui [8], and is shown in figure 3. The bottom layer can be entirely devoted to exploration, the intermediate layer is a compromise between exploitation and exploration and the top layer concentrates on refining solutions. Following Hansen and Ostermeier [15], the HAPEA method uses a covariance matrix adaptation mutation operator that gives second order estimation of the problem topology, and has some similarities with most deterministic descent methods. The algorithm has been tested and parallellized [16] using the Parallel Virtual Machine (PVM) framework [17] and shows good performance in several multi-objective test problems [18].

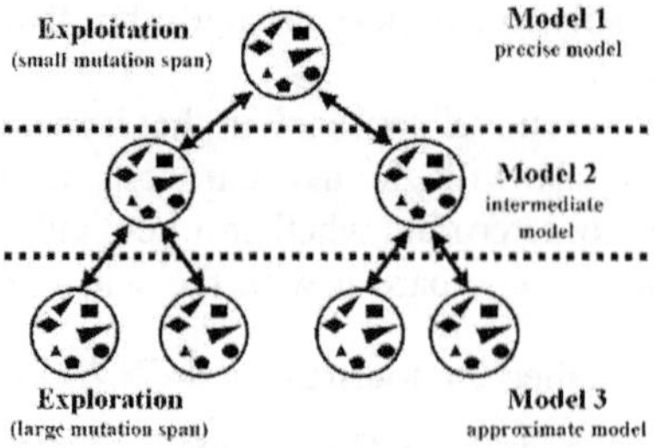

Figure 3: Hierarchical Topology

4 Test Problems

This section will describe the design tools used in the calculations and will present two test cases related to aircraft conceptual design, namely a one objective subsonic transport design and a two objective Unmanned Combat Air Vehicle (UCAV) design comparison using the concepts of Pareto optimality and Nash equilibrium.

4.1 Solution Method

The solver used to evaluate the aircraft configuration is FLOPS (Flight OPtimisation System). FLOPS [19] is a workstation based code with capabilities for conceptual and preliminary design and evaluation of advanced design concepts. The sizing and synthesis analysis in FLOPS are multidisciplinary in nature. It has a numerous modules for noise, detailed takeoff, performance, structures, control, aerodynamics and other capabilities; it is used in some universities for MDO development as well as aerospace firms and government. The aerodynamics module uses a modified version of the EDET (Empirical Drag Estimation Technique), and modifications to include smoothing of the drag polars. Different hierarchical levels could be adapted for drag build up using higher fidelity models that evaluate full Navier-Stokes flow and wake. FLOPS has capabilities for optimisation but in this case it was used only for analysis and adapted to the EA design optimiser. Details on the solver can be found in [19].

4.2 Case I: One Objective Subsonic Transport Design

4.2.1 Problem Definition

The objective in this case is to find an optimum set of design variables for a subsonic medium size transport aircraft as given in [19]. The aircraft has two wing-mounted engines, and the number of passengers and crew is fixed to 200 and 8 respectively. The design requirements are specified in Table 1. The aircraft is designed to cruise at 40000 ft and Mach 0.8.

Table 1. Design Requirements

Description	**Value**
Range [R, Nm]	2500
Ultimate Load Factor [n_{ULT}]	4.2
Maximum Takeoff Field Length [s_{TO}, ft]	6000

4.2.2 Design Variables and Constraints

The design variables for optimisation and their upper and lower bounds are represented in Table 2.

Table 2. Design variables for a subsonic medium size transport aircraft.

Description	Low Bound	Upper Bound
Aspect Ratio [AR_w]	7.0	13.1
Engine Thrust [T, lbf]	30500	50000
Wing Area [S_w, sq ft]	1927	2872
Sweep [Λ_w, deg]	25	40
Thickness [t/c]	0.091	0.235

Constraints in this case are minimum takeoff distance, moment coefficient for stability and control and range required. Violation of these constraints is treated with an adaptive penalty criterion.

4.2.3 Fitness Functions

This case is a single objective minimisation problem, the fitness function devised for this problem is toward minimum fuel weight required to complete the mission.

$$\min(f_1): \quad f = W_f \qquad (4)$$

4.2.4 Implementation

The solution to this problem has been implemented using a single population and parallel asynchronous evaluation, with the optimiser only considering a single objective. After an empirical study, it was found that a small population size of 10 and buffer size of 30 produced acceptable results.

4.2.5 Results

The algorithm was allowed to run for 1500 functions evaluations. Table 3 shows the design variables and results for the best configuration found as compared to a conventional Broyden-Fletcher-Goldfarb-Shano (BFGS) algorithm and a conventional conjugate gradient (CG) based (Polak-Ribiere) algorithm [19]. Through our approach, the EA gives a 3.5% and a 2.4% improvement in gross weight respectively.

Table 3. Design variables and results for best configuration found after optimisation.

Description	EA Best	BFGS	CG
Aspect Ratio [AR_w]	13.1	13.0	12.8
Engine Thrust [T, lbf]	34,770	38,852	39,021
Wing Area [S_w, sq ft]	1,929	2,142	2,218
Sweep [Λ_w, deg]	27.0	28.4	27.32
Thickness[t/c]	0.091	0.112	0.096
Taper Ratio [λ_w]	0.267	0.267	0.267
Fuel Weight [W_f, lbs]	**34,337**	*37,342*	*36,092*
Gross Weight [Wg , lbs]	**216,702**	*222,154*	*224,618*

4.3 Case II: Two Objective Air Superiority Unmanned Combat Air Vehicle (UCAV), a Pareto Optimality - Nash Equilibrium Comparison

4.3.1 *Problem Definition*

Unmanned Aerial Vehicles (UAVs) and Unmanned Combat Aerial Vehicles (UCAV) are ever increasingly becoming important topics for aerospace research. There are difficulties in these concepts because of the compromising nature of the missions that can be performed. The goal in the following is to address this issue from multidisciplinary aircraft conceptual design optimisation. Aerodynamic performance, cost minimisation and range might be improved if a multi-objective multi point optimisation can be developed that consider numerous separate design points. In this case the objectives are maximisation of the cruise efficiency $M_\infty \times (L/D)_{cruise}$ and minimisation of gross weight. . Two different approaches were run for comparison, one using the concepts of Pareto optimality and Nash equilibrium. The design requirements are specified in Table 4. The mission profile is represented in Figure 4.

Table 4. Design Requirements for UCAV.

Variable	**Requirement**
Range [R, Nm]	1000
Cruise Mach Number [M_{cruise}]	1.6
Cruise Altitude [h_{cruise}, ft]	40000
Ultimate Load Factor [n_{ult}]	12
Takeoff Field Length [s_{to}, ft]	7000

4.2.1 *Design Variables and Constraints*

The design variables after optimisation and the upper and lower bounds are represented in Table 5.

Table 5: Design Variables for optimisation of the UCAV.

Description	**Lower Bound**	**Upper Bound**	**Nash Best**
Aspect Ratio [AR_w]	3.1	5.3	5.14
Engine Thrust [T, lbf]	32000	37000	32923
Wing Area [S_w, sq ft]	600	1400	625
Sweep [Λ_w, deg]	22	47	30.7
Thickness [t/c]	0.02	0.09	0.023
Taper Ratio [λ_w]	0.15	0.55	0.19

The performance constraints are: 6 G's at Mach 0.6 at 10000 feet, 5 G's at Mach 0.9 at 30000 feet, accelerate from Mach 0.9 to 1.5 at 20000 feet in 30 seconds, maintain a turn rate of 15 deg/sec at Mach 0.9 at 20000 feet and have an excess energy of 50 ft/sec at Mach 0.9, 30000 feet and 4 G's. Violation of these constraints is treated with adaptive penalty criteria.

4.2.2 Fitness Functions

This case is a multi-objective problem where the fitness functions to be optimised are gross weight and cruise efficiency:

$$\min(f_1): \quad f_1 = W_g \qquad (5)$$

$$\min(f_2): \quad f_2 = \left(M_\infty \left(\frac{L}{D} \right)_{cruise} \right)^{-1} \qquad (6)$$

4.2.3 Implementation

In the Nash approach we split the variables between two players; player one maximises $M_\infty \times (L/D)_{cruise}$ using variables $S_1 = (AR_w, t/c, \lambda_w)$ while player two minimises W_g using variables $S_2 = (S_w, T, \Lambda_w)$. For both players we have chosen ('split') variables most relevant to its design objective. We implement this using two hierarchical EAs for each player, information is exchanged between the EAs after an epoch has terminated, and in this case it is equal to five times the population size in evaluations. The Pareto optimality approach involves one level and population size 40.

4.2.4 Results

The algorithm was allowed to run for 600 functions evaluations but converged after 300. Table 5 shows the parameters for the best configuration found. Figure 5 shows an example of convergence history of one of the Players. The final population (including the Pareto optimal front) and comparison with the Nash equilibrium result are shown in figure 6, although the computation expense for the Pareto approach is two times the Nash approach we can see the point obtained by the Nash equilibrium is a suboptimal in the Pareto sense.

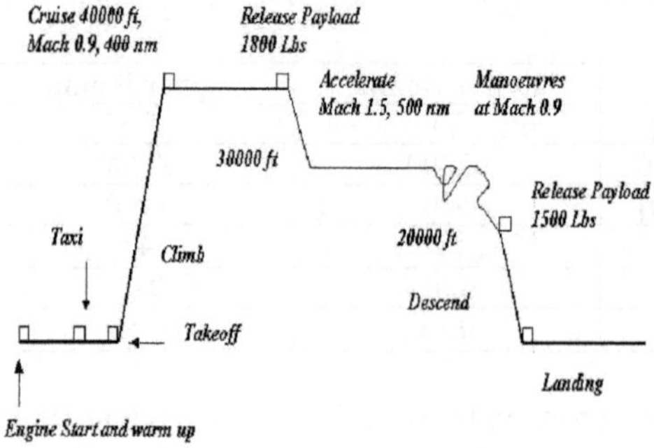

Figure 4: Mission profile for the UCAV.

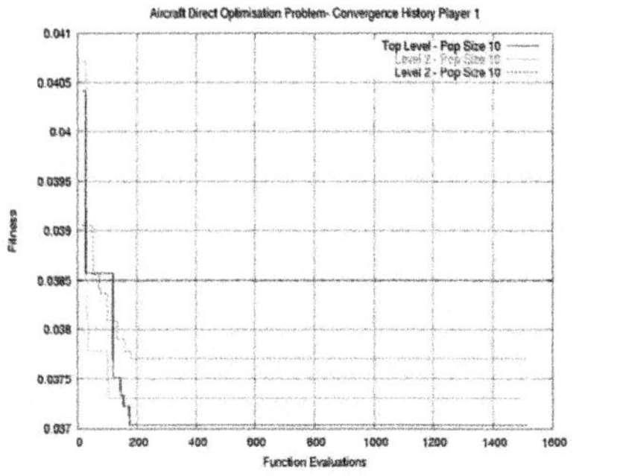

Figure 5: Optimisation progress player one

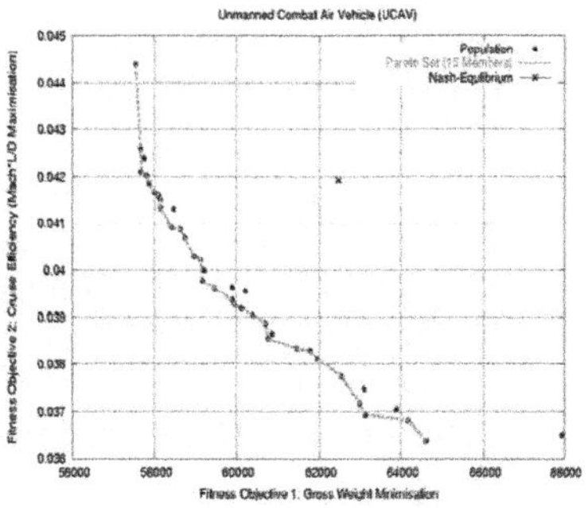

Figure 6: Pareto front and Nash equilibrium obtained for UCAV.

5 Conclusions

In this study a methodology and formulation for multidisciplinary optimisation problems using a parallel evolutionary algorithm has been presented and applied to realistic practical aircraft conceptual design problems. Two different aircraft configuration problems had been analysed presenting the applicability of the method. The use of the Pareto optimality and Nash equilibrium for conceptual aircraft design problems has been investigated enables the splitting of the optimisation problem and provides feasible results. When compared to traditional optimisation methods the evolutionary algorithm approach is capable of exploring the solution space better and provides feasible alternative solutions for single and multi-objective concepts. The MDO-EA technique can reduce weight and cost of an aircraft concept by minor changes in the design variables. Research is now underway to apply the methodology with hierarchical, multiple fidelity models and CAD integration.

Acknowledgments

The authors would like to thank Arnie McCullers at NASA LARC who kindly provided the FLOPS code and Mourad Sefrioui of Dassault Aviation for cooperation on hierarchical methods.

References

1. Dasgupta D, Michalewicz Z, (1997) Evolutionary Algorithms in Engineering Applications. Springer-Verlag, Berlin, Heidelberg.
2. Obayashi S, (1998) Multidisciplinary Design Optimization of Aircraft Wing Planform Based on Evolutionary Algorithms. Procs. IEEE International Conference on Systems, Man, and Cybernetics 1998, La Jolla, California.

3. Parmee IC and Watson AH, (1999) Preliminary Airframe Design Using Co-Evolutionary Multiobjective Genetic Algorithms. Procs. Genetic and Evolutionary Computation Conference, 2:1657-1665. W. Banzhaf, J. Daida, A. E. Eiben, M. H. Garzon, V. Honavar, M. Jakiela and R. E. Smith, editors, Orlando, Florida, USA, Morgan Kaufmann.
4. Raymer D, (2002) Enhancing Aircraft Conceptual Design Using Multidisciplinary Optimization. Department of Aeronautics, FLYG2002-2, Stockholm, KTH.
5. Deb K, (2003) Multi-Objective Optimization Using Evolutionary Algorithms, Wiley.
6. Pareto V, (1896) Cours d'Economie Politique, Rouge, Lausanne, Switzerland.
7. Nash JF, (1950), Equilibrium Points in N-Person Games. In Nat. Acad. Sci, USA, 36:46-49.
8. Sefrioui M, (1998) Algorithmes Evolutionnaires pour le calcul scientifique. Application à l'electromagnetisme et à la mécanique des fluides numériques, PhD thesis University Pierre et Marie Curie, Paris.
9. Sobieski J, Haftka RT, (1996) Multidisciplinary Aerospace Design Optimization: Survey of Recent Developments, AIAA Paper No. 96-0711.
10. Alexandrov NM, Lewis RM, (2000). Analytical and Computational Properties of Distributed Approaches to MDO, AIAA 2000-4718.
11. Wakunda J, Zell A, (2000) Median-selection for parallel steady-state evolution strategies. In Marc Schoenauer, Kalyanmoy Deb, Günter Rudolph, Xin Yao, Evelyne Lutton, Juan Julian Merelo, and Hans-Paul Schwefel, editors, ParallelProblem Solving from Nature – PPSN VI, pages 405–414, Berlin, Springer.
12. Goldberg D, (1989) Genetic Algorithms in Search, Optimization and Machine Learning, Addison-Wesley.
13. Sefrioui M, Périaux J, Ganascia J-G, (1996) Fast Convergence Thanks to Diversity, L. J. Fogel and P. J. Angeline and T. Back editors, Procs. Fifth Annual Conference on Evolutionary Programming 1996, San Diego, California, IEEE Computer Society Press, MIT Press.
14. Coello C, Christiansen A, (1998) Two New GA-Based for Multiobjective Optimisation, In Civil Engineering Systems.
15. Hansen N, Ostermeier A, (2001) Completely Derandomized Self-Adaptation in Evolution Strategies, In Evolutionary Computation, 9(2): 159—195.
16. Cantu-Paz E, (2000) Efficient and Accurate Parallel Genetic Algorithms. Kluwer Academic Pub.
17. Geist A, Beguelin A, Dongarra J, Jiang W, Manchek R and Sunderam V, (1994). PVM: Parallel Virtual Machine. A User's Guide and Tutorial for Networked Parallel Computing. Massachusetts Institute of Technology.
18. Whitney EJ, (2003) A Modern Evolutionary Technique for Design and Optimisation in Aeronautics. PhD Thesis, The University of Sydney.
19. McCullers A, (2003) FLOPS User's Guide, Release 6.02, NASA Langley Research Center.

CSAA: a Constraint Satisfaction Ant Algorithm Framework

Koenraad Mertens, Tom Holvoet

AgentWise, Distrinet
Department of Computer Science, K.U.Leuven
Celstijnenlaan 200A
B-3001 Leuven, Belgium
Koenraad.Mertens@cs.kuleuven.ac.be
Tom.Holvoet@cs.kuleuven.ac.be

Abstract

In this paper the Constraint Satisfaction Ant Algorithm (CSAA) framework is presented. The underlying infrastructure and the ants behavior are described in detail. The CSAA framework is an ant-based system for solving discrete Constraint Satisfaction Problems (CSP) and Partial Constraint Satisfaction Problems (PCSP). CSPs and PCSPs are used among others to design facility layouts and schedule workflow and repairs. Ant-based systems use stochastic decision making and positive feedback processes to reach their goal. Ant algorithms have already proven their value in solving various optimization problems. In this paper we show that they are also useful for more general constraint reasoning. We combined the strengths of ant-based systems – flexibility, the ability to adapt to changes – with heuristics from traditional constraint reasoning in order to obtain a flexible, yet efficient algorithm. The flexibility is used to continuously improve on the solution. This aspect of the framework gives the algorithm a great advantage over traditional solving methods when constraints and/or variables are added or removed at run-time. This becomes important when for example workflow should change dynamically according to user demands.

1 Ant-based Systems and Optimization Problems

An Ant-based System can be tought of as a special kind of multi-agent system, where each agent is modelled after a biological ant (therefore, agents are called *ants* in ant-based systems). Each ant-based system consists of an *environment* and a number of ants. The environment is the topology of the system, the structure wherein ants are situated.

Ants can move around in their environment and manipulate the objects that are placed inside it. Moving around (*walking*) allows the ants to find a solution to the problem at hand. Each ant continuously tries to improve the solutions found by other ants. The ants are not able to communicate directly with each other, but they put objects (named *pheromones*, after the biological chemicals) in the environment, which can be observed by other ants. Pheromones *evaporate*, thereby limiting their influence over time. In ant-based systems, a large number of ants act in the environment simultaneously (a *swarm*), each dropping a small amount of pheromones, but enough to influence other ants when a number of similar pheromones are dropped at the same locations. These influences are exploited while the ants walk in the environment. At first, each ant walks around randomly, but the ants who find good approximations to the solution drop pheromones at the paths they walked on. In a next step, other ants (slightly) prefer these edges. This process is known as *positive feedback*[2]. Stochastic processes and evaporation of pheromones allow for further exploration of the graph, but in a guided manner, eventually causing ants to prefer the best solution that one of them found. The ants continuously try to improve the currently best solution, in a manner similar to local search.

In general, ant-based systems tend to perform well on a number of optimization problems (*e.g.* the quadratic assignment problem [3], the travelling salesman problem [10], telecommunication networks [9], and other [6]). In [7], we presented a system for constraint satisfaction problem (CSP) solving. Now, by changing the structure of the system environment, we are able to incorporate a number of heuristics that are widely used in traditional solving methods (while maintaining the flexibility of ant-based systems), and to allow ants to solve partial constraint satisfaction problems (PCSPs). Our ultimate goal is to solve dynamic constraint satisfaction problems (DCSP). These are problems where the problem instance itself changes while it is being solved (addition/removal of variables, values and constraints). Therefore, maintaining the flexibility and adaptability of ant-based systems was a very important aspect for the system. We call the new graph structure, in combination with an improved solving mechanism the *Constraint Satisfaction Ant Algorithm* (CSAA).

Section 2 describes the design of the environment of our ant-based system and the behavior of the pheromones. A detailed description of the behavior of the ants is given in Section 3: how they choose their path, when they check constraints, when and where they drop pheromones. Section 4 elaborates on the heuristics of traditional algorithms that are mimicked in the CSAA algorithm. Section 5 gives experimental results for different configurations of the algorithm. The conclusion and future work on handling online changes is described in Section 6.

2 Environment and Pheromones

Figure 1 shows the construction of a graph for a problem with three variables: $A \quad 4, 5, 6$, $B \quad 2, 3$ and $C \quad 2, 3$. There are two constraints: $A = B + C$ and $B > C$. The graph consists of main nodes, selection nodes, selection edges and value edges. Each main node represents a variable. The number of main nodes is the same as the number of variables. This is depicted in Figure 1(a). In Figure 1(b), we see that each main node is connected through a number of selection edges to selection nodes. The selection edges determine the sequence the main nodes are visited in: from a selection node, only one main node can be reached. In a problem with N variables,

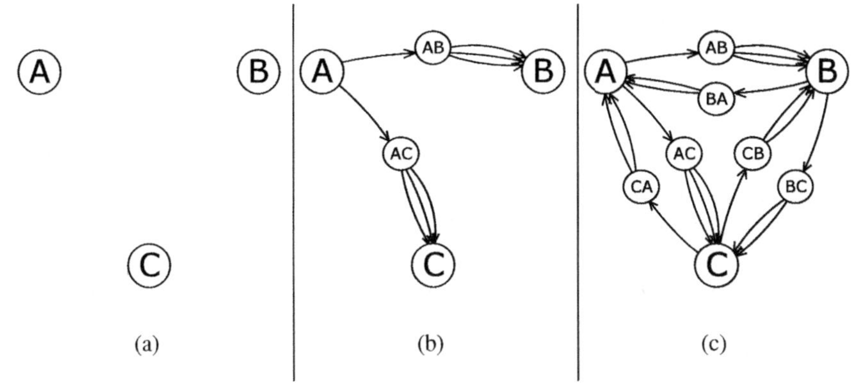

Fig. 1. Construction of a graph for a problem with 3 variables: $A \in \{4\ 5\ 6\}$; $B \in \{2\ 3\}$; $C \in \{2\ 3\}$ and 2 constraints: $A = B + C$; $B > C$ (a) Each main node represents a variable. (b) Going to the next main node is a two-step process: first the selection edge is chosen, then the value for the current variable (A has 3 possible values, thus there are 3 value edges from AB). (c) The complete graph for the problem.

every main node has $N \quad 1$ selection nodes. The selection nodes are connected to the next main node through value edges. Each value edge is associated with one possible value for the variable that is associated with the previous main node. Why the graph is constructed like this will be explained in Section 4, where we discuss the simulation of heuristics.

After building the graph, the constraints of the problem are inserted into the graph. Each main node keeps a reference of all constraints that apply to its variable. For the example of Figure 1, this means a reference to $A = B + C$ is kept in nodes A, B and C. A reference to $B > C$ is kept in nodes B and C.

Pheromones are dropped on the edges by the ants (how many pheromones, which sort of pheromones and when they are dropped is discussed in Section 3). When an ant arrives in a node (main node or selection node), it has to choose which edge to take next. That decision is based upon the constraints that hold in that node and the value of the pheromones that are present in the edges that start in that node. In order to limit the influence of pheromones over time, the pheromones *evaporate* and *diffuse*. In [1], mathematic formulas for how fast pheromones must evaporate and how sensitive ants must be to them in order to obtain a good search behavior can be found. There are two important parameters for the sensitivity of the ants: the *osmotropotaxic sensitivity* (β) and the *sensor capacity* ($1/\delta$). As indicated in [1], a good combination for those two parameters is $\beta = 3.5$ and $\delta = 0.2$, so we performed our experiments using these values. In the CSAA framework, we chose to let the pheromones evaporate on discrete intervals: each time an ant arrives in a node, the pheromones that are present on the edges that start in that node evaporate. The rate of evaporation is chosen such that after N (the number of variables in the problem) evaporation steps 10% of the pheromones has evaporated.

3 Behavior of the Agents

The algorithm uses a swarm of ants. The size of the swarm depends on the size of the problem. For our experiments (see Section 5), we found a swarm of 250 ants to give good results. Each ant is placed at a randomly chosen main node of the graph. The ant walks the graph, thereby remembering the path it walked upon. The ants are also responsible for checking the constraints. Dropping pheromones is only done after each ant of the swarm has finished. Only the best ants (those that found the best approximated solutions for a PCSP) and ants that failed are allowed to drop pheromones. We first give a detailed overview of the ant behavior in the CSP case. Differences for the PCSP case are described afterwards.

3.1 CSP Case

When an ant arrives in a main node (suppose: main node A in Figure 1), it has to choose which selection edge to take next. The possibilities are limited by the path the ant has already walked upon: an ant should not return to a main node it has already visited (because in that case, it would have to assign a value to a variable that is has already found a value for). The ant has to choose between the remaining edges. The choice is influenced by the pheromones that are present on the edges (suppose it chooses the edge that leads to selection node AB).

From each selection node, there are a number of value edges, each leading to the same main node (node B in our example). The number of edges is equal to the number of values in the domain of the previous main node (node A). Each value edge is associated with one of those values (so there are three value edges between AB and B, because the domain of A consists of three values: A $4, 5, 6$). Choosing a value edge means choosing a value for the variable, associated with the previous main node. Only value edges whose values do not create a conflict with the values that have already been assigned can be allowed. This implies that the ant has to check the appropriate constraints at this point. Appropriate constraints are those that involve the variable that is being given a value (variable A). As stated in Section 2, a reference to these constraints is kept in the previous main node. If no value exists that does not violate any constraints, no solution can be found with the currently chosen values: the variable associated with the previous main node *fails*. When there are edges whose values can be allowed, the ant chooses between them. The choice is again influenced by the pheromones that are present on the edges. When all variables have been assigned a value, the ant has constructed a solution by combining all collected values from the value edges it walked upon.

When a variable fails, the ant returns to all edges it walked upon in order to drop *pheromones* on them. A different type of pheromones is dropped on value edges and on selection edges: value edges receive *negative* pheromones. This indicates to other ants that this edge is probably a bad choice. The more negative pheromones that are present on a value edge, the less likely it is that that edge will be chosen by another ant. Selection edges receive *positive* pheromones. They indicate that the next main node should receive a value as soon as possible. The more positive pheromones that are present on a selection edge, the more likely it is that that edge will be chosen by another ant. Heuristics can be mimicked by making this distinction between positive and negative pheromones. Those heuristics and how the pheromones simulate them are described in Section 4.

The amount of pheromones that is dropped depends on the number of variables that are assigned a value: the more variables that are assigned a value, the more positive pheromones are dropped on selection edges and the less negative pheromones are dropped on value edges. The amount of pheromones that is dropped by one ant is not the same for all edges of the ant's path. The edges that the ants walked upon at the end of its path, receive more pheromones than the edges it walked upon when it started. We found that the best results were obtained when using a linear function for calculating the amount of pheromones: when an ant has walked upon N selection edges, the last selection edge receives N units of pheromones, the one before that $(N \quad 1)$ units,... and the first one receives 1 unit of pheromones.

An example: an ant starts in B, goes to BC and chooses the value 2 for the variable B. In C, the ant is forced to go to CA, where no valid value can be found for C $2, 3$ so that $B > C$: variable C *fails*. One amount of negative pheromones is dropped on the value edge $BC \quad C$. Positive pheromones are dropped on the selection edges $B \quad BC$ and $C \quad CA$. $C \quad CA$ receives two amounts and $B \quad BC$ receives one amount of positive pheromones.

3.2 PCSP Case

When solving a PCSP, it is assumed that no solution can be found that satisfies all constraints. Instead of searching for a complete solution, *approximated solutions* (variable-value assignments that satisfy part of the constraints) are allowed. A ranking between these approximated solutions has to be made. Our framework is well suited for a special case of such an ranking, one that assigns priorities to constraints. The rank of an approximated solution is then defined as the sum of the priorities of the violated soft constraints: the lower this sum, the better the approximated solution. We distinguish between *hard* constraints (with an infinite priority) and *soft* constraints (with a limited priority). While walking the graph, ants must satisfy hard constraints, while violations of soft constraints can be allowed. This means a variable can only *fail* when all possible values for that variable violate at least one hard constraints.

In contrast to the CSP case, where ants can stop searching when one of them has found a solution, ants should try to improve an approximated solution in the PCSP case. Therefore, next to the ants that found a failing variable, the *best* ants of a swarm also drop pheromones. The best ants are those that found approximated solutions with the lowest ranks. The number of ants in the group of best ants can be chosen freely: the more ants, the more feedback, but that feedback will also be less accurate. We found a group of 1 to 5% of the swarm of ants to reach satisfying results. After all ants of a swarm have found an approximated solution or a variable that failed, a central algorithm constructs the group of best ants and these drop their pheromones. In our experiments, the same sort of pheromones were dropped by ants that found a failing variable and by ants that belonged to the best ants. Because the number of pheromones depends on the number of assigned values, ants that found a failing variable drop more negative pheromones and less positive pheromones than ants that found an approximated solution. The behavior could be further differentiated in order to optimize the algorithm for a specific class of PCSPs.

4 Simulation of Heuristics

Pheromones are used to influence choices: positive pheromones on an edge increase the probability that that edge is chosen by an ant, negative pheromones decrease this probability. The use of pheromones, in combination with the structure of the graph, mimics the behavior of aspects of traditional algorithms:

Arc-consistency: Values for a specific variable that always result in a conflict with values of other variables, should no longer be chosen. The same goes for values that cause a soft constraint with a high priority to be violated. In CSAA, there is no hard restriction on such values, but the probability that they are chosen decreases because the ants drop negative pheromones on the value edges that lead to a failed main node. In PCSPs with large differences in soft constraint priorities, a group of *worst* ants can be formed, next to a group of *best* ants. Those ants should indicate values that violate high-priority soft constraints with negative pheromones.

Min-conflict: The min-conflict heuristic [8] says that the number of violations with variables that do not have a value yet should be minimized. Traditionally, all variables start with a (random) value and changes are based on the values of the other variables. In our ant-based approach, the choice of values is influenced by the pheromone trails on the value-edges. The same negative pheromones that were used for arc-consistency, increase the probability that a chosen value will not conflict with variables that will be assigned a value later.

First-fail: A variable that has a lot of constraints associated with it, receives a value as soon as possible [4]. Failed variables are approximations for such highly constrained variables. The ants drop positive pheromones on the selection edges that lead to the failed main node. This way, the other ants have a greater probability of reaching the failed variable (and thus assigning it a value) sooner. This is the reason we explicitly model value edges a well as selection edges: the negative pheromones on a value edge are only valid for one value, but the positive pheromones on a selection edge are valid for the whole of the variable.

Hill-climbing: The hill-climbing heuristic is only applicable for PCSPs. When using hill-climbing, an ant chooses the value that violates 'the least' constraints (the sum of the priorities of the violated soft constraint must be as low as possible). In our stochastic ant-based system, we increase the probability that an ant chooses a value that violates few constraints. Values that violate at lot of high-priority soft constraints, have a low probability of being chosen. This heuristic is not achieved by pheromones, but by checking constraints: when an ant has to choose an edge, it checks which values violate which constraints and it assigns probabilities to each edge. These probabilities are combined with the probabilities that are associated with the pheromones on each edge. At the beginning of the algorithm,

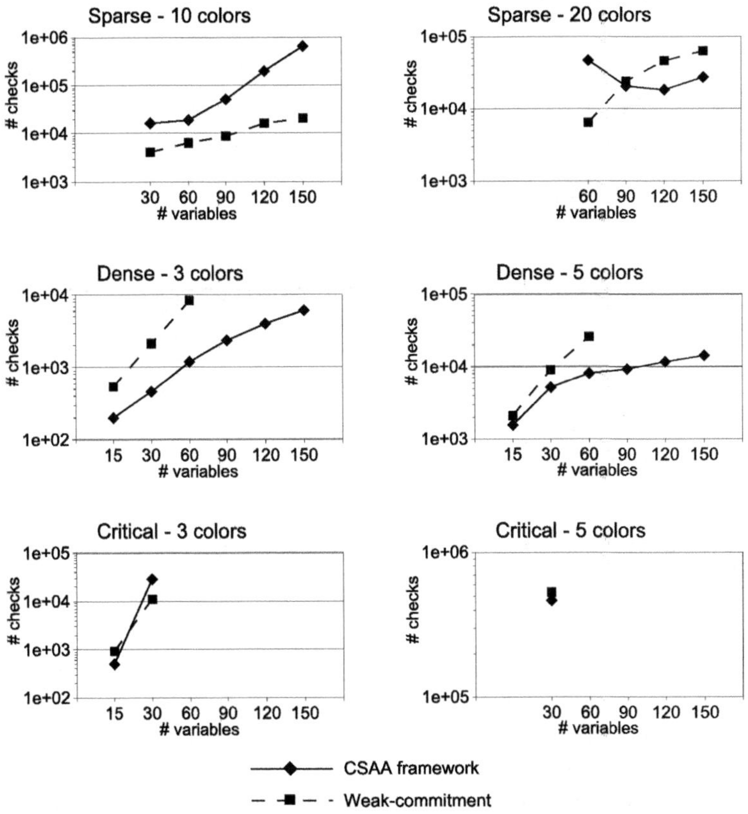

Fig. 2. Average number of needed constraint checks for different types of CSPs. The solid lines with diamonds are the results of the CSAA framework, the dashed lines with squares are the results of the weak-commitment algorithm. The different types of problems are, with n the number of variables and m the number of constraints: Sparse problems: $m = 2n$; Dense problems: $m = n(n - 1)\ 4$; Critical problems: $m = 2\ 7n$ for 3 colors, $m = 6\ 7n$ for 5 colors

the pheromone trails do not contain much information, so the impact of the hill-climbing probabilities is large compared to those of the pheromones. The longer the algorithm is working, the larger becomes the impact of the pheromones.

5 Experimental Results for CSPs and PCSPs

First we tested the system on some CSPs. Following the guidelines in [8], we created graph coloring problems (where n is the number of variables and m the number of constraints) that could be described as *sparse* (few constraints: $m = 2n$), *dense* (a lot of constraints: $m = n(n \quad 1)/4$) and *critical* ($m = 2.7n$ for 3 colors, $m = 6.7n$ for 5 colors). Sparse problems with 15 variables and 10 colors, sparse problems with 15 or 30 variables and 20 colors, critical problems with 15 variables and 5 colors cannot

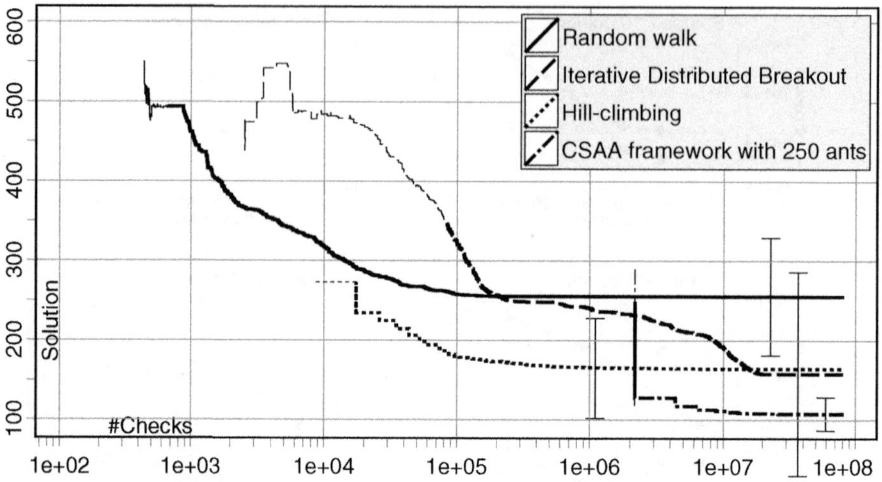

Fig. 3. Any-time curves for PCSP problems with 30 variables and 20 colors, using only random walk (solid line), using hill-climbing (dashed line) and using the CSAA framework with a swarm of 250 ants (dotted line). Error bars indicate 95% confi dence intervals around the mean.

be constructed due to the values for m and n. For each tested number of variables and colors, 10 different problems were generated and each solved 10 times, using both the CSAA framework and the weak-commitment algorithm, described in [11]. To be able to compare the two algorithms, we counted the number of constraint checks that each algorithm needed to find a solution. The average number of constraint checks for those 100 runs is displayed in Figure 2.

For dense problems our ant-based system outperforms weak-commitment (problems with 90 or more variables could not be solved by the weak-commitment algorithm). The high degree of random search used in ant-based systems is responsible for this. Sparse problems on the other hand can be solved faster by the weak-commitment algorithm, due to less duplicated checks. For critical problems (the hardest ones), both systems were unable to compute solutions for larger problems (problems with 60 variables or more). For the critical problems that could be solved, a solution was found almost equally fast with both systems.

We also computed solutions for particular PCSPs. Again, we tried to solve 10 different graph-coloring problems, with 10 runs for each problem. Each problem consisted of a number of hard constraints (in the first problem 3% of all constraints were hard, in the second 6%,..., in the last one 30%), all other possible edges between nodes in the graph-coloring problem were considered a soft constraint and given a random weight between 1 and 50.

We tested four cases: the first one was a CSAA variant where no pheromones were dropped and the ants only walked in a random manner, one by one. In the second test, we used the Iterative Distributed Breakout algorithm (introduced in [5]). In the third test, we used a CSAA variant with only the hill-climbing heuristic: the probability of an edge being chosen by an ant increased when few constraints were violated by the value of that edge. In the last test, a swarm of 250 CSAA-ants walked the graph simultaneously. They used all heuristics described in Section 4. An average any-time curve of the solution of problems with 30 variables and 20 colors (10 runs for 10 instances each) is depicted in Figure 3. This Figure also shows the variances (as 95% confidence intervals) of the different results.

The hill-climbing heuristic and the IDB algorithm were chosen because they lead quickly to a solution. This is an important characteristic when dynamic problems have to be solved. The CSAA framework is also designed to exhibit this behavior.

When the ants walk randomly, a first solution is found rather quickly (mostly after the first ant completed a run, at about 7e+02 constraint checks). The hill-climbing ants found a first solution at a later time: preferring the best edge, means that all possible edges have to be checked for constraints before the best one can be chosen. It took the hill-climbing ants approximately 20 times longer to find a first solution than it did for the random ants (about 2e+04 checks). However, the first solution of the hill-climbing ants was already better than the solution that was found by the random ants at that time. The final solution was better also (an average solution of 255 for the random search, compared to 165 for the hill-climbing algorithm). The IDB algorithm ended with a solution that was about as good as the solution of the hill-climbing algorithm, but it took longer to get there. The variance of IDB was a lot larger than the variance of the other algorithms: it found the best solutions, but also the worst. The CSAA swarm found a first solution approximately 250 times later than the hill-climbing ants (after about 2e+06 constraint checks). This is because only after a complete swarm of 250 ants completed one tour of the graph, a first solution was formed. Again, the first solution as well as the final solution are better than the random and the hill-climbing algorithms (an average solution of 108 for the CSAA framework). As a drawback, it takes longer to reach that final solution.

6 Conclusion and Future Work

The improvements we made to the system in [7] allow for the incorporation of traditional heuristics, more flexibility and higher performance in solving (Partial) Constraint Satisfaction Problems. This should make the CSAA framework well suited to solve complex industrial tasks, such as designing layouts and scheduling workflows.

The high level of flexibility is an important step towards our main goal: to cope with change in optimization problems. The logical next step in our research process is to evaluate and test the effectiveness and efficiency of the way ant systems can deal with change, *i.e.* additions, deletions and modifications of constraints, variables and values. The pheromone trail that is present in the system at the time of the update is the instrument that plays an essential role in this phase.

References

1. D. R. Chialvo and M. M. Millonas. How Swarms Build Cognitive Maps. In Luc Steel, editor, *The Biology and Technology of Intelligent Autonomous Agents*, volume 144, pages 439–450. Nato ASI Series, 1995.

2. M. Dorigo, V. Maniezzo, and A. Colorni. Positive Feedback as a Search Strategy, Technical Report 91016, Dipartimento di Elettronica e Informatica, Politecnico di Milano, Italy, 1991.
3. L. Gambardella, E. Taillard, and M. Dorigo. Ant Colonies for the QAP. *Journal of the Operational Research Society*, 50:167–176, 1999.
4. Robert M. Haralick and Gordon L. Elliot. Increasing Tree Search Effi ciency for Constraint Satisfaction Problems. *Artificial Intelligence*, 14(3):263–313, 1980.
5. Katsutoshi Hirayama and Makato Yokoo. Distributed Partial Constraint Satisfaction Problem. In *Proceedings of the Third International Conference on Principles and Practice of Constraint Programming (CP'97)*, pages 222–236, 1997.
6. V. Maniezzo and A. Carbonaro. Ant Colony Optimization: an Overview. In *Proceedings of MIC'99, III Metaheuristics International Conference, Brazil.*, 1999.
7. Koenraad Mertens, Elke Steegmans, and Tom Holvoet. Cyclic Path-Based Environment: an Ant Environment for Solving Distributed Constraint Satisfaction Problems. In *Proceedings of the Third International Workshop on Distributed Constraint Reasoning*, pages 94–103, 2002.
8. Steven Minton, Mark D. Johnston, Andrew B. Philips, and Philip Laird. Minimizing Conflicts: A Heuristic Repair Method for Constraint Satisfaction and Scheduling Problems. *Artificial Intelligence*, 58(1-3):161–205, 1992.
9. Ruud Schoonderwoerd, Owen E. Holland, Janet L. Bruten, and Leon J. M. Rothkrantz. Ant-Based Load Balancing in Telecommunications Networks. *Adaptive Behavior*, 5(2):169–207, 1996.
10. Thomas Stützle and Marco Dorigo. ACO Algorithms for the Traveling Salesman Problem. In K. Miettinen, M. Makela, P. Neittaanmaki, and J. Periaux, editors, *Evolutionary Algorithms in Engineering and Computer Science*. Wiley, 1999.
11. Makoto Yokoo. Weak-Commitment Search for Solving Constraint Satisfaction Problems. In *Proceedings of the 12th National Conference on Artificial Intelligence (AAAI-94); Vol. 1*, pages 313–318, Seattle, WA, USA, July 31 - August 4 1994. AAAI Press, 1994.

Adaptive Computing in Support of Traffic Management

Kalin Penev

Technology Research Centre, Faculty of Technology, Southampton Institute, Southampton, S014 0YN
Kalin.Penev@solent.ac.uk

Abstract

The article presents an exploration of a novel optimisation method, called Free Search. Free Search is population-based and can be classified as an evolutionary computational method.

Free Search is examined by using a hard non-linear constrained optimisation problem. The experimental results of twenty and fifty dimensional variants of the test problem are presented and discussed.

The algorithm is also applied to a traffic management optimisation model. It explores how adaptive computing can support air traffic dispatchers, who attempt to satisfy requirements for safety and efficiency constrained by the environmental impacts. The results suggest that the Free Search can provide decision-making with optimised traffic information.

1 Introduction

A recently proposed optimisation method called Free Search (FS) is subject to evaluation and examination [20]. FS is a population-based evolutionary computational method. Free Search demonstrates better performance by comparison with Genetic Algorithm BLXα modification [11], Particle Swarm Optimisation [8] and Differential Evolution [22], applied to five heterogeneous numerical problems [21]. In the current study FS is tested with the so-called "bump problem", with additional constraints that define the feasible space for the objective function. The test problem is widely discussed in the literature. The optimum is unknown, and the best-achieved results are published [13, 17, 18, 24].

Free Search models the animals' behaviour. The individuals of the population are called animals. The term "animal" is an abstraction. It denotes characteristics and relationships, essential for the search purposes. Considering the animals' behaviour in nature, it is assumed that the sense and the mobility support the search within the natural environment. It is assumed, also, that the relation between them supports the search process. The sense, the mobility and the relation between them are abstracted and modelled in an optimisation algorithm. In the model, each animal has the essential peculiarities - sense and mobility. The sense gives to the animal a capability for orientation within the search space. The animals consider the experience of the previous generations as knowledge but not as an obligation or a restriction. The animals, following their own sense, can select any search area, even, the previous experience (the best achieved from the previous generation value of an objective function) suggests exploration around a certain location. This ability is accepted as fundamental for the algorithm.

The second peculiarity is a capability for action. Each animal can operate either with small precise steps (local search) or with large steps (global exploration). Each animal decides how to search: with small or with large steps. The previous experience can be taken into account, but it is not compulsory. The sense and the action are related. The relationship is also fundamental for the algorithm.

The aim of the study is to explore whether the probabilistic approach is suitable for unknown constrained optimisation problems, to clarify whether the new concepts lead to better performance and how they reflect on the optimisation process. In particular, the aim is to examine whether uncertainty, individualism or freedom to make own decisions support both global exploration and precise local search. Dealing successfully with hard tests, similar to the real-word problems, can be a good indicator for reliability of the search method. The excellent examination results encourage explorations with real problems. FS is applied to a model of traffic management. The results from the experiments demonstrate the benefits, which adaptive computing can bring to decision-making.

This paper has 7 sections. Section 2 "Free Search essential properties" presents the FS concept and its architecture. Section 3 "The optimisation problem" defines the test problem. Section 4 "Free Search applied to the bump problem" summarises the achieved experimental results. Section 5 "Discussion" comments on the results. Section 6 "Free Search support for Traffic Management" focuses on the FS application to the optimisation of traffic data for decision-making purposes. Section 7 "Conclusion" makes a summary and proposes direction for further work.

2 Free Search essential properties

Free Search [20] can be applied to the optimisation of wide range non-discrete, non-linear problems and to search within continuous data space as well as other population based methods - Genetic Algorithm (GA) [5, 11, 14], Differential Evolution (DE) [22, 23] and Particle Swarm Optimisation (PSO) [1, 8]. Free Search relates to Ant Colony Optimisation (ACO) [6]. The difference is that ACO is applied to discrete optimisation problems [7]. Free Search has similarities with ACO modified for continuous space search [2, 3], real-coded GA [11], PSO and DE. However, Free Search is not ant, swarm or genetic algorithm. The ACO, PSO, GA and DE concepts cannot describe the operation of the algorithm. In particular, they cannot explain a free, uncertain, individual's behaviour.

In the FS algorithm, the animals have a freedom to take a decision where and how to search. The Free Search concept is suitable to explain the functioning of the algorithm and the behaviour of the individuals. The difference from all other population based optimisation methods is that the behaviour of the animals in FS is not defined. It is not strictly controlled and can be described by some probability, only. The success of this approach is based on the appropriate regulation of the probability for access to any location within the search space. It is based on the assumptions: *With an uncertainty can cope other uncertainty. With infinity can cope other infinity.*

2.1 Free Search a conceptual model

In the model the animals can move step by step (discrete movement) through multi-dimensional non-discrete search space. During the search process each animal takes exploratory walks. The aim is to find a favour. In terms of numerical optimisation, the aim is to find better satisfaction of an objective function. The walks are for a determinate period. The period is limited by a number of steps. The first walk starts from a random or a certain (for example the middle) location within the search space and the directions are random. During the exploration each animal achieves some favour (an objective function solution) and distributes a pheromone in amount proportional to the amount of the found favour (the objective function satisfaction). The pheromone is fully replaced with a new one after each walk. The animals abstract, essential for the search, entities from the information about marked with pheromone locations. Abstraction is memorised as cognition about the search space. The knowledge from all animals can be used during the selection of start locations for the next walk. The process continues until satisfaction of the termination criterion. If the criterion is satisfied, search is terminated.

How do the animals use the distributed pheromone? Each animal has a sense and a propensity for pheromone. The animal uses its sense for a selection of a location for the next walk. The decision is a function of two variables the pheromone amount and the sensibility. The animal can select, for a search, any location marked with pheromone, which suits its sense. The sense is a conceptual improvement of population-based optimisation methods. It has no analogue in GA, DE, PSO and ACO. In terms of numerical optimisation the sense is a tool for regulation of the divergence and the convergence within the search process and a tool for guiding space exploration. During the exploration walk the animals make steps within the neighbouring space. The neighbouring space is limited and can vary during the walk and throughout the whole search process. It varies also for the different animals. The search space borders restrict the limitation of the neighbouring space only. The probability for access to any location of the search space is non-zero. The steps are a model of the ability for motion and action. The steps can be large or small, and can vary. In terms of numerical optimisation the neighbouring space is a tool for the tuning of rough and precise searches. The concept for a neighbouring space has a similarity to the concept of neighbour nodes in ACO [6] and to the concept of search radius in ACO applied to continuous search space [2, 3]. In ACO the neighbour nodes are discrete locations. The search radius in ACO for continuous space is defined [3]. The difference is that the neighbouring space in FS is variable continuous or partially discontinuous area appropriate for numerical optimisation.

The animal in FS can explore any area no matter whether this is the area around the best own solution found, or the best solution found from another animals, or the best solution found from all the population. By enhancement of the sensibility the animal can be forced to search to the area of the best-found solution from all animals – local search. By reducing the sensibility, the animal can be allowed to explore around locations found from other animals but not the best for the population – global exploration. The sensibility is not determined. The different animals can have different sensibility. It varies also during the optimisation process, and one animal can have different sensibility before different walks. In Free Search the sense, the step limit, which define an action potential, and the pheromone trail are subject to adaptive self-regulation during the stochastic search process on a random principle. In FS the sensory perceptions and the action are related. This is considered to represent a model of artificial thinking [20].

The algorithm architecture is designed, based on the conceptual model.

2.2 Free Search algorithm architecture

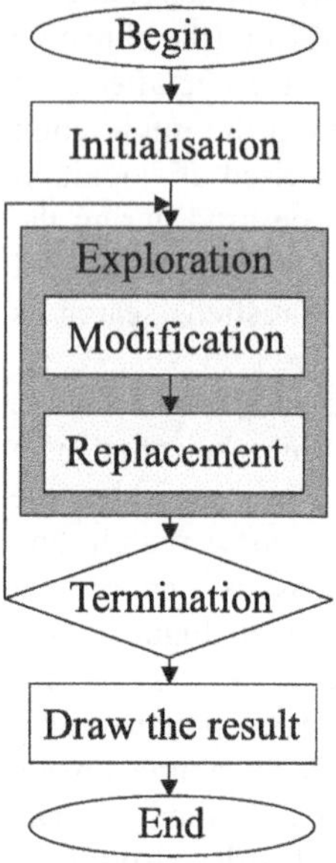

Free Search{
    Initialisation{
        initialise $X_{max\,i}$, $X_{min\,i}$, n, m, G, x_{0ji}, T, R_{ji}
        take initial walks, f_{tj} ($x_{0ji}+\Delta x_t$)
        generate an initial pheromone P_k
        distribute the initial pheromone P_k->x_{kP}
        learn the initial achievements P_k ->x_{jkP}
    } // end initialisation
    do{ // **exploration**
        generate sensibility S_j
        select start locations for a walk $x'_{0ji} = x_{ji}(S_j,P_k)$
        take exploration walks f_{tj} ($x'_{0ji}+\Delta x_t$)
        generate a pheromone P_k
        distribute the pheromone P_k->x_{kP}
        learn the achievements P_k ->x_{jkP}
    } while ($f(x_j)<f_{opt}$ && g<G) // **termination**
    Output the result
} // end Free Search

Figure 1: Free Search – algorithm architecture and an example in pseudo code.

On Figure 1 $Xmin_i$ and $Xmax_i$ are the search space borders, m is the population size, $j = 1,..,m$, $k = 1,..,m$, $i = 1,..,n$, n is the number of dimensions. T is the step limit, t -the current step. Rji is the variable neighbouring space $Rji = [Rmin, Rmax]$.

Acceptable initialisation strategies are:

- random values: $x_{0ji} = Xmin_i + (Xmax_i - Xmin_i)*random_{ji}(0,1)$
- certain values: $x_{0ji} = a_{ji}$, $a_{ji} \in [Xmin_i, Xmax_i]$
- one location: $x_{0ji} = c_i$, $c_i \in [Xmin_i, Xmax_i]$

Where $random(0,1)$ is a random value between 0 and 1, a_{ji} and c_i are constants.

The exploration walk generates $x_{tji} = x_{0ji} + \Delta x_{tji}$. The modification strategy is:

$\Delta x_{tji} = R_{ji} * (Xmax_i - Xmin_i) * random_{tji}(0,1)$

$i = l$ for uni-dimensional step, $i = 1,..,n$ for multi-dimensional step, $t = 1,..,T$.

The modification strategy is independent from a current or the best achievements. The strategy allows nonzero probability for access to any location of the search space and highly encourages escaping from trapping in local maxima.

The individual behaviour, during the walk, is modelled and described as:

$f_{tj} = f(x_{tji})$, $f_j = \max(f_{tj})$. f_j is the, only, location marked with pheromone from an animal after the walk. The pheromone generation is: $P_j = f_j / \max(f_j)$. The sensibility generation is: $S_j = Smin + \Delta S_j$, $\Delta S_j = (Smax - Smin) * random_j(0,1)$

Where $Pmax = Smax$, $Pmin = Smin$.

The selection of a start location x'_{0ji} for an exploration walk is:

$x'_{0ji} = x_{ji}\ (P_k \geq S_j)$, where $j = 1,..,m$, $k = 1,..,m$, k is the marked locations number.

Acceptable criteria for termination are:

- satisfaction of optimisation criteria: $fmax \geq fopt$, where $fmax$ is maximal achieved solution, $fopt$ is an acceptable value of the objective function.
- expiration of generation limit: $g \geq G$, where G is a limit and g - current values.
- complex criterion: $((fmax \geq fopt) \,||\, (g \geq G))$.

The Free Search structure is similar to the general description of the evolutionary algorithms given by Corne [4] and Eiben & Smith [9]. The algorithm architecture is reduced and consists of generalised events initialisation, exploration and termination. An example, of possible implementation in pseudo code, is presented on Figure 1. The algorithm is implemented as a computer program and it is applied to the bump test.

3 The optimisation problem

The test problem used is hard, non-discrete, non-linear and constrained [15, 16]. It is generalised for multidimensional search space. Large research efforts have been directed towards its exploration [2, 10, 13, 15, 16, 17, 18, 24]. The search space is continuous. It has many peaks. The objective function and the conditions are:

$$\text{Maximise: } f(x_i) = \left| \sum_{i=1}^{n} \cos^4(x_i) - 2\prod_{i=1}^{n} \cos^2(x_i) \right| / \sqrt{\sum_{i=1}^{n} i x_i^2}$$

$$\text{subject to: } \prod_{i=1}^{n} x_i > 0.75 \quad (1), \qquad \text{and} \quad \sum_{i=1}^{n} x_i < 15n/2 \quad (2),$$

for: $0 < x_i < 10$, and $i=1,\ldots,n$, starting from $x_i = 5$, $i=1, \ldots, n$, where x_i are the variables (expressed in radians) and n is the number of dimensions [15, 16, 17].

4 Free Search applied to the bump problem

FS is applied to the bump problem with a common general configuration. It does not require special settings of the optimisation parameters for any particular problem. The configuration of the optimisation parameters in FS can be considered as a desirable frame or as an advice to the individuals of the population rather than as a restriction or strict deterministic rule. FS has to adapt to the problem.

The optimisation problem is implemented for $n = 20$ and $n = 50$. The number of iterations is limited to 20000 for $n = 20$ and to 50000 for $n = 50$. These limitations are accepted according to the earlier published results [15, 16]. The population size is 10 (ten) for all experiments. The explored start conditions are: (1) the optimisation starts from $x_i = 5$, $i = 1,..,n$; (2) start from random points in the middle of the search space $x_i = 4 + r_i$, $r_i \in (0, 2)$; and (3) with start points near to the edge of the feasible region $x_i = 3.3 - i{*}0.165$. A start from the random or certain locations in the middle of the search space requires global exploration, and the experiments with such start are classified as global optimisation. The experiments with a start from points near to the constraint border, where, suppose to be, the optimum, are similar to a local search, and can be classified as local search tasks. Let us clarify that for the animals in Free Search there is no restriction to search near to the border. They can explore any location of the search space. The probability for access to any location within the search space is non-zero. However, the probability, an animal to be attracted from a location with low level of pheromone, is low. On Table 1 are presented the maximal values achieved from 320 experiments, for 20 dimensional search space.

N	start from	iterations	constraint (1) > 0.75	f_{max20}
20	$x_i = 5$	20000	0.750022060021	0.80015448156
20	$x_i = 4 + r_i$	20000	0.750000636097	0.803558343297
20	$x_i = 3.3 - i{*}0.165$	20000	0.750015669223	0.8035972606
20	$x_i = x_{max}$	20000	0.7500013889841	0.80361832569

Table 1: Maximal values achieved for $n = 20$.

x[0]=3.162532907780581	x[1]=3.1281253167718561	x[2]=3.0943772167322559
x[3]=3.0616174215392533	x[4]=3.0279646795850135	x[5]=2.9929794640414609
x[6]=2.9582035131302815	x[7]=2.9209071521426648	x[8]=0.49574861849745588
x[9]=0.48803013906648524	x[10]=0.4839089601325185	x[11]=0.4761613052301595
x[12]=0.4706140151519968	x[13]=0.46570991858146327	x[14]=0.4620819047541968
x[15]=0.45667780906709243	x[16]=0.45266314887859582	x[17]=0.4484183223424154
x[18]=0.44359479201658231	x[19]=0.44018809287300481	

Table 2: The variables values for $f_{max20} = 0.80361832569$.

N	start from	iterations	constraint (1) > 0.75	f_{max50}
50	$x_i = 5$	50000	0.750064071725	0.830462857742
50	$x_i = x_b$	50000	0.750000512903	0.835238805797361
50	$x_i = x_{max}$	50000	0.750000001829	0.83526232992710514

Table 3: Maximal values achieved for $n = 50$.

On Table 1, n is the number of dimensions. x_i are initial start locations. x_{max} are variables of an achieved local maximum. f_{max20} is a maximal achieved value for the objective function. $i = 1,\ldots,n$, r_i is random and $r_i \in (0, 2)$.

Table 3 presents the maximal values achieved from 320 experiments, for 50 dimensional search space. n is the number of dimensions. x_i are initial locations for start. i =1,...,n,. x_b are locations near to the border. x_{max} are the variables values of an achieved local maximum. f_{max50} is a maximal achieved value.

x[0]=6.2835817832628997	x[1]=3.1700104901760713	x[2]=3.1560884495602837
x[3]=3.1424421833129745	x[4]=3.1287732473796535	x[5]=3.1152068177611656
x[6]=3.1018610108965325	x[7]=3.0885301727729457	x[8]=3.0751715298137889
x[9]=3.0617448579952153	x[10]=3.0483956917288917	x[11]=3.035068161813653
x[12]=3.0216324553508298	x[13]=3.0079998110016151	x[14]=2.9943671520531048
x[15]=2.9806315071240381	x[16]=2.9666712361989416	x[17]=2.9523764516033806
x[18]=2.9379527576886626	x[19]=2.9232048550828269	x[20]=0.4881832042511519
x[21]=0.4859448863788553	x[22]=0.4836864969794788	x[23]=0.4814337527395312
x[24]=0.4793552624096046	x[25]=0.4771169704099637	x[26]=0.4750639516411976
x[27]=0.4732043079365083	x[28]=0.4712006454175081	x[29]=0.4692825428554116
x[30]=0.4671621886338754	x[31]=0.4653838695069296	x[32]=0.4635002813169994
x[33]=0.4617420094117654	x[34]=0.4600035377086197	x[35]=0.4583661842728538
x[36]=0.456602073521977	x[37]=0.4549844310375236	x[38]=0.4531609308605502
x[39]=0.4513627730229094	x[40]=0.4499850692495182	x[41]=0.4483422436194472
x[42]=0.4470344599615064	x[43]=0.4453551192557063	x[44]=0.444098810871133
x[45]=0.4422113211889435	x[46]=0.4408906125000003	x[47]=0.4393163619905571
x[48]=0.4380135616854353	x[49]=0.436532488971716	

Table 4: The variables values for f_{max50} = 0. 0.83526232992710514.

Maximal value	Constraint (1)	Maximal value	Constraint (1)
0.83526232474344719	0.750000000896	0.83526232464164241	0.750000002134
0.83526232992710514	0.750000001829	0.83526232800561506	0.750000000161
0.83526232471460016	0.7500000005	0.8352623279412521	0.750000000163
0.83526232459059935	0.750000000704	0.8352623246002131	0.750000000944
0.83526232376123954	0.750000001384	0.83526232561135005	0.750000004843

Table 5: The best results for n = 50.

The variables' values for the best achieved objective function value for n = 20 and for n = 50 are presented respectively on Table 2 and Table 4. The best-achieved values of the objective function for n = 50 and the corresponding values of the constraint parameter (1) are presented on Table 5. The constraint parameter values are an indicator whether the found maximum belongs to the feasible region. They indicate also expected possible improvement and can be valuable for further research. The best results presented on the Table 5, suggest that for n = 50 with the precision of seven decimal digits (0.0000001) the optimum is f_{opt} = 0.8352623.

For $n = 50$, $x_i \in (0, 10)$, $i = 1,\ldots,50$ there are 10^{400} solutions with a precision of seven decimal digits. FS achieves the maximum with such precision after exploration of less than 10^9 solutions. The results point more than one maximum with the precision of seven digits. Let us note that: (1) these results are achieved on probabilistic principle; (2) the search space is continuous and the results can be clarified to an arbitrary precision.

For a measurement of the convergence speed, and the computational cost for the achieving of an acceptable optimum for n = 50, are made, two additional series of 640 experiments each. The termination criterion is complex, satisfaction of an optimal value or expiration of a certain number of iterations. For the first series the criterion is reaching of the optimal value $f_{crtr} = 0.8$ or expiration of 50000 iterations. Measured are the minimal number of iterations g_{min} and the average number of the iterations g_{av}, used to reach $f_{crtr} = 0.8$. The average value is calculated from the sum of the iterations used to reach the optimisation criterion, divided on the number of the experiments. In case of expiration of the iterations' limit, the value of the limit is used for the calculations.

For the second series the criterion is satisfaction of a higher optimal value $f_{crtr} = 0.83$ or expiration of 50000 iterations. Measured are the minimal number of iterations g_{min} used to reach $f_{crtr} = 0.83$. For these experiments the population is 10 individuals, starting from $x_i = 5$, $i = 1,\ldots,n$, neighbouring space varies from 0.1 to 2.1 with step 0.1.

n	start from	f_{crtr}	f_{max}	constraint (1)	g_{min}	g_{av}
50	$x_i = 5$	0.8	0.800106225496685	0.80279775699	5335	36650
50	$x_i = 5$	0.83	0.830013250356044	0.75038942747	44610	-

Table 6: Convergence speed.

On Table 6, n is the number of dimensions. f_{crtr} is the satisfactory value of the objective function. The criterion is reaching of f_{crtr} or expiration of 50000 iterations. f_{max} is the maximal value reached. g_{min} is the minimal number of iterations used. g_{av} is the average number of iterations used for reaching of f_{crtr} from 640 experiments.

5 Discussion

How the new concept reflects on the performance? Free Search outperforms other methods discussed in the literature [10, 13, 15, 16, 17, 18, 24]. How the new concept affects the search process and the individuals within the population? The own sensory perception creates an element of individualism. This individualism supports personal and social creativity within the population. The individuals can escape easily, from unessential stereotypes, expired acceptances and hypotheses, than whole population in GA, ACO, PSO and DE. In particular, the experiments with a start from a local maximum, illustrates that animals can escape easily from trapping in local optima. The results suggest that a modelling of the individual behaviour contributes to the better performance. The individuals in FS can adapt their own behaviour flexibly during the optimisation process for global exploration, local search or search near to the constraints edge. The results demonstrate that the individual abilities for adaptation support the performance.

On conceptual level there are several aspects of the Evolutionary Computation [9, 12, 18], which are affected from the new method and can be reconsidered. One of the aspects is a requirement for special optimisation operators for any particular task. For example, according to the approach for search, exists a conclusion: "Searching boundaries of a feasible part of the search space isn't easy. It requires specialized operators that are implemented for just this purpose, on just this problem." [18: p.15]. The results achieved from Free Search indicate that the boundaries can be explored successfully without specialised operators.

Another aspect is according to the selection and determinism of the search space: "Choosing the right search space is of paramount importance." [18: p.36]. For uncertain unknown problems the "right search space" is subject of clarification during the whole optimisation process. Contrary changes are possible after any iteration. Free Search method does not rely on a choosing of a "right search space". And achieved results demonstrate good performance without determination or discrimination of the search space.

6 Free search support for Traffic Management

The good results on previous [20, 21] and current tests encourage Free Search exploration with real problems. And the algorithm is applied to the optimisation of a highly uncertain traffic management model, which has additional constraints.

The main issues of the Traffic Management are safety, effectiveness and environmental impact [19]. The decision-makers often face a dilemma trying to satisfy both safety and effectiveness, and to find an optimal solution [25]. From other hand modern Evolutionary Computational Methods offer a powerful potential in optimisation of wide range hard problems, which require satisfaction of contrary criteria and constraints [4]. Therefore, the Free Search method is applied to a model of the air traffic dispatch. The aim is to explore how the adaptive computing can support the air traffic dispatchers, who attempt to satisfy the safety requirements, the criteria for effectiveness and the constraints of the changeable weather. A grooving traffic demands from the dispatchers to take a decision for a short time using large amount of uncertain information [25]. The adaptive computational methods can support decision-making with a quick optimisation of the existing data. A generalised simplified model of the air traffic dispatch attempts to maximise the profit from the flights and at the same time to keep high level of flight safety. The profit is defined by an objective function $f(x_i, w_i)$ and the safety is described by a constraint condition $S(x_i, w_i)$ as follows:

Maximise: $f(x_i, w_i) = \sum_{i}^{n} (a_{1i} - a_{2i}x_i - a_{3i}w_i)$,

subject to: $S(x_i, w_i) = a_{4i} + a_{5i}x_i - a_{6i}w_i > SR$, for $i = 1,\ldots,n$

where in relative units x_i present the impact of the human factor, $x_i \in (1, 10)$, w_i present the weather impact, $w_i \in (2, 10)$, a_{1i} - the profit from the flight, a_{2i} - the losses from delays, a_{3i} - the losses from the weather impact, a_{4i} - the airplane safety, a_{5i} - the possible safety improvement, a_{6i} - the negative weather impact on the safety, SR is the required flight safety, n is a number of airplanes. The algorithm optimises possible dispatchers behaviour for particular number of airplanes, uncertain weather condition and for particular safety requirements. The explored example is: The requirement for a minimal flight safety is SR = 5. The airplanes are 10 $n = 10$. The profit from each flight for normal operations is $a_1[10]$ = {0.25, 0.45, 0.10, 0.30, 0.75, 0.30, 0.30, 0.60, 0.35, 0.50}. The losses from dispatch delay per unit time for each flight are a_2 [10] = {0.05, 0.05, 0.01, 0.01, 0.10, 0.03, 0.03, 0.06, 0.05, 0.05}. The losses from weather impact for each flight are $a_3[10]$ = {0.02, 0.02, 0.02, 0.01, 0.03, 0.01, 0.01, 0.01, 0.01, 0.01}. The airplane safety for each airplane is a_4 [10] = {5.50, 5.40, 5.40, 5.30, 5.70, 5.30, 5.30, 5.60, 5.50, 5.50}. A possible safety improvement for each flight is a_5 [10] = {0.01, 0.01, 0.01, 0.01, 0.10, 0.03, 0.03, 0.01, 0.02, 0.02}. The negative weather impact on the safety is a_6 [10] = {0.02, 0.02, 0.02, 0.01, 0.05, 0.01, 0.01, 0.05, 0.01, 0.01}. Three series, of then experiments, are made. Each series is limited to the g= 100, g= 1000 and g= 10000 iterations. The achieved results are presented on Table 7 and Figure 2.

	experiment number										
g	1	2	3	4	5	6	7	8	9	10	A
100	1.48	2.57	1.86	1.91	1.83	1.96	1.96	2.00	1.99	2.30	1.9908
1000	2.57	2.75	2.60	2.78	2.78	2.83	2.65	2.68	2.79	2.79	2.7278
10000	2.65	2.69	2.76	2.73	2.78	2.74	2.79	2.80	2.87	2.81	2.7667

Table 7: Optimal profit - experimental results.

Table 7 presents the achieved optimal profit from the flights, constrained by the weather impact and requirements for the flight safety. g is the iterations limit, A is the average result.

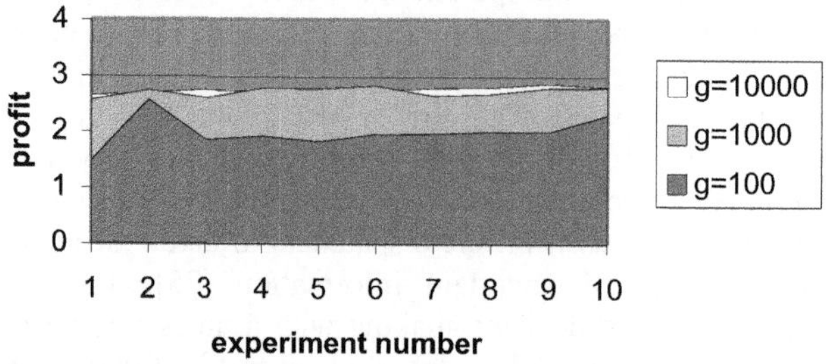

Figure 2: Optimal profit - experimental results.

With a precision of two decimal digits for $x_i \in (1, 10)$ $w_i \in (2, 10)$ i = 1,.,10 all possible situations are around 10^8. The exhaustive exploration of such number of situations requires large efforts and time. In that case, the algorithm plays a role of an intelligent search system, which achieves optimal solutions after heuristic exploration of small number of iterations. The search potential is illustrated with the results on Table 7. For 10 experiments 100 iterations (g=100) after exploration of 10^3 situations the found average profit is f_{100a} = 1.9908, after exploration of 10^4 situations (g=1000, 10 experiments) f_{1000a} = 2.7278 is the average profit, and after exploration of 10^5 situations (g=10000, 10 experiments) the average profit found is f_{10000a} = 2.7667. The results present the possible optimal profit for the considered traffic condition. This information can extricate the dispatchers from a search of impossible profit level and can support them in management of the traffic dispatch, for particular weather condition, keeping required flight safety.

7 Conclusion

The research work presented in this paper explores an original population-based optimisation method, called Free Search. It implements novel concepts for the behaviour of the individuals during the optimisation process. FS is examined with a hard constraint non-linear test problem [15, 16]. The results suggest that: (1) FS outperforms other methods and has better exploration and search potential, than the methods tested with the bump problem and discussed in the literature; (2) the individuals in FS adapt effectively their behaviour during the optimisation process, taking into account the constraints on the search space; (3) FS can be reliable in solving real-world, non-linear, constraint, optimisation problems.

Free Search is applied to a generalised traffic management, optimisation model. The aim of the model is to find the maximal profit and to satisfy the requirements for safety, both constrained by the environmental impact. FS demonstrates quick abstraction of optimised information for traffic management purposes. The exploration highlights the adaptive computing method as an intelligent search tool. The optimised information can support the decision-making in Traffic Management, and a development of a generalised traffic model for ship and road traffic management can be a subject of further research.

The new concepts are a contribution to the research efforts in the domain of population-based search methods, and, also, can contribute in general to the Computer Science in modelling of natural and uncertain behaviour. A Free Search assessment requires further exploration with other real-world problems.

References

1 Angeline P., (1998), Evolutionary Optimisation versus Particle Swarm Optimisation: Philosophy and Performance Difference, The 7-th Annual Conference on Evolutionary Programming, San Diego, USA

2 Bilchev G., I. Parmee, (1996), Constrained Optimisation With an Ant Colony Search Model, *Proceedings of ACED'96,* PEDC, University of Plymouth, UK.

3 Bilchev G., I. Parmee, (1995), The Ant Colony Metaphor for Searching Continuous Design Space, *Proceedings of the AISB Workshop on Evolutionary Computation,* University of Sheffield, UK, April 3-4.

4 Corne D.,M. Dorigo, and F. Glover, (1999), *New Ideas in Optimization.* ISBN 007 7095065, McGraw-Hill International.

5 Goldberg D., (2001), Genetic Algorithms in Search, Optimisation, and Machine Learning, ISBN 0-201-15767-5, Addison-Wesley.

6 Dorigo M.,G. Agazzi, G.Di Caro, L.Gambardella, R. Michel, M. Middendorf, T. Stutzle, E. Taillard, (1999) Part One, Ant Colony Optimization, in Editors Corne D.,M. Dorigo, and F. Glover, *New Ideas in Optimization.* ISBN 007 7095065, McGraw-Hill International. (pp. 9-76).

7 Dorigo M.,G.Di Caro, L.Gambardella, (1998), Ant Algorithms for Discrete Optimisation, *TR 98-10*, IRIDIA, University Libre de Bruxelles.

8 Eberhart R. and J.Kennedy, (1995), Particle Swarm Optimisation, *Proceedings of the IEEE International Conference on Neural Networks*, vol.4, 1942-1948.

9 Eiben, A.E., and J.E. Smith, 2003, *Introduction to Evolutionary Computing*, Springer, ISBN 3-540-40184-9, (pp 15 – 35).
10 El-Beltagy M.A., and A.I. Keane, (1998), Optimisation for Multilevel Problems: A Comparison of Various Algorithms, In I.C. Parmee editor, *Adaptive computing in design and manufacture,* ISBN 3-540-76254-X Springer – Verlag London Limited. (pp. 111-120).
11 Eshelman, L.J., & Schaffer, J.D., (1993), Real-coded genetic algorithms and interval-schemata, Foundations of Genetic Algorithms 2, Morgan Kaufman Publishers, San Mateo, pp. 187-202.
12 Fogel G., (2000), Evolutionary Computation: Towards a New Philosophy of Machine Inteligence, Second Edition, IEEE Press, ISBN: 0-7803-5379-X
13 Ghasemi M.R., E. Hinton and S. Bulman, (1998), Performance of Genetic Algorithms for Optimization of Frame Structures, In I.C. Parmee editor, *Adaptive computing in design and manufacture*, ISBN 3-540-76254-X Springer - Verlag London Limited. (pp. 287-299).
14 Holland J., (1975), Adaptation In Natural and Artificial Systems, University of Michigan Press.
15 Keane A. J., (1995), Genetic algorithm optimization of multi-peak problems: studies in convergence and robustness, *Artificial Intelligence in Engineering* 9(2) (pp. 75-83).
16 Keane A. J., (1996), A Brief Comparison of Some Evolutionary Optimization Methods, In V.Rayward-Smith, I. Osman, C. Reeves and G.D. Smith, J. Wiley (Editors), *Modern Heuristic Search Methods*, ISBN 0471962805 pp 255-272.
17 Michalewicz, Z. and Schoenauer, M., (1996), Evolutionary Algorithms for Constrained Parameter Optimization Problems, *Evolutionary Computation*, Vol.4, No.1, (pp.1-32).
18 Michalewicz, Z. and Fogel, D., (2002), How to Solve It: Modern Heuristics, ISBN 3-540-66061-5 Springer-Verlag, Berlin, Heidelberg, New York.
19 Penev, K., (2001), GIS in Support of Traffic Management, MPhil thesis submitted in partial fulfilment of the requirements of The Nottingham Trent University, UK, August, (pp 5-23).
20 Penev, K., and Littlefair, G., (2003), Free Search – a Novel Heuristic Method, *Proceedings of the PREP 2003*, 14-16 April, Exeter, UK, (pp 133-134).
21 Penev, K., and Littlefair, G., (2003), Free Search – A Comprative Analysis, Submitted to *Information Sciences*, Special Issue on Genetic and Evolutionary Computing, Elsevier.
22 Price K., and R. Storn, (1997), Differential Evolution, *Dr, Dobb's Journal* 22 (4), (April), (pp. 18-24).
23 Price K., K Chisholm, J. Lampinen, R. Storn,, I. Zelinka, (1999), Part Two Differential Evolution, in Editors CorneD.,M.Dorigo, and F.Glover, *New Ideas in Optimisation.* ISBN 007 7095065, McGraw-Hill International (pp 77-158).
24 Schoenauer, M. and Michalewicz, Z., (1996), Evolutionary Computation at the Edge of Feasibility, *Proceedings of the 4th Parallel Problem Solving from Nature*, H.M. Voigt, W. Ebeling, I. Rechenberg, and H.P. Schwefel (Editors), Springer-Verlag, Lecture Notes in Computer Science, Vol.1141 (pp.245-254).
25 Smith K., (2001), Incompatible goals, uncertain information and conflicting incentives: the dispatch dilemma, *Human Factor and Aerospace Safety*, Ashgate Publishing 1(4), (pp. 361-380).

Chapter 6

Control and Robotics

An Online Hierarchical Fuzzy Rule Based System for Mobile Robot Controllers

Antony Waldock[1], Brian Carse[2], Chris Melhuish[2]

[1]Advanced Technology Centre
BAE SYSTEMS
Bristol, BS34 7QW, UK
antony.waldock@baesystems.com

[2]Intelligent Autonomous Systems Laboratory,
University of the West of England
Bristol, BS 16 1QY
{brian.carse, chris.Melhuish}@uwe.ac.uk

Abstract

The introduction of automated robots has revolutionised the manufacturing industry. The further development of autonomous mobile robots capable of functioning in unstructured and dynamic environments is highly desirable. This paper outlines a novel method for the online development of an interpretable mobile robot controller using supervised learning. An information theoretic approach is used to control the rate of expansion in a Hierarchical Fuzzy Rule Based System (FRBS). Experimental results, on a simulated mobile robot, are provided to demonstrate how the uncertainty tolerated can be used to control the trade-off between accuracy and interpretability.

1 Introduction and Related Work

Over the last 30 years, autonomous robots have revolutionised the manufacturing industry by allowing mass production of a wide range of products from cars to highly complex printed circuit boards. These robots remain relatively unintelligent and operate in sanitised and caged environments. The development of industrial robots capable of coping with unstructured dynamic environments, where humans may be present, could revolutionise the industry again. This paper outlines a novel method for the construction of an interpretable controller for mobile service robots through interaction with a human supervisor.

The construction of a mobile robot controller for a real world application is a complex and time-consuming process. Classical Control Theory (Ogata, 1997) allows the generation of control rules from accurate analytical models of the environment, platform and control laws. The construction of models that accurately represent the dynamics of an uncertain and complex environment can be a long and difficult, if not impossible, process. Soft Computing (Zadeh, 1994) provides a collection of methods that aim to exploit a tolerance for imprecision and approximation to achieve robustness and low cost solutions e.g. Neural Networks and Fuzzy Logic. Fuzzy Logic has been demonstrated to cope well with uncertain and imprecise data when used in a mobile robot controller (Hagras et al., 2001). Using Fuzzy Rule Based Systems (FRBS) to build mobile robot controllers (Tunstel et al., 1996, Hoffman and Pfister, 1997) has become increasing popular due to their ability to automatically generate rules from large data sets while maintaining a degree of human interpretability (Holve and Protzel, 1996).

A FRBS divides the input space into a number of linguistic symbols, each associating a fuzzy set with a natural language meaning (Zadeh, 1998); e.g. very small, small, large etc. A fuzzy set links an input variable with a membership function, to represent its applicability with regard to the current environmental state, and hence, determine its influence in the decision process. The decision space is partitioned using a series of IF...THEN... rules. The combination of meaningful linguistic symbols and well-structured rules provides a high degree of human interpretability. For example, a service robot that is required to avoid collisions might use a fuzzy control rule such as:

IF (wall is close) AND (wall_direction is infront-left) THEN (turn right)

The way in which linguistic symbols are represented can directly affect the accuracy of the approximated function (Bastian, 1994). If a greater degree of accuracy is demanded then the structure (i.e. size, shape and position) of the membership functions, which represent the linguistic symbols, must be altered. Increasing the granularity of the linguistic symbols can facilitate improvements in accuracy, but as the complexity of a system increases, our ability to make precise and yet significant statements about its behaviour diminishes. As the number of linguistic symbols increases human interpretability declines. Hence within this contribution, the number of rules is used as a measure of interpretability. This relationship is referred to as the trade off between accuracy and interpretability (Alcala et al., 2001).

A new variant of a FRBS has been developed in an attempt to improve the accuracy of the approximation and is referred to as an Approximate FRBS (also known as "non-grid-oriented" or "free semantics" FRBS). An approximate FRBS increases the accuracy of the model by manipulating fuzzy sets directly instead of the linguistic symbols. By allowing the fuzzy sets to vary in number, size and position, they can be mapped directly to the data giving significant improvements (Carse et al., 1996). At the end of rule generation, the fuzzy variables are semantically free and can tend to be excessively specialised. This limits the human interpretability of an approximate FRBS. However, it must be noted they often remain more interpretable than other models such as neural networks.

The Hierarchical Fuzzy Rule Based System (HFRBS) has been developed, in recent years, as an attempt to improve accuracy while maintaining interpretability (Cordón et al., 2001b). A HFRBS divides the input space into a fixed number of linguistic symbols each corresponding to a natural language meaning e.g. very small, small etc. Training data is then used to automatically generate rules as in a standard FRBS. The key extension of a HFRBS is the use of an *expansion policy* to determine inaccurate areas of the decision space and the corresponding rules. When an inaccuracy is identified, the rule representing that area is specialised into a set of more specific rules. This involves partitioning the rule's decision space into smaller areas each represented by a separate rule. This process of specialisation continues until a desired level of accuracy is satisfied. Figure 1 shows an example of a partitioned decision space (a) and its corresponding hierarchical representation (b). This concept of increasing the granularity to fit the underlying decision space has been applied to classifier systems for a number of years (Melhuish and Fogarty, 1994).

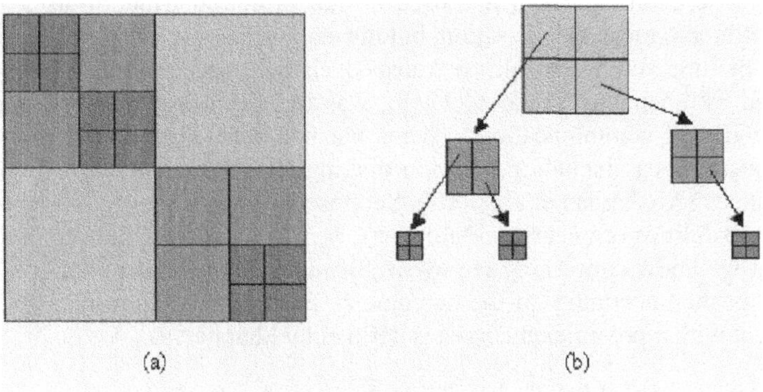

Figure 1: Hierarchical FRBS

This type of linguistic symbol expansion increases the accuracy depending on the complexity of the data modelled and is commonly referred to as "ad-hoc data driven learning" (Alcala et al., 2001). The expansion can be controlled depending on the accuracy desired, human interpretability required and the complexity of the function to be approximated. Some of these properties are mutually exclusive, for example, high human interpretability on a complex, accurate model may be impossible.

Holve outlines a method for specialisation by carefully pre-processing training data such that, when a conflict is encountered, the linguistic symbol is expanded (Holve, 1998). Although demonstrated to approximate complex functions, pre-processing limits its use to applications where all the training data is available and certain. Cordón, Herrera and Zwir (Cordón et al., 2001b) outline a hierarchical FRBS that uses expansion techniques to specialise linguistic symbols with a large degree of error. The error of each rule is calculated by the percentage of the Mean Square Error (MSE) associated with the rule over the MSE for the entire training set. A rule with bad performance is determined by comparing this error with a tuneable parameter α, which dictates the rate of expansion and hence the accuracy of the function approximated. This method relies on the complete training set and test set being available during the expansion phase. For an online application, such as learning behaviour in a mobile robot, access to the entire training set is not possible requiring an Online or Anytime Algorithm (Grefenstette and Ramsey, 1982).

The next section outlines the proposed algorithm with a new method of expansion based on Information Theory. The performance of the proposed algorithm is compared to existing function approximators in section three followed by a simulated mobile robot application in section four. Section five and six discuss the mobile robot results and possible areas of future work.

2 The Information Theoretic Hierarchical Fuzzy Associative Memory (IT-HFAM)

The rule generation algorithm proposed in the current contribution uses a similar method of hierarchical specialisation, but the expansion policy is determined by the amount of uncertainty within a rules decision space. Information Theory, developed by Shannon (Shannon, 1948), was initially concerned with modelling the efficiency of communication systems but has been applied to a multitude of other research areas including decision making (ID3 (Quinlan, 1990)) and fuzzy logic systems (Al-sharhan et al., 2001). Suppose we have a set of n possible events whose probabilities of occurrence are $p_1,p_2...p_n$. These probabilities are known but that is all we know concerning the event. Shannon defines the amount of 'choice' involved or the uncertainty of the outcome as Entropy. The entropy, $H(p_1,p_2...p_n)$ of an event with n possible outcomes is defined by Shannon as

$$H(p_1,p_2..p_n)=\sum_{i=1}^{n}(p_i \times \log(p_i))$$

This measure of information can be used to calculate the uncertainty within each rule (Zadeh, 1965) and to determine the expansion of a HFRBS. A FRBS can be viewed as a Fuzzy Associative Memory (FAM) where the linguistic symbols M_i, for each input n, produce an n-dimensional decision space partitioned by an i-dimensional grid. Each cell represents an IF-THEN rule with n linguistic inputs corresponding to a single output linguistic symbol. The proposed Information Theoretic Hierarchical Fuzzy Associative Memory (IT-HFAM) extends the traditional FAM approach by including an applicability distribution over all the output linguistic symbols. Figure 2 displays a FAM partitioned into four cells (rules) each with an applicability distribution over all possible output symbols.

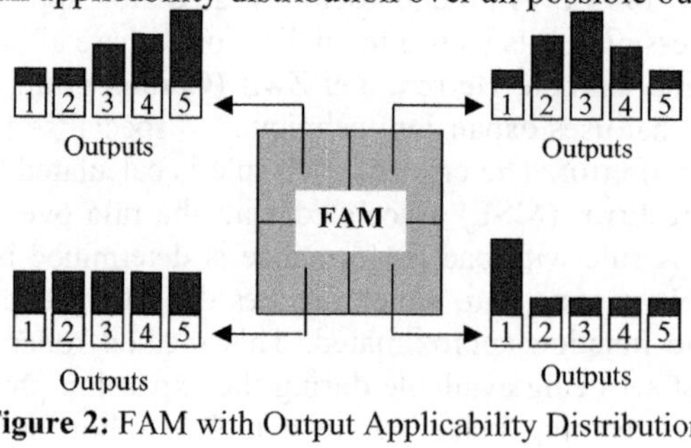

Figure 2: FAM with Output Applicability Distributions

Using Shannon's measure of uncertainty on the output applicability distribution gives the amount of choice experienced by this cell. If the applicability distribution is flat i.e. A_i are equal, (Figure 2: Bottom Left Rule) then the uncertainty of the cell is at a maximum. This indicates that the cell at this granularity cannot effectively model the decision space it represents. In order to model this area of the decision space more effectively it is necessary to divide the cell into a number of smaller cells. The algorithm can be initialised with a single linguistic symbol covering each input or with a set of linguistic symbols if input space partitions are known.

Training of an IT-HFAM consists of updating the applicability distribution when exposed to a new training pattern. T_p. Each pattern is a $(n+1)$ dimensional vector, $T_p = <x_1, x_2, x_3...x_n, y>$ where n is the number of inputs; x is the current input values and y the target output value. The algorithm is trained on all patterns until the training set has been exhausted. The algorithm is trained by first identifying all the cells with an activation greater than zero when presented with the inputs $(x_1..x_n)$ from the training pattern. The applicability distribution for each active cell is updated depending on the target value y. For each output membership function, M_j^y, the applicability of the output symbol $a(M_j^y)$ is increased by the level of activation for the target value y (see Figure 3).

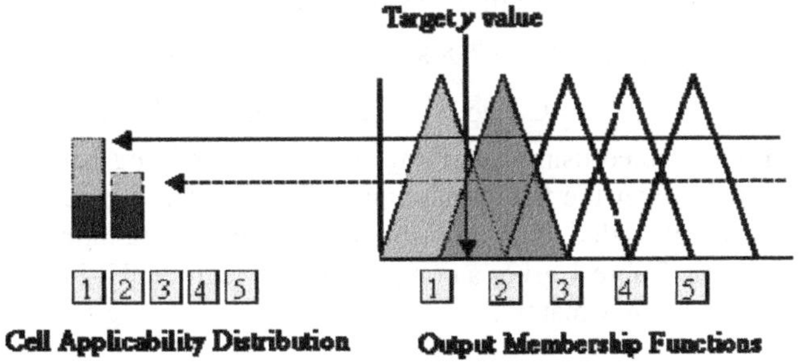

Figure 3: Cell applicability update procedure

Centre of Sums is used on the applicability distribution to calculate the current output symbol for this cell in the FAM. After each training pattern, the entropy (uncertainty) of each cell is compared with a tuneable parameter $E_{\max}$, which represents the maximum amount of uncertainty tolerated within each cell. If the uncertainty of the cell exceeds $E_{\max}$ and the applicability distribution covers more than two consecutive output symbols then the cell is divided into four smaller specialised cells (assuming the FAM consists of only two dimensions – see Figure 4).

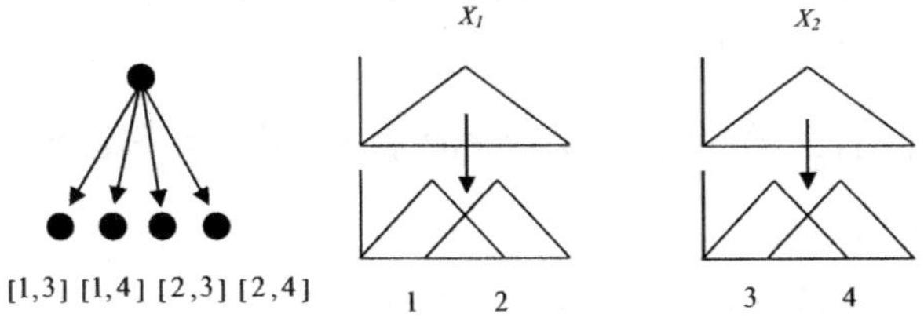

Figure 4: Membership Function Specialisation

The cell expansion is halted when only two consecutive output symbols are active in the applicability distribution. A final stage of rule reduction is applied which merges neighbouring cells with the same output fuzzy set. Rule reduction can be performed after a number of training phases or after training has been completed.

3 Comparison with Existing Function Approximations

This section describes the performance of the IT-HFAM algorithm, on an intermediate complexity function, compared with existing function approximators. Within the experiment below, the accuracy is determined using the Root Mean Squared Error (RMSE) equation using a test set of 100 evenly distributed points over the input space.

Cordón, Herrera and Zwir (Cordón et al., 2001a) demonstrate the ability to train a hierarchical FRBS using a Mean Squared Error (MSE) expansion policy on an intermediate complexity function defined as:

$$f(x_1,x_2) = e^{x_1} \times \sin^2(x_2) + e^{x_2} \times \sin^2(x_1)$$

$$x_1, x_2 \in [-8,8], f(x_1,x_2) \in [0,5836]$$

The system was trained using a training set of 1156 training patterns, which were evenly distributed over the function, compared to Cordón's 1089 training patterns. The RMSE was calculated after exposure to the complete training set and then after a second iteration. Table 1 gives a comparison between the IT-FAM algorithm and existing function approximators.

Table 1: Comparison with existing function approximators

Function Approximator	No Rules	RMSE
Static Weight Counting Algorithm (S-WCA) (Bardossy and Duckstein, 1995)	9	0.0868
Hierarchical S-WCA (Cordón et al., 2001b)	316	0.0134
Fuzzy C-Means (FCM) (Bezdek, 1992)	6	0.1124
Hierarchical FCM (Cordón et al., 2001b)	9	0.0403
IT-HFAM (E_{max} 100%)	7	0.0880
IT-HFAM (E_{max} 45%)	336	0.0291

The results shown in Table 1 demonstrate that the on-line IT-HFAM performance is comparable to the other off-line methods.

4 Learning Behaviour on a Mobile Robot

In this section the IT-HFAM algorithm will be used to construct a set of control rules for a mobile robot though interaction with a supervisor. The supervisor presents examples of correct operation (training patterns) to the algorithm when the robot's behaviour is incorrect. The task taught is inspired by a collision avoidance problem demonstrated by Furuhashi, Nakaoka and Uchikawa (Furuhashi et al., 1996). The implementation of the task differs from that addressed by Furuhashi et al. by using simulated robot sensors to detect obstacles instead of the exact range and bearing information. The performance and rules generated by the IT-HFAM algorithm over a number of E_{max} values are compared.

The task involves navigating a mobile robot to reach a goal while avoiding two other robots (Figure 5). The controlled robot moves with a constant speed with the two crossing robots moving at half the speed of the controlled robot. Varying the angular velocity, ω, alters the direction of the controlled robot. The controlled robot starts from the bottom line and attempts to navigate towards the goal at the top. The two other robots appear from the left and right hand side and cross between the robot and the goal. The aim of the task is to reach the goal without hitting either of the two crossing robots. The task is performed using the Pioneer simulation environment, Stage[1]. The controlled, and crossing robots, are simulated Pioneer robots, while a red box is used to represent the goal. The controlled robot can detect an obstacle using eight sonar sensors evenly distributed over the front of the robot. Each sonar sensor returns a value indicating the distance to the nearest obstacle. The sensors are divided into two sets, left and right, and the minimum distance for each set is normalised ([0,1]) before being used as an input to the controller (Figure 6).

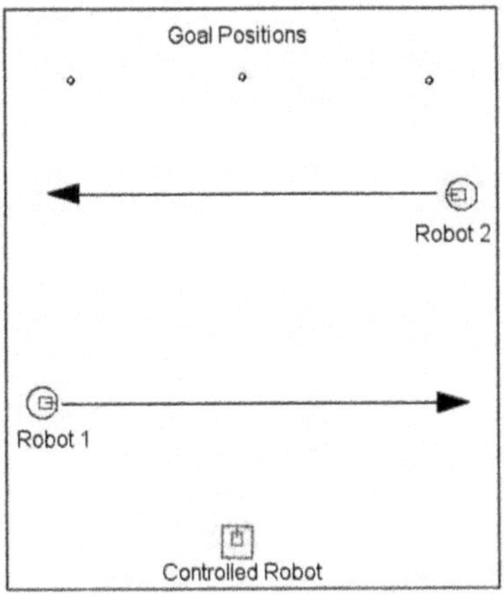

Figure 5: Robot Task

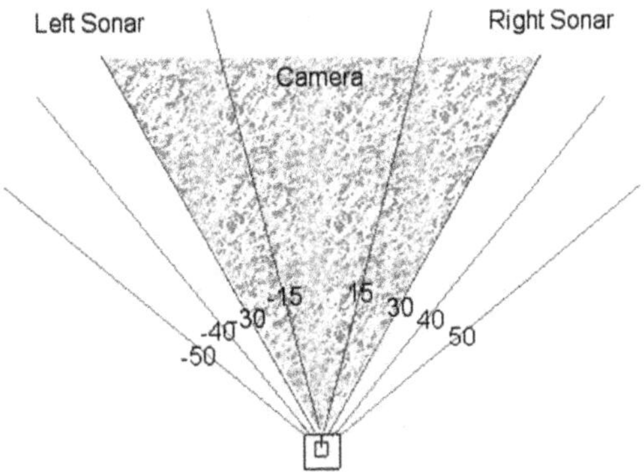

Figure 6: Robot Platform

A bearing to the goal is acquired by simulating a front mounted camera to identify the position of the red box representing the goal. The goal is detected in the camera's field of the view by extracting the red areas of the image (red blob detection). The pixel column with the highest concentration of red within the image is marked as the centre of the goal. A normalised bearing to the goal is found by dividing the pixel column with the highest concentration by the total pixel width of the image. The input to the controller is a continuous value in the range [0,1] indicating the bearing to the goal. In summary, the inputs to the controller are a bearing to the centre of the goal and the distance to the nearest object on the front left and right of the robot.

The controlled robot is trained using a human supervisor to present examples of correct operation during three scenarios. The starting position of the controlled robot is varied in each scenario giving the controlled robot exposure to a

---

[1] http://playerstage.sourceforge.net

large proportion of the decision space. The supervisor uses a joystick to indicate the correct angular velocity (continuous value over the range [0,1]) by monitoring the inputs to the algorithm (left/right sonar and the position of the goal). The training examples are only provided when the supervisor deems the robot's behaviour is incorrect.

The IT-HFAM algorithm is initialised with three membership functions evenly distributed over each of the input variables and five membership functions over the output variable. The task performance and characteristics of the IT-HFAM algorithm are compared over a number of E_{max} values: 100, 90, 75, 50. Three test scenarios, where the position of the goal and crossing robots are varied, are used to determine the performance of the trained controller. Task performance is obtained by studying the amount of deviation from the shortest path to the goal. The three test scenarios are displayed in Figure 7. The shortest path for each scenario is a straight line between the start position and the goal. The amount of deviation is calculated by comparing the distance travelled by the controlled robot during the test scenario and the shortest path.

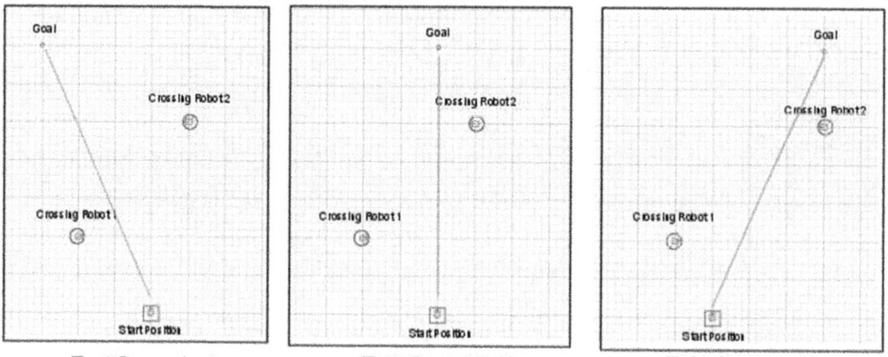

Test Scenario 1 Test Scenario 2 Test Scenario 3

Figure 7: Test Scenarios

4 Robot Results and Discussion

Table 2 outlines the average deviation for three test runs on each scenario (D1, D2, D3 in millimetres) and the number of rules (C_r) created for values of E_{max}.

Table 2: IT-HFAM Simulated Robot Task Results

E_{max}	C_r	D_1	D_2	D_3	D_{total}
100	15	171	249	190.5	610.5
90	33	156	223.5	157	536.5
75	183	165.5	193.5	143	502
50	609	141.5	170	62.5	374

Table 2 demonstrates that E_{max} can be used to control the trade off between the number of rules generated (C_r) and the accuracy of the solution (D_{total}). With no specialisation (E_{max} = 100), the controlled robot was capable of navigating to the target in all test scenarios but deviated from the optimal path by 203mm on average. The algorithm produced only 15 rules that remain highly interpretable to the supervisor. Some of the rules are listed in Table 3. The selection of rules presented resembles control rules a human may have used to solve the problem and hence demonstrates that the algorithm can produce interpretable solutions. Only 6 out of the 15 rules are displayed helping to demonstrate that a human designer may not have designed a rule base that captures all situations and the arbitration between the two tasks.

Table 3: Rules Generated for E_{max} = 100

Task	*Sonar Left*	*Sonar Right*	*Vision*	*Action*
Track	*Far*	*Far*	*Left*	*Left*
Track	*Far*	*Far*	*Middle*	*Forward*
Track	*Far*	*Far*	*Right*	*Right*
Avoid	*Close*	*Far*	*Left OR Middle OR Right*	*Left*
Avoid	*Inrange*	*Far*	*Left OR Middle OR Right*	*Forward*
Avoid	*Close OR Inrange OR Far*	*Close*	*Left OR Middle OR Right*	*Right*

As the amount of uncertainty tolerated (E_{max}) is decreased, the deviation from the optimal path is reduced with the side affect that the number of rules generated increases significantly. For example when E_{max} is 50, the deviation from the optimal path is 38.7% better than without specialisation while the number of rules generated has increased to 609. The hierarchical structure used to perform specialisation can now be used to increase the interpretability of a complex rule base containing 609 rules. As the hierarchical structure is maintained, a representation containing the initial 15 rules can be extracted. For example when E_{max} = 50, a representation containing the initial rules (as in Table 3 when E_{max} = 100) can be extracted. This indicates that this hierarchical structure could provide a means of increasing the interpretability of complex rule bases. Work is continuing on how this hierarchical structure could allow a supervisor to explore and understand a complex collection of rules.

These results demonstrate that adjusting the amount of uncertainty tolerated within the decision making process can be used as a trade off between the interpretability and the performance of the solution. Figure 8 demonstrates how the deviation from the optimal path is reduced in test scenario 3 over E_{max} values 100, 75 and 50. The reduction in the deviation from the optimal path is achieved through specialisation of the decision space by more accurately modelling the behaviour demanded by the supervisor. For example, specialisation adjusts the distance at which the controlled robot performs obstacle avoidance. These results also indicate that specialising the input space in all dimensions limits the scalability of the proposed algorithm. A method of specialising the most uncertain dimension could be employed to improve scalability.

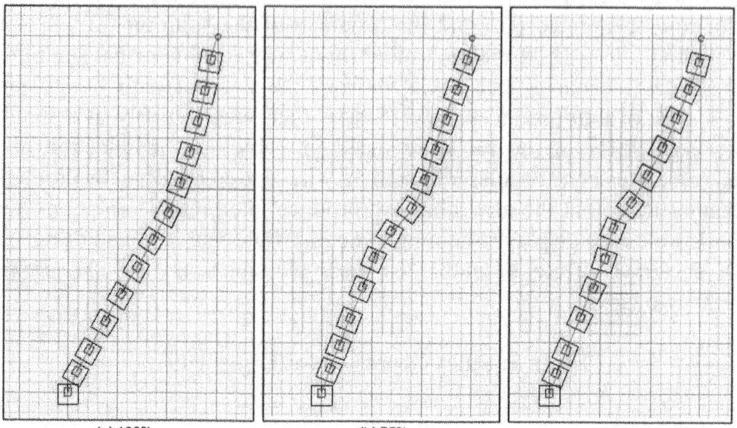

Figure 8: Robot path in test scenario 3 for Emax = 100,75,50

5 Conclusions and Further Work

This paper has demonstrated a novel online method to construct a hierarchical Fuzzy Rule Based System that can achieve levels of accuracy comparative with previous offline methods. The algorithm uses an information theoretic approach to determine if any rule does not effectively model the underlying decision space within a specified degree of uncertainty. The algorithm has been demonstrated to approximate an intermediate complex function and learn a simulated mobile robot task. The results demonstrate that accuracy and interpretability can be controlled using a single tuneable parameter (E_{max}) but more intelligent specialisation is required to improve scalability.

Future work will focus on investigating limitations of the algorithm when applied to a real mobile robot using a human supervisor. Another important aspect of constructing a human interpretable controller is the visualisation of complex controllers with large numbers of rules. When a high degree of certainty in the decision is required the algorithm described within this paper produces a solution with a large number of rules. A possible method of interpreting a large number of rules could arise from the hierarchical structure of the rule base. This may provide a means of allowing the supervisor to explore different parts of the decision space (rules) at varying levels of complexity.

Acknowledgements

This work was funded by the Advanced Technology Centre Filton, BAE SYSTEMS Ltd.

References

Alcala, R., Casillas, J., Cordón, O. and Herrera, F. (2001) *Building Fuzzy Graphs: Features and Taxonomy of Learning for Non-grid-oriented Fuzzy Rule-based Systems, Journal of Intelligent and Fuzzy Systems,* 11, pp 99-119.

Al-sharhan, S., Karray, F., Gueaieb, W. and Basir, O. (2001) In Fuzzy Entropy: A Brief Survey at The 10th IEEE International Fuzzy Systems Conference, 2001, pp 1135-1139

Bardossy, A. and Duckstein, L. (1995) *Fuzzy Rule-based Modelling with Application to Geophysical, Biological and Engineering Systems,* CRC Press.

Bastian, A. (1994) *How to handle the flexibility of linguistic variables with applications, Intl. Journal of Uncertainty, Fuzziness and Knowledge-Based Systems,* 2, pp 463-484.

Bezdek, J. C. (1992) *Fuzzy Models For Pattern Recognition: Methods that search for structures in Data,* IEEE Press.

Carse, B., Fogarty, T. C. and Munro, A. (1996) *Evolutionary Learning of Fuzzy Rule Based Controllers using Genetic Algorithms, Fuzzy Sets and Systems,* 80, pp 273-293.

Cordón, O., Herrera, F., Hoffmann, F. and Magdalena, L. (2001a) *Genetic Fuzzy Systems: Evolutionary Tuning and Learning of Fuzzy Knowledge Bases,* World Scientific.

Cordón, O., Herrera, F. and Zwir, I. (2001b) *Fuzzy Modeling by Hierarchically built Fuzzy Rule Bases, International Journal of Approximate Reasoning,* 27, pp 61-93.

Furuhashi, T., Nakaoka, K. and Uchikawa, Y. (1996) *A Study on Fuzzy Classifier Systems For Finding a Control Knowledge of Multi-input Systems, Genetic Algorithms and Soft Computing*, pp 489-502.

Grefenstette, J. J. and Ramsey, C. L. (1982) In An Approach to Anytime Learning at Ninth International Machine Learning Workshop, San Mateo, CA, Morgan Kaufmann, pp 189--195

Hagras, H., Callaghan, V. and Colley, M. (2001) *Outdoor Mobile Robot Learning and Adapation, IEEE Robotics and Automation Magazine,* 8, pp 53-69.

Hoffman, F. and Pfister, G. (1997) *Evolutionary Design of a Fuzzy Knowledge Base for a Mobile Robot, International Journal of Approximate Reasoning,* 17, pp 447-469.

Holve, R. (1998) In Investigation of Automatic Rule Generation for Hierarchical Fuzzy Systems at IEEE World Congress on Computational Intelligence, FUZZ IEEE, Anchorage, Alaska, May 4-9 1998, pp 971-978

Holve, R. and Protzel, P. (1996) In Generating Fuzzy Rules by Learning from Examples at Biennial Conference of the North American Fuzzy Information Processing Society - NAFIPS, Berkeley, CA, USA, June 19th-22nd 1996, pp 451-455

Melhuish, C. and Fogarty, T. (1994) In Applying Restricted Mating Policy to Determine State Space Niches Using Immediate and Delay Reinforcement at Evolutionary Computing, AISB Workshop, Leeds, UK, 1994, pp 224-237

Ogata, K. (1997) *Morden Control Engineering,* Tim Robbins.
Quinlan, J. R. (1990) *Induction of Decision Trees,* Morgan Kaufmann.
Shannon, C. E. (1948) *A mathematical theory of communication, Bell System Technical Journal,* 27, pp 379-423 and 623-656.
Tunstel, E., Lippincott, T. and Jamshidi, M. (1996) In Introduction to Fuzzy Logic with Application to Mobile Robotics at First National Students Conference of the National Alliance of NASA University Research Centres, NC A&T State Univ, Greensboro, NC, Marc 1996, pp
Zadeh, L. A. (1965) *Probability measures of Fuzzy Events, Journal Math. Anal. Appl.,* 23, pp 421-427.
Zadeh, L. A. (1994) *Soft Computing and Fuzzy Logic, IEEE Software*, pp 48-56.
Zadeh, L. A. (1998) *Fuzzy Logic, Computer,* 21, pp 83-93.

Improving Robot Manipulator Performance with Adaptive Neuro-Control

A. G. Pipe, A. Lenz

Faculty of Computing, Engineering and Mathematical Sciences,
University of the West of England
Bristol, BS 16 1QY
Anthony.Pipe@uwe.ac.uk, alex@lenz-alexander.de
www.ias.uwe.ac.uk

Abstract

An adaptive controller that modifies its characteristics to deal with new situations, in a timely and accurate way, would be a valuable improvement on many existing industrial plant controllers. Furthermore, if such a controller could effectively be "strapped around" the existing controller then there could be many opportunities for performance enhancement of systems already "out in the field". The new situations that this adaptive controller would need to deal with could arise from time variations in the plant's characteristics due to wear and tear, or from learning, on-line, about unforeseen new parts of the plant's operational envelope. The universal approximation abilities of neural networks, combined with permanently active on-line learning, yield powerful features that can be used to great advantage in creating adaptive controllers for such applications. Together they allow accurate control strategies to be developed for these types of plant without the need for a mathematical model. Furthermore, neuro-control algorithms can learn about these dynamical features using signals from the plant that are normally easily obtained. Multi-axis revolute-jointed robot manipulators are good examples of this class of plant. In order to illustrate some of the important advantages of these methodologies, and to inform the reader about some of the important characteristics of these adaptive structures, we review some of our recently reported experiments in applying an on-line learning neuro-control approach to joint level trajectory control of two different industrial robots. In each case, the neuro-controllers are used to enhance performance of the existing PID controllers. This paper is mainly concerned with highlighting these features via the experimental results. However, when a controller learns on-line whilst acting as part of a plant's closed-loop controller, it is crucial that a careful and rigorous approach is adopted. A strict theoretical basis that guarantees the whole system's stability is required. To set the experimental work in context therefore, we briefly review our on-line learning neuro-control method, which is used for both sets of experiments.

1 Introduction

In recent years, much attention has been paid to artificial neural network based controllers. Their abilities to perform non-linear mappings and their learning potential have led to great improvements in the performance of open and closed loop controllers for a variety of complex applications, including robotics. The work reported on here is a contribution to this area. One of the important problems of dynamic control of a multi-jointed robot arm is generating the appropriate joint-motor commands so that the end-effector follows a desired trajectory as precisely as possible, even under extreme speed and payload conditions. To reach this goal using traditional advanced control schemes requires a thorough knowledge of the system behaviour, usually represented as a highly cross-coupled non-linear mathematical model. There are some, commonly occurring, circumstances in which some of the parameters of this model are very hard to obtain [3,8].

Neural identification and control techniques are very well suited to the problem of controlling robot dynamics at joint level. Further, the on-line learning capabilities of some categories of neural networks make them very efficient for adaptive control architectures. On-line learning refers to a situation where the neural network is trained, often in conjunction with a linear controller, while the system (robot manipulator) is in operation. In such a set-up the neural network is not only able to acquire knowledge about unmodelled dynamics of the system but also to adapt the controller to a change in plant dynamics. A change in dynamics can be, for example, the result of a sudden change in payload or slow drifts of model parameters due to wear. Clearly, if the controller is active whilst its internal parameters are changing, then stability of the complete system (controller and plant) will be of particular interest in an on-line learning situation.

The work presented in this paper is concerned with the application of these techniques to trajectory control of a Yaskawa Motaman industrial manipulator extracted from daily use as a manufacturing welding robot in a local company, and a Puma 500 industrial robot manipulator. The two pieces of work have subtly different objectives. The first is a practical demonstration of these controllers' abilities to accurately learn complex non-linear dynamics without need of any complex *a priori* mathematical model; the second illustrates their performance in the face of disturbances in the dynamics of the controlled system. In the first set of experiments a Cerebellar Model Articulation Controller (CMAC) neural network is used, whilst in the second a "placement on demand" Radial Basis Function (RBF) neural network is used. The first of these experiments was described in 2001 [12], whilst the second was described in 2003 [6], but this paper is the first time that these two pieces of work have been described together with a wider emphasis.

2 Related Work

Over the years much research effort has been put into the design of neural network applications for manipulator control. For example, Albus [1] used the Cerebellum Model Articulation Controller (CMAC) to control manipulators as early as 1975.

Typically, the earlier work did not make a link between neural network based approaches to control, and learning convergence or stability guarantees. However, later examples where this link is made explicit are Narendra [10,11] in 1990 and 1992, Sanner & Slotine [14] in 1991, Suykens, Van de Wall & De Moor [15] in 1995, and Lewis *et al* [8] in 1999; in addition to our own laboratory's work published between 1993 and 2001 (e.g. [2,3,4]). Lewis *et al* have produced an excellent advanced textbook (as opposed to edited collections of research papers) [8] covering the relevant control theory, dynamics of robot manipulators, and the use of neural networks in these control applications. It is a strong sign of the increasing maturity of this field that such a textbook has become available.

3 Neural Network On-Line Learning Control

The work presented here is based on a conceptual neural network set; besides their basic features there are three special characteristics.

- First, they possess a universal approximation feature; i.e. any continuous function may be approximated as long as the function inputs are *within a compact set.*
- Second, within our conceptual set there is a sub-class of special interest, which we call Linear-Equivalent. This term is referred to as "Linear-In-the-Parameters" in the textbook by Lewis *et al* [8]. The adjustable weights are linear with the neural network output, i.e. $\boldsymbol{y}=\boldsymbol{G}^{\mathrm{T}}(\boldsymbol{u})\boldsymbol{W}$, where $\boldsymbol{y}$ is the output, $\boldsymbol{u}$ is the input vector, $\boldsymbol{W}$ is the adjustable weight vector, and $\boldsymbol{G}$ is the vector of non-linear mapping functions. For a particular neural network, $\boldsymbol{G}$ is fixed. However, $\boldsymbol{G}$ varies according to neural network type and parameter settings. If there are multiple outputs, each could have its own non-linear mapping vector and weight vector, i.e. for an n output neural network,

$$\begin{pmatrix} y_1 \\ y_2 \\ \vdots \\ y_n \end{pmatrix} = \begin{pmatrix} g_1(u) \\ g_2(u) \\ \vdots \\ g_n(u) \end{pmatrix}^{\mathrm{T}} \begin{pmatrix} W_1 \\ W_2 \\ \vdots \\ W_n \end{pmatrix} \equiv G^T(u)\, W \tag{1}$$

  in most cases however, all the non-linear mapping vectors ($\boldsymbol{g}_i$) are the same.
- Third, for all neural networks in our conceptual set $\boldsymbol{g}_i$ have bounded output values. Of course this is the normal case for most neuron mapping functions.

Specific paradigms include - but are not limited to - the Radial Basis Function (RBF) neural network [14] and the Cerebellum Model Articulation Controller (CMAC) [1]. We have previously shown [3,4] that the convergence and stability results presented in this paper are valid for either of these neural networks.

4 Robot Manipulator Dynamics

One could consider the three major axes (base, shoulder & elbow) of a revolute jointed manipulator, such as the Yaskawa Motaman or Puma 500, as a coupled system of links and revolute joints driven by electric motors. The Yaskawa is fitted with a shaft encoder and tachometer on each of these joints so as to provide position and velocity feedback, whilst the Puma robot provides only position.

A general equation of motion for a rigid manipulator is

$$H(q)\ddot{q} + C(q,\dot{q})\dot{q} + \overline{g}(q) + F(q,\dot{q}) = \tau(q,\dot{q},\ddot{q}) \tag{2}$$

where $\boldsymbol{q}$ is the n x 1 vector of joint displacements, $\boldsymbol{\tau}$ is the n x 1 vector of applied joint torques, $H(q)$ is the n x n symmetric positive definite manipulator inertia matrix, $C(q,\dot{q})\dot{q}$ is the n x 1 vector of centripetal and Coriolis torques, $\overline{g}(q)$ is the n x 1 vector of gravitational torques, and $F(q,\dot{q})$ are the unstructured uncertainties of the dynamics including friction and other disturbances.

By exploiting the neural network universal approximation feature we can assume that the left-hand side of the above equation can be approximated by a neural network, i.e.

$$H\ddot{q} + C\dot{q} + \overline{g} + F = G(q,\dot{q},\ddot{q})^T W + \xi(q,\dot{q},\ddot{q}), \tag{3}$$

where $\boldsymbol{G}$ is the non-linear mapping matrix of an RBF or CMAC neural network, $\boldsymbol{W}$ is its weight vector, which is initially unknown, and ξ is its approximation error, which should be as small as possible by careful design. This process of careful design can be approached with some confidence because of the availability of both existence proofs of the universal function approximation abilities of these neural network types, and of published guidelines on how to achieve the appropriate structure (e.g. [4,8,14].

5 On-Line Learning for Neural Networks

The main objective of on-line learning, in this context, is to force the manipulator to follow the desired trajectory, i.e. $\boldsymbol{q}_d$. In order to simplify the final control structure, we will use only the *desired* joint values as the neural network inputs. In this way we can satisfy one of the conditions of the neural network universal approximation feature, i.e. that all inputs must be bounded. Although this control structure could be viewed as a feedforward arrangement, the *actual* joint values are used for *training* the neural network on-line, thus completing an adaptive on-line learning feedback path.

During on-line learning, each neural network is used as a part of the manipulator controller. Its output forms part of the control signals, which are used to drive manipulator joints. The joint values, in turn, are used to train the neural network. Therefore the on-line learning algorithm must guarantee that these two coupled-systems (neural network & manipulator) work in a converging fashion. The fact that the neural network inputs differ from the manipulator joint values makes this task complicated. Since learning convergence and system stability are such crucial issues for an on-line learning system, we review our main theoretical research results below. The control architecture combines a simple linear controller with the neuro-controller.

In each of the two applications covered in this paper we had to make a copy of part, or all, of the original controller in order to add our neuro-control. However, this was only due to advanced age of the existing controller in each case. We were not able to obtain complete circuit diagrams for either of them at a sufficiently low level to establish where to inject the neuro-control signals. Care was taken, in each case, to make accurate copies of parts of the original controller where necessary. For the Yaskawa robot, this is a PD controller. The velocity control was used unaltered, implemented in analogue electronic hardware (we were able to "reverse engineer" the circuit diagram for this). The position control was originally implemented in software on a simple microcomputer; we recreated this in the software of our replacement control computer. For the Puma robot, we had to replace the entire PD controller with one of similar performance.

Let us consider the neuro-control structure for the Yaskawa robot in the remainder of this section; the Puma robot's control arrangements were very similar to this structure, except for the type of neural network used. The control diagram is shown in fig. 1. The desired joint values (displacements, velocities, and accelerations) are its inputs. In the feedback loop, there is the (partly copied) existing PD controller. A linear combination of joint displacement errors and velocity errors is used to train the neural network.

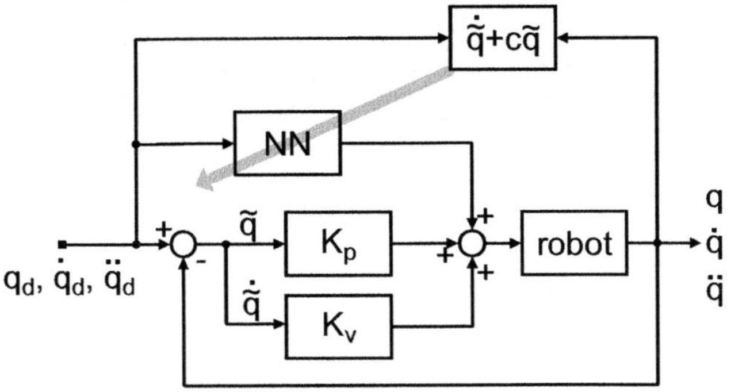

Fig. 1: Robot manipulator neuro-control architecture

One can see from figure 1, that we have implemented the system defined by equation 2 with the following neuro-control law, defined by the combination of a linear PD controller and a neural network

$$\tau = \mathrm{K}_p \tilde{\mathrm{q}} + \mathrm{K}_v \dot{\tilde{\mathrm{q}}} + \mathrm{G}(\mathrm{q}_d, \dot{\mathrm{q}}_d, \ddot{\mathrm{q}}_d)^T \hat{\mathrm{W}} \tag{4}$$

and that we have used the following on-line learning algorithm

$$\dot{\hat{\mathrm{W}}} = \Gamma \mathrm{G}(\mathrm{q}_d, \dot{\mathrm{q}}_d, \ddot{\mathrm{q}}_d)(\dot{\tilde{\mathrm{q}}} + c\tilde{\mathrm{q}}) \tag{5}$$

where the neural network matrix of non-linear mapping functions, $\boldsymbol{G}$, is the same as the one in equation 1, $\boldsymbol{\Gamma}$ is a constant positive definite matrix and is called the learning rate. We have already shown in [3,4] that the use of both the principal error signal *and* its first time-derivative, where they are both available, can greatly improve neural network learning rate. Further, if $\mathbf{K}_p$ and $\mathbf{K}_v$ are positive definite matrices and chosen to be sufficiently large, then the closed-loop manipulator system will be asymptotically stable and therefore $\lim_{t\to\infty} q = q_d$, $\lim_{t\to\infty} \dot{q} = \dot{q}_d$.

The proof of these statements is rather complicated. We use the Lyapunov method in that proof, to show that the whole system is asymptotically stable. Those interested in the details are referred to [3,4,12].

Fig. 2: Yaskawa robot

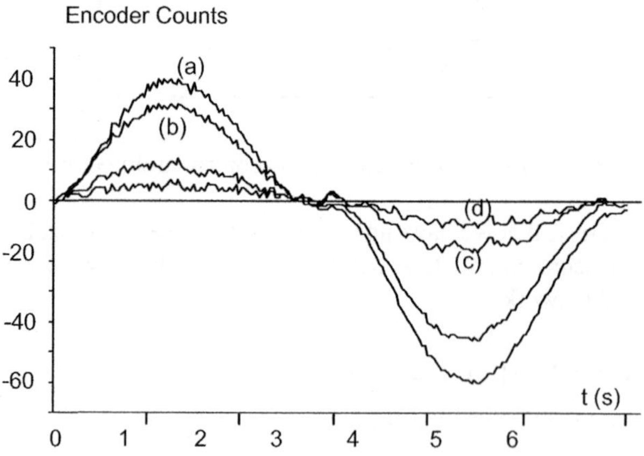

Fig. 3: The Trajectory Errors of the base joint. (a) Using the original controller; (b) Using the neuro-controller in the first trial; (c) Using the neuro-controller in the third trial; (d) Using the neuro-controller in the fifth trial

6 Neuro-Control of the Yaskawa Manipulator

The Yaskawa robot is shown in fig. 2. Three CMAC (Cerebellum Model Articulation Controller) neural network [1] controllers are used, one for each of the three major axes.

Each neural network utilises the same demand and feedback information available to the existing controller, and the output signals from the existing and neuro-controllers are combined by simple analogue voltage addition just before the power stages of the motor drive circuit. A neural network therefore simply adds its control effort to that of the existing controller in order to help reduce errors.

The control arrangement for one joint is as defined by equations 4 and 5, and depicted in fig. 1. A single PC-resident Texas Instruments DSP board, based around the 40MHz 320C40 processor, generates desired values for each joint. Each composite joint controller (linear PD and neural network) receives new position, velocity and acceleration demands every 2ms. As may be seen from the equation above and from fig. 1, the two term linear controller operates on only the position and velocity error signals to derive control actions. The position errors $(q_d - q)$ are calculated within the computer and added to the desired velocity.

Each CMAC has 32,768 adjustable weights. For a particular input set, 120 weights are selected for each output. Between each 2 millisecond sampling period and the next, the 320C40 must therefore complete one pass of the learning algorithm and a forward pass of the neural network for each CMAC. Over repeated trials, performance improves; though the bulk of this is obtained in the first five trials of a given trajectory.

Fig. 3 shows the base joint trajectory tracking errors. Line (a) shows the errors without the neural network. Line (b) shows the errors of the neuro-controller in the first trial. Lines (c) and (d) show the errors in the third and fifth trials. The tracking errors are finally reduced by more than six times. This improvement was the worst of the three axes, in general an improvement of six to ten times was observed across the base, shoulder and elbow joints.

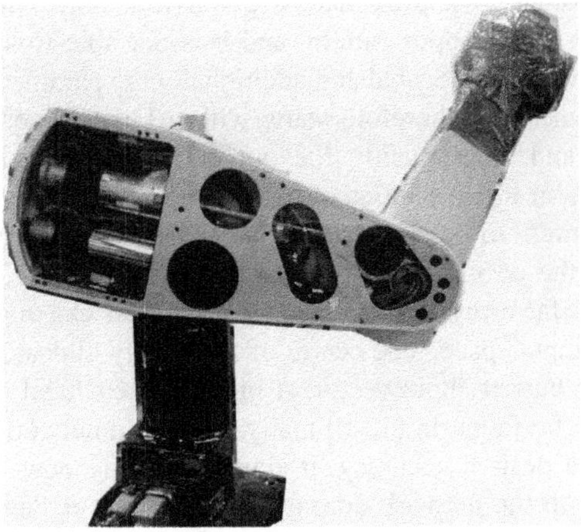

Figure 4: Puma 500

7 Neuro-Control of the Puma Manipulator

Figure 4 shows a photograph of a, partly dismantled, PUMA 500 manipulator. This "famous" robot arm, used in many industrial applications and research laboratories, was employed as a test-rig for the development of the second neuro-controller. In order to carry out this work, the robot was not used with its original control electronics from the manufacturer. Instead, a new control system, based on some pre-developed units and a 16-bit microcontroller, was designed and built. This was necessary because we did not have access to circuit diagrams for the controller, and so could not establish where to inject the neuro-controller's correcting signals. However, care was taken to ensure that dynamic performance of the new PD controller was similar to the original system. All control algorithms were written in the language C. Since the microcontroller was not very powerful (20MHz clock), with no maths co-processor, all algorithms were written using integer arithmetic. For these experiments we were interested in evaluating the other main representative of the class of Linear-Equivalent neural networks, the RBF network. In its standard form, unlike the CMAC network, this network suffers from a "curse of dimensionality" as the number of inputs increases. However, this is because the number of neurons increases as a power of the number of inputs if neurons are placed evenly across all input domains. A "placement on demand" algorithm can result in an appropriately sparse RBF structure, with neurons only placed where they are required.

8 A Dynamically Sized RBF Neural Network

John Platt [13] described a Resource-Allocating Network (RAN) for function interpolation. It is essentially a RBF neural network that starts off with no (Gaussian) nodes in the hidden layer when performing a supervised learning algorithm in order to approximate a given function. Whenever the network is presented with a new input pattern, and it is not able to approximate the desired output, a new node is allocated and additionally the parameters of existing units are adjusted. The network therefore starts with a blank slate. No nodes exist in the hidden layer and consequently the output is zero. When the network is then presented with an input, the output of the neural network y (which is always zero for the first pattern to be presented) is compared with the desired output y_d. If the magnitude of the error $E = (y_d - y)$ exceeds a threshold e_{min}, and additionally the input vector is far away from the center of nearest existing nodes, a new node is placed in the input space. The centre of this newly allocated node is placed at the position of the current input vector. A more mathematical expression of these two conditions can be found in [13]. Finally, when the network has learned the entire function to the desired accuracy, it stops allocating new nodes altogether. If no node is added to the network during a learning phase, "normal" gradient descent can be used to train the existing nodes in order to decrease the error.

Some extensions to this algorithm have been made, [5, 7,16]. Although these extended algorithms do enhance the performance of the RAN algorithm, they are not suitable for our research. There are two main reasons for this. Firstly, the reader has to bear in mind that the computations to implement a linear controller, a feed-forward pass of the neural network, *and* the training (including node allocation) of the neural network have to be performed in a short time-period, for the Puma controller 4ms or less. This constitutes a very high computational load for a 16-bit integer-only microcontroller, and does not allow implementation of the more computationally intense extended algorithms within the given timing constraints. Secondly, and more importantly, a RBF neural network that is trained by adjusting the centres and the width of Gaussian nodes does not belong to the class of Linear-Equivalent (Linear-In-the-Parameter) artificial neural networks. Consequently, proofs of stability and convergence [4,8] would no longer be valid. These proofs only allow adjustments in the weight values of the output layer. For these reasons the RAN specified by Platt was adopted, and designed into an on-line learning structure.

The inputs to the neural network (the RBF RAN) are the desired position and the desired velocity of the joint. As in the case of the Yaskawa control structure, this provides the network with bounded inputs. The position error and its time-derivative are used in linear combination as an error signal, again like the Yaskawa control structure.

The output of the RBF RAN injects a (potentially) correcting signal into the inner velocity-control loop of the cascaded controller. Although there are detailed implementational differences, the resulting overall control structure is very similar to figure 1 above.

The following experiment was designed to test whether the RBF RAN algorithm can be employed to compensate for predictable disturbances in the trajectory of the shoulder joint introduced by regular 'rapid' movements of the elbow joint. To

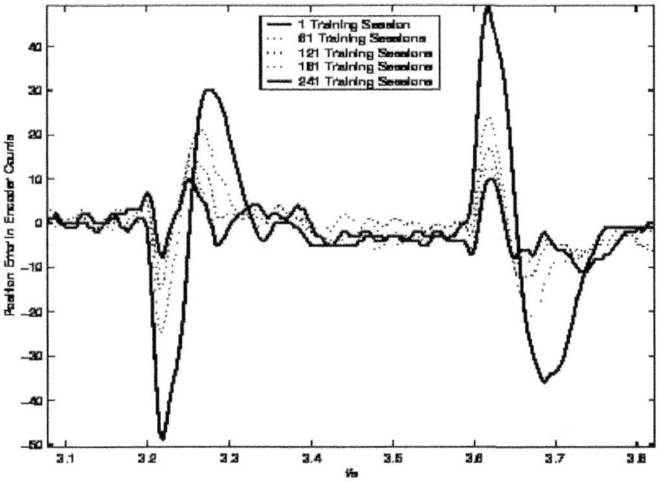

Figure 5: Position error of shoulder joint due to rapid acceleration of elbow, compensated by RBF RAN

further increase the impact of elbow joint movement on the shoulder joint, the 'wrist' area of the manipulator was loaded with approximately 7 kg of lead, wrapped around the manipulator and secured with packaging tape. First, we observed the impact of this on the position error of the shoulder joint when only the linear controller was used.

Two pairs of spikes were visible at approximately 3 seconds into the trajectory (see figure 5). A closer look at the spikes revealed that their duration is quite short (approximately 150 ms). In all other regions the position error was quite low. Testing of the RBF RAN architecture added to this control structure indicated a requirement for temporal adjustment in the learning algorithm. Clearly, when observing a position error at time t_n, this error has to have been caused by an event in the past; let us say at time t_{n-k}. Consequently, it makes much more sense to train a neural network with the error observed at time t_n on an input vector from the past (t_{n-k}). How far back in the past is obviously dependent on the overall system dynamics.

For this experiment the following RBF RAN parameters were used:

- 120 nodes per axis
- learning rate = 0.4
- spacing threshold = 275
- width = 160
- error threshold = 15
- learning delay "k-parameter" = 4 steps at 4ms per step

Figure 5 shows how the RBF RAN is able to reduce the position error of the shoulder-joint with the help of the training architecture. The peak magnitude of the position error is reduced from approximately 50 encoder counts to 10 encoder counts. The plot also shows that it took more than 200 training sessions to reach

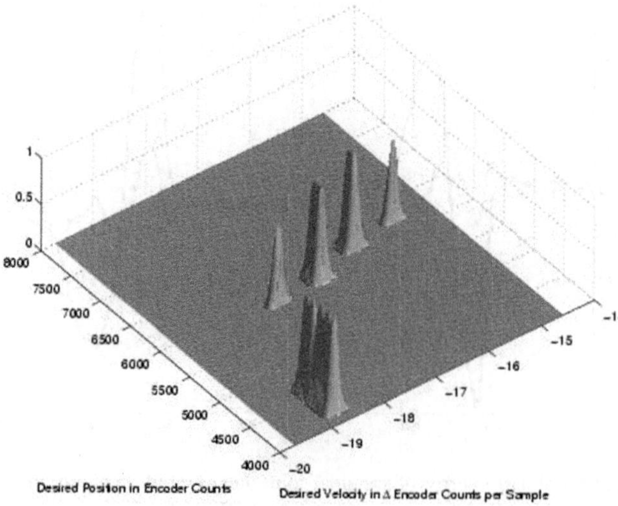

Figure 6: RBF node placement for RAN network

this result. This is mainly due to the conservative learning rate and the severity of the disturbances. Altogether, the system placed only 61 nodes in the input space. Compared to the number of nodes the system would need to place to cover the complete 2-dimensional input space with the same resolution ($160^2 = 25600$), this is remarkably low. This is illustrated graphically in figure 6.

9 Conclusions and Further Work

This experimental work illustrates that adaptive on-line learning neuro-control using Linear-Equivalent neural networks can achieve impressive improvements in accuracy; in both cases a five- to ten-fold improvement in trajectory tracking accuracy was achieved, compared with the robot's original joint controllers (or copies of them in this case). This is accomplished without using any detailed mathematical models of the plant, and using feedback signals already available. Further, it has been demonstrated that it is possible to implement such algorithms on low to medium performance computer architectures. The neuro-controller can normally be "strapped round" the existing control system. We believe that these developments could have an important impact on performance improvement for robot manipulator control systems, and for control systems generally, that are already "out in the field" in Manufacturing (and other) settings. Although the emphasis of this article is on experimental results from real plant, a crucial step in achieving acceptance of these Intelligent Control techniques is the firm establishment of proofs for learning convergence and system stability. Our approach to this is briefly described here, and covered in more depth in [3,4,12].

A robot manipulator is just one example of a complex system requiring accurate control. This research is therefore only a contribution towards the wider venture of applying neuro-control techniques in challenging new areas.

This research is already in a quite mature state, representing 10 years of work by our group overall. However, there is a clear item of detailed further work. The learning delay parameter k used in the Puma experiments reported on here, which is critical for performance and stability of the system, is not currently part of our neuro-controller design process; it needs to be incorporated into this process so that it also can also benefit from a sound theoretical background.

References

1 Albus J S, 1975, A new approach to manipulator control: the Cerebellar Model Articulation Controller (CMAC), J. Dynamics Systems, Measurement, & Control, Vol.97, pp.220-227

2 Jin Y, Pipe A G, Winfield A, 1994, Robot Trajectory Control using Neural Networks - Theory & Puma Simulations, Procs. IEEE Int. Symp. on Intelligent Control, Columbus Ohio, USA

3 Jin Y, 1994, Intelligent Neural Control & its Applications in Robotics, Ph.D. Thesis, University of the West of England

4 Jin Y, Pipe A G, Winfield A, 1997, Chapter 5: Stable Manipulator Trajectory Control Using Neural Networks, Neural Systems for Robotics, Academic Press, ISBN 0-12-526280-9
5 Kadirkamanathan, Niranjan, 1992, A function estimation approach to sequential learning with neural networks, Cambridge University, Eng. Dpt
6 Lenz A, Pipe A G, 2003, A Dynamically Sized Radial Basis Function Neural Network for Joint Control of a PUMA 500 Manipulator, Procs. IEEE Int. Symp. Intelligent Control, ISBN 0-7803-7892-X, pp.170-175
7 Leong, Saratchandran, Sundararajan, 2002, Real-time performance evaluation of the minimal radial basis function network for identification of time varying nonlinear systems, Computers and Electrical Engineering, vol. 28, pp.103-117
8 Lewis F L, Jagannathan S, Yesildirek A, 1999, Neural Network Control of Robot Manipulators & Nonlinear Systems, London, Taylor & Francis
9 Lutz, Wendt, 1998, Taschenbuch der Regelungstechnik, 2. Aufiage Verlag Harri Deutsch
10 Narendra K S and Parthasarathy K, 1990, "Identification and control of dynamical systems using neural networks", IEEE Trans. Neural Networks, vol. 1, pp.4-27
11 Narendra K S, 1992, Adaptive Control of Dynamical Systems Using Neural Networks, Handbook of Intelligent Control, pp.141-183, Van Nostrand Reinhold, New York
12 Pipe A G, Randall M J, Cherian R P, Jin Y, 2001, Experiments In On-Line Learning Neuro-Control, International Journal of Computer Research, 10(2), pp.269-295, Nova Science
13 Platt J, 1991, A Resource-Allocating Network for Function Interpolation, Neural Computing, vol. 3, no. 2, pp.213-225
14 Sanner R M and Slotine J J E, 1992, "Gaussian Networks for Direct Adaptive Control", IEEE Trans. Neural Networks, Vol.3, pp.837-863
15 Suykens, Van de Wall & De Moor, 1995, Artificial Neural Networks for Modelling & Control of Non-Linear Systems, Kluwer Academic
16 Yingwei, Sundararajan, Saratchandran, 1997, A sequential learning scheme for function approximation by using minimal radial basis function neural networks, Neural Computing, vol. 9,pp.462-478.

Chapter 7

Algorithm Development

New Strategies in Differential Evolution
Design Principle

Vitaliy Feoktistov and Stefan Janaqi

Laboratoire de Génie Informatique et d'Ingénierie de Production,
Site EERIE – l'École des Mines d'Alès,
Parc Scientifique Georges Besse, 30035, Nîmes, France.
{Vitaliy.Feoktistov, Stefan.Janaqi}@ema.fr

Abstract

Differential Evolution, a quite recent evolutionary optimization algorithm, is gaining more and more popularity among evolutionary algorithms. Proposed as a method for the global continuous optimization, Differential Evolution has been easily modified for mechanical engineering purposes and for handling nonlinear constraints. In this paper we introduce a new type of strategies which increase stability of the algorithm reducing its computational expenses. Also we propose a new principle of strategies' design. Theoretical discussions lead us to a tradeoff that helps to choose the better strategy. The strategies are illustrated, tested and compared on a classical test suite. We present a part of the testing results.

1 Introduction

Differential Evolution (DE) is a recently invented global optimization technique. It can be classified as an *iterative stochastic method*. Enlarging the evolutionary algorithms' group, DE turns out to be one of the best *population-based* optimizers [1]. Originally it has been proposed for continuous optimization [2]. Later, thanks to the work of J. Lampinen and I. Zelinka, DE becomes a universal method that can handle both mixed (continuous-integer-discrete) variables [3, 4] and nonlinear constraints [5, 6]. These authors used DE in order to minimize the steel wires volume necessary for manufacturing a spring with given characteristics. Among evolutionary algorithms DE shows the best results for mechanical engineering design problems [7, 8]. Also DE was applied for finding an optimal schedule of an Earth observing satellite [9].

In Section 2 we introduce DE in its typical form. A thorough analysis of the Differentiation operation is given in Section 3. This analysis helps to understand the differentiation principle leading us to a better choice of the strategy. The next Section 4 is devoted to the strategy design principle. Then, in Section 5 we examine in detail some examples of strategies. In Section 6 we test these strategies on the De Jong's functions' suite and compare them with the typical one. For a comparison of the typical DE with other evolutionary algorithms see the paper of R. Storn and K. Price [1].

2 Differential Evolution

In the following lines we give a brief description of the DE algorithm.

An optimization problem is represented by a set of variables. Let these variables form a *D*-dimensional vector in continuous space $X = (x_1, \ldots, x_D) \in \Re^D$. Let there be some criterion of optimization $f(X)$: $\Re^D \propto \Re$, usually named *fitness* or *cost* function. Then the goal of optimization is to find the values of the variables that minimize the criterion, *i.e.*, to find

$$X^* : f(X^*) = \min_X f(X) \tag{1}$$

Often, the variables satisfy boundary constraints

$$L \leq X \leq H : \quad L, H \in \Re^D \tag{2}$$

As all evolutionary algorithms, DE deals with a *population* of solutions. The population P of a generation *g* has *NP* vectors, so-called *individuals* of population. Each such individual represents a potential optimal solution.

$$\mathrm{P}^g = X_i^g, \quad i = 1, \ldots, NP \tag{3}$$

In turn, the individual contains *D* variables, so called *genes*.

$$X_i^g = x_{i,j}^g, \quad j = 1, \ldots, D \tag{4}$$

The population is *initialized* by randomly generating individuals within the boundary constraints,

$$\mathrm{P}^0 = x_{i,j}^0 = rand_{i,j} \cdot (h_j - l_j) + l_j \tag{5}$$

where *rand* function generates values uniformly in the interval [0,1].

Then, for each *generation* the individuals of a population are updated by means of a *reproduction* scheme. The reproduction of each individual *ind* needs a set of other individuals π randomly extracted from the population. To produce a new one the operations of *Differentiation* and *Recombination* are applied one after another. Next, *Selection* is used to choose the best. We briefly consider these operations.

Here, we show a typical model of *Differentiation*, others can be found in [2]. For that, three different individuals $\pi = \{\xi_1, \xi_2, \xi_3\}$ are randomly extracted from the population. So, the result, a *trial* individual, is

$$\tau = \xi_3 + F \cdot (\xi_2 - \xi_1), \tag{6}$$

where F > 0 is the constant of differentiation.

After, the trial individual τ is recombined with updated one *ind*. *Recombination* represents a typical case of a genes' exchange. The trial one inherits genes with some probability. Thus,

$$\omega_j = \begin{cases} \tau_j & \text{if} \quad rand_j < Cr \\ ind_j & \text{otherwise} \end{cases} \tag{7}$$

where j=1, …, D and $Cr \in [0,1)$ is the *constant of recombination*.

Selection is realized by comparing the cost function values of updated and trial individuals. If the trial individual better minimizes the cost function, then it replaces the updated one.

$$ind = \begin{cases} \omega & \text{if} \quad f(\omega) \le f(ind) \\ ind & \text{otherwise} \end{cases} \tag{8}$$

Notice that there are only three control parameters in this algorithm. These are *NP* – population size, *F* and *Cr* – constants of differentiation and recombination accordingly. As for the terminal conditions, one can either fix the number of generations g_{max} or a desirable precision of a solution *VTR* (*value to reach*).

The pattern of DE algorithm is presented in the following way (see Alg.1).

Algorithm 1 Typical Differential Evolution

Require: *F*, *Cr*, *NP* — control parameters

initialize $\mathrm{P}^0 \leftarrow \{ind_1, \ldots, ind_{NP}\}$
evaluate $f(\mathrm{P}^0)$
while (not terminal condition) **do**
    for all $ind \in \mathrm{P}^g$ **do**
        $\mathrm{P}^g \rightarrow \pi = \{\xi_1, \xi_2, \ldots, \xi_n\}$
            $\tau \leftarrow Differentiate(\pi, F)$
            $\omega \leftarrow Recombine(\tau, Cr)$
        $ind \leftarrow Select(\omega, ind)$
    end for
    $g \leftarrow g + 1$
end while

3 Analysis of Differentiation

Differentiation occupies a quite important position in the reproduction cycle. So, we try to analyze it in detail.

Geometrically, Differentiation consists in *two* simultaneous operations: the first one is the choice of a Differentiation's direction and the second one is the calculation of a step length in which this Differentiation performs. From the optimization point of view we have to answer the next two questions:

1. How to choose the optimal direction from all available ones?
2. What step length is necessary in order to better minimize the cost function along the chosen direction?

Let us remind that the principle of Differentiation is based on a random extraction of several individuals from the population and the geometrical manipulation of them.

Possible directions of Differentiation entirely depend on the disposition of extracted individuals. Also, their disposition influences the step length. Furthermore by increasing either the size of population or the number of extracted individuals we augment the diversity of possible directions and the variety of step lengths. Thereby we intensify the exploration of the search space. But on the other hand, the probability to find the best combination of extracted individuals goes considerably down.

Example.

We take the typical differentiation strategy $u = x_1 + F \cdot (x_2 - x_3)$, where for each current individual three other individuals are randomly extracted from the population.

- In the first case we suppose that the population consists only of *four* individuals (Fig. 1). So there are (4 - 1)(4 - 2)(4 - 3) = 3 ·2 ·1 = 6 possible directions and 6 possible step lengths. Imagine then that only one combination gives the best value of the cost function. Therefore the probability to find it, is 1/6.
- In the second case the population size is equal to *five* individuals. It gives (5 - 1)(5 - 2)(5 - 3) = 4 · 3 · 2 = 24 directions and as many step lengths. But, in this case, the probability to find the best combination is much less – 1/24.

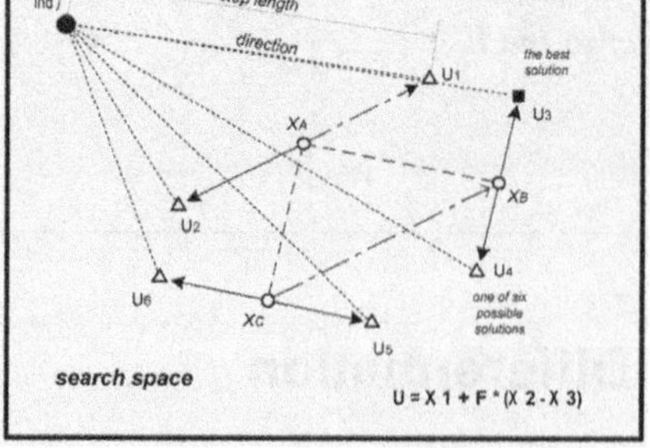

Figure 1. The illustration of the example

If we choose another strategy consisting of two randomly extracted individuals, $u = x_1 + F \cdot (x_2 - x_1)$ for example, then for the population size of five individuals the diversity of possible directions and step lengths is equal now to (5 - 1)(5 - 2) = 12 (two times less then in the previous case).

As we can see only two factors control the capability of the search space exploration. These are the population size *NP* and the number of randomly extracted individuals *k* in the strategy. In the case of the consecutive extraction of individuals the dependence of the potential individuals' diversity from both the population size and the number of extracted individuals is shown in the Formula 9 (Fig. 2.).

$$f(NP,k)=\prod_{i=1}^{k}(NP-i) \tag{9}$$

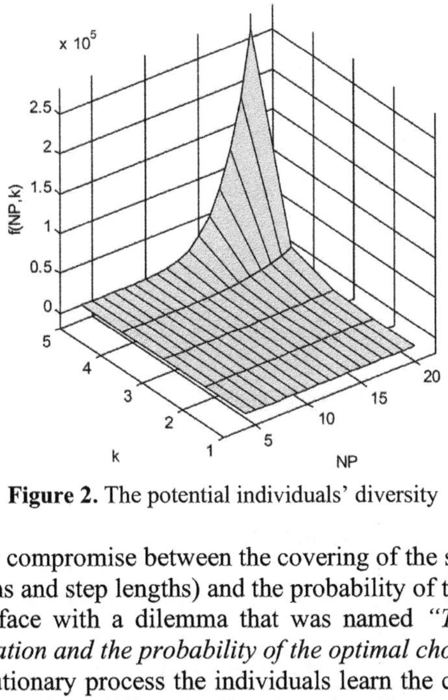

Figure 2. The potential individuals' diversity

But, where is the compromise between the covering of the search space (*i.e.* the diversity of directions and step lengths) and the probability of the best choice? This question makes us face with a dilemma that was named *"The Dilemma of the search space exploration and the probability of the optimal choice"*.

During the evolutionary process the individuals learn the cost function surface [10]. The step length and the difference direction adapt themselves accordingly. In practice, the more complex the cost function is, the more exploration is needed. The balanced choice of *NP* and *k* defines the efficiency of the algorithm.

4 New Principle of the Strategy Design

We introduce a new principle of the strategy design consisting of the strategies which use the values of the cost function of chosen individuals to determine a *"good"* differentiation direction. These strategies are called *directed*. By this way the probability of the optimal direction choice increases at least twice.

Let randomly extracted individuals X_i be divided into two classes C_+ and C_- with n_+ and n_- elements so that for each element from the class C_+ its cost function value would be less than the cost function value of any element from class C_-. That is,

$$\begin{aligned} &(\forall X_i \in C_+) \wedge (\forall X_j \in C_-) : f(X_i) \le f(X_j) \\ &i = 1, \ldots, n_+, \quad j = 1, \ldots, n_- \end{aligned} \tag{10}$$

We find then the *maximal* and *minimal* elements of each of the classes.

$$\begin{aligned} X_{C_\pm}^{\max} &: \forall X_i \in C_\pm \quad f(X_i) \le f(X_{C_\pm}^{\max}) \\ X_{C_\pm}^{\min} &: \forall X_i \in C_\pm \quad f(X_i) \ge f(X_{C_\pm}^{\min}) \end{aligned} \tag{11}$$

Next, we calculate the positive and negative *shifts* inside of the classes.

$$\begin{aligned} &V_s^{\pm} = \lambda \cdot (X_{C_\pm}^{\min} - X_{C_\pm}^{\max}) \\ &\lambda - 1/2 - \textit{influence constant} \end{aligned} \tag{12}$$

So the *average* is equal to

$$V_S = (V_s^+ + V_s^-)/2 \tag{13}$$

Thus, the new principle of strategy design, so called *RAND/DIR* principle, is

$$\omega = V_{C_+} + F \cdot (V_{C_+} - V_{C_-} + V_S), \tag{14}$$

where V_{C_+} and V_{C_-} are barycenters of the C_+ and C_- classes accordingly. They are calculated by Formula 15.

$$V_{C_\pm} = \frac{1}{n_\pm} \sum_{i=1}^{n_\pm} X_i \tag{15}$$

This principle is illustrated in Figure 3.

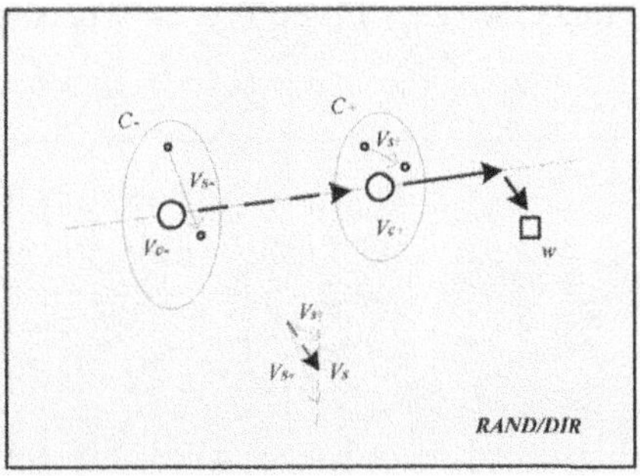

Figure 3. The new principle of the strategy design

5 Examples of the Strategies

Let us show some examples of the strategies designed with *RAND/DIR* template. Firstly we introduce a notation:

- *ind* – current individual;
- $\{X_i\}$ – set of extracted individuals;
- $V^* \in \{X_i\}$ is an individual which has the minimal value of the cost function among extracted individuals and the current one;
- V' and V'' are other individuals, if there are more than three extracted individuals we denote them V_1, V_2, V_3, ...;
- δ – difference vector;
- φ – particular constant of differentiation;
- ρ – base vector, the point of the difference vector application;

So the differentiation formula is represented as

$$\omega = \rho + \varphi \cdot \delta \tag{16}$$

Rand3/Dir2 strategy.
Three randomly extracted individuals X_1, X_2 and X_3 are used in this strategy. Taking into account that $f(V^*) \le f(V')$ and $f(V^*) \le f(V'')$, where V^*, V', $V'' \subseteq \{X_1, X_2, X_3\}$ are sorted according to their values of the cost function. The formula of the strategy is

$$\omega = V^* + \varphi \cdot (2V^* - V' - V'') . \tag{17}$$

Notice that the superposition of two conjugate directions (difference vectors) $(V^* - V')$ and $(V^* - V'')$ appears here. To adjust the differentiation formula with Template (14) imagine that the individuals form two barycenters $V_{C_-} = (V' + V'')/2$ and $V_{C_+} = V^*$. Thus, $\omega = V^* + 2\varphi \cdot (V^* - (V' + V'')/2) = V_{C_+} + 2\varphi \cdot (V_{C_+} - V_{C_-})$. There is no average shift vector in this case ($V_S = 0$). So $F = 2\varphi$. The strategy is shown on the Figure 4.

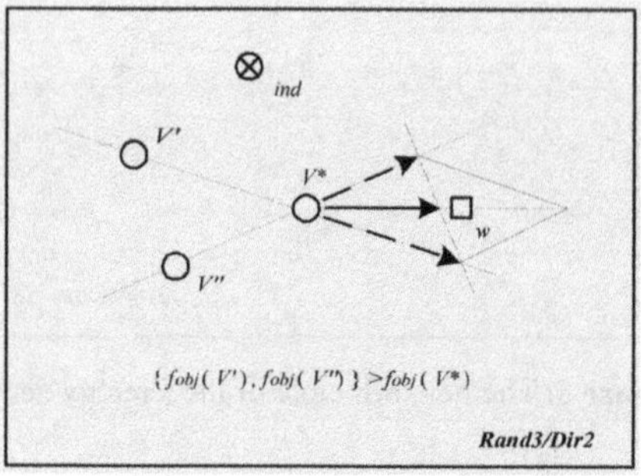

Figure 4. *Rand 3/Dir2* strategy

Rand3/Dir3 strategy.

This strategy completely repeats the previous one. In addition to it the average shift vector V_S is formed by comparing the cost function values of two other individuals V' and V''. Assume that $f(V^*) \le f(V') \le f(V'')$, then the differentiation formula is

$$\omega = V^* + \varphi \cdot \left(2V^* - \frac{1}{2}V' - \frac{3}{2}V'' \right). \tag{18}$$

Supposing $V_{C_-} = (V' + V'')/2$ and $V_{C_+} = V^*$ the shift vectors for each of the barycenters are equal to $V_s^+ = 0$ and $V_s^- = (V' - V'')/2$. Then the average shift $V_S = (V_s^+ + V_s^-)/2 = (V' - V'')/4$. By substituting it in the Template (14) we obtain the next: $\omega = V_{C_+} + F \cdot (V_{C_+} - V_{C_-} + V_S) = V^* + F \cdot \left(V^* - \frac{(V' + V'')}{2} + \frac{(V' + V'')}{4} \right)$ $= V^* + \frac{F}{2} \cdot \left(2V^* - \frac{1}{2}V' - \frac{3}{2}V'' \right)$. Thus, $\varphi = F/2$. This strategy is shown on Figure 5.

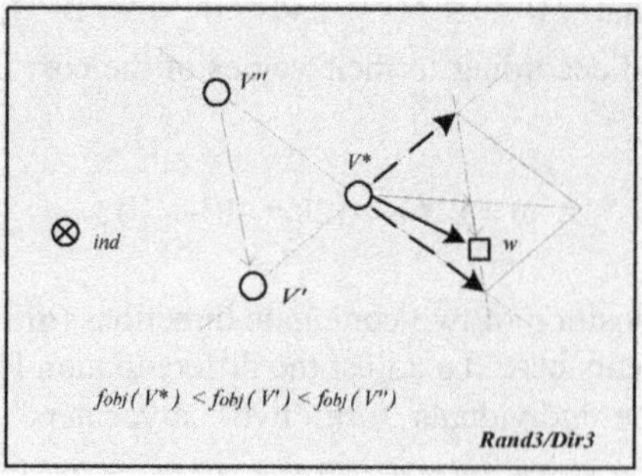

Figure 5. *Rand3/Dir3* strategy

Rand4/Dir2 strategy.

This strategy is based on the choice of four random individuals X_1, X_2, X_3 and X_4. Let us denote these individuals as V_1, V_2, V_3 and V_4 so that $f(V_1) \leq f(V_2)$ and $f(V_3) \leq f(V_4)$. Hence the differentiation formula is

$$\omega = V_1 + \varphi \cdot (V_1 - V_2 + V_3 - V_4). \tag{19}$$

It is obvious that $V_{C_+} = (V_1 + V_3)/2$ and $V_{C_-} = (V_2 + V_4)/2$. The only distinction is that the base point $\rho = V_1$ but not V_{C_+} as in the Template. Such a little difference allows to simplify the strategy without losing quality. Moreover it is easy to verify that $\varphi = F/2$ and $V_S = 0$. This strategy is shown on the Figure 6.

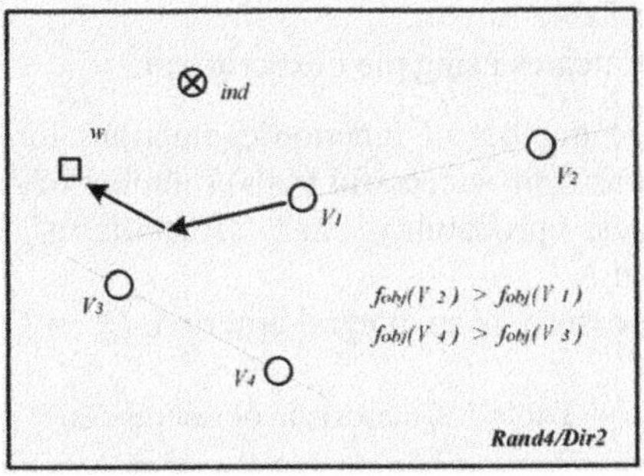

Figure 6. *Rand4/Dir2* strategy

6 Results

In order to test the strategies we choose four test functions (20) from a standard test suite for Evolutionary Algorithms [11]. These functions are in the De Jong's testbads [12]. The first one f_1, Sphere, is a "dream" of every optimization algorithm. It is smooth, unimodal and symmetric function. The performance on the Sphere function is a measure of the general efficiency of the algorithm. Whereas the second one f_2, Rosenbrock's function, is too complex. It has a very narrow ridge. The tip of the ridge is very sharp and it runs around a parabola. The Step function f_3 is a representative of the problem of flat surfaces. Flat surfaces are obstacles for optimization algorithms, because they do not give any information as to which direction is more favourable. Unless an algorithm has variable step sizes, it can get stuck at one of the flat plateaus. The fourth function f_4 is named Shekel's function or Foxholes. It has many (in this case 25) local optima. Many standard optimization algorithms get stuck in the first peak they find.

$$f_1(X)=\sum_{i=1}^{3}x_i^2 \qquad f_2(X)=100\left(x_1^2-x_2\right)^2+\left(1-x_1\right)^2$$

$$f_3(X)=\sum_{i=1}^{5}\lfloor x_i\rfloor \qquad f_4(X)=\frac{1}{0.002+\sum_{j=1}^{25}\frac{1}{c_j+\sum_{i=1}^{2}\left(x_i-a_{i,j}\right)^6}}\,. \tag{20}$$

We fixed the differentiation F and recombination Cr constants to be the same for all functions: $F = 0.5$; $Cr = 0$ (there is no recombination) in order to make the DE algorithm rotationally invariant [13, 10]. The population size $NP = 10 \cdot D$, where D is the dimension of the test problems. The terminal condition of the algorithm is a desirable precision of the optimal solution *VTR* (*value to reach*). It is fixed for all tests as $VTR = 10^{-6}$, except Foxholes ($VTR_4 = 0.998005$). The maximal number of generations g_{max} is also fixed to 10000. The average results of 10 runs are summarized in the Table 1.

We compare the strategies using the next criterions:

1. *Conv* – average number of function evaluations for successful trials, *i.e.*, (total evaluations from successful trials)/(number of successes).
2. *Prob* – success probability, *i.e.*, (fraction of successful trials) = successes)/(trials).
3. Q_m – quality measure[1] is an integral criterion, $Q_m = Conv/Prob$.

Table 1. Comparison of test results

function	*Strategy*	*Conv*	*Prob*	Q_m
Sphere f_1	typical	1088	1.0	1088.70
	Rand3/Dir2	849	0.7	1213.47
	Rand3/Dir3	939	1.0	939.7
	Rand4/Dir2	742	1.0	742.4
Rosenbrock's f_2	typical	525	0.6	875.00
	Rand3/Dir2	943	1.0	943.00
	Rand3/Dir3	448	1.0	448.50
	Rand4/Dir2	361	0.9	401.11
Step f_3	typical	6143	0.6	10238.89
	Rand3/Dir2	6053	0.3	20178.89
	Rand3/Dir3	10424	0.3	34746.67
	Rand4/Dir2	5817	0.9	6463.83
Shekel's f_4	typical	975	0.5	1950.40
	Rand3/Dir2	–	–	–
	Rand3/Dir3	477	0.3	1591.11
	Rand4/Dir2	392	0.1	3920.00

The tests show that the new strategies outperform the typical one. The augmentation of the number of conjugate directions increases the stability of the

[1] Q-measure was introduced thanks to personal communication with K. Price

algorithm (see Sphere function). Although the directed strategies reduce the search space exploration capability we can easily compensate it by simple increasing either the number of extracted individuals (conjugate directions) or the population size (see Rosenbrock's function). Furthermore, the use of these strategies may be efficient for some specific functions such as Step and Foxholes function and others.

We have made tests on a great number of functions with a large variety of control parameter's values. These tests empirically provide us the way of choosing the best strategy when solving a given problem. Since there is no theoretical result for the dilemma's solution (Section 3), we have adopted the preceding experimental approach.

7 Conclusion

The new principle of strategy design (RAND/DIR) permits to generate a range of strategies with directed Differentiation. These strategies increase the stability of the algorithm by imitating the gradient descent. The gradient, in this case, is constructed from randomly extracted individuals of the population. The analysis of Differentiation reveals the interrelation between the population size and the number of individuals participating in the strategy. The formulated Dilemma gives the possibility to catch the needed extent of the search space exploration. Further work is carried out to find a universal formula generalizing other types of strategies. Nevertheless, a theoretical work is necessary in order to derive convergence conditions of these strategies.

References

1 Storn R, Price K, (1996) Minimizing the Real Functions of the ICEC'96 Contest by Differential Evolution. In Nagoya, IEEE International Conference on Evolutionary Computation, IEEE, New York, NY, USA; 842-844.

2 Storn R, Price K, (1995) Differential Evolution – a Simple and Efficient Adaptive Scheme for Global Optimization over Continuous Spaces. Technical Report TR-95-012, International Computer Science Institute, Berkeley, CA.

3 Lampinen J, Zelinka I, (1999) Mixed Integer-discrete-continuous Optimization by Differential Evolution, Part 1: The Optimization Method. In Osmera P, (ed.), Proceedings of MENDEL'99, 5th International Mendel Conference on Soft Computing, Brno University of Technology, Faculty of Mechanical Engineering, Institute of Automation and Computer Science, Brno (Czech Republic) (9-12 June 1999); 71-76.

4 Lampinen J, Zelinka I, (1999) Mixed Integer-discrete-continuous Optimization by Differential Evolution, Part 2: A Practical Example. In Osmera P, (ed.), Proceedings of MENDEL'99, 5th International Mendel Conference on Soft Computing, Brno University of Technology, Faculty of Mechanical Engineering, Institute of Automation and Computer Science, Brno (Czech Republic) (9-12 June 1999); 77-81.

5 Lampinen J, (2001) Solving Problems Subject to Multiple Nonlinear Constraints by the Differential Evolution. In Matousek R, Osmera P, (eds.), Proceedings of MENDEL 2001, 7th International Mendel Conference on Soft Computing, Brno University of Technology, Faculty of Mechanical Engineering, Institute of Automation and Computer Science, Brno (Czech Republic) (6-8 June 2001); 50-57.

6 Lampinen J, (2002) A Constrain Handling Approach for the Differential Evolution Algorithm. In Fogel DB, EL-Sharkawi MA, Yao X, Greenwood G, Iba H, Marrow P, Shackleton M, (eds.), Proceedings of the 2002 Congress on Evolutionary Computation CEC2002, IEEE Press; 1468-1473.

7 Lampinen J, Zelinka I, (1999) Mechanical Engineering Design Optimization by Differential Evolution. In Corne D, Dorigo M, Glover F, (eds.), New Ideas in Optimization, McGraw-Hill, London; 127-146.

8 Lampinen J, (1999) Differential Evolution – New Naturally Parallel Approach for Engineering Design Optimization. In Topping BH, (ed.), Developments in Computational Mechanics with High Performance Computing, Edinburgh (Scotland), Civil-Comp Press; 217-228.

9 Feoktistov V, Janaqi S, (2003) Gestion de Mission des Satellites d'Observation avec l'Évolution Différentielle. In 5-ème Congrès de la Société Française de Recherche Opérationnelle et d'Aide à la Décision – ROADEF 2003, Université d'Avignon et des Pays de Vaucluse; 228-230

10 Price K, (2003) New Ideas in Optimization, Part 2: Differential Evolution, McGraw-Hill, London, UK.

11 Whitley D, Rana SB, Dzubera J, Mathias KE, (1996) Evaluating Evolutionary Algorithms. Artificial Intelligence, **85**; 245-276.

12 Jong KAD, (1975) Analysis of the Behavior of a Class of Genetic Adaptive Systems. PhD Thesis, University of Michigan.

13 Salomon R, (1996) Re-evaluating Genetic Algorithm Performance under Coordinate Rotation of Benchmark Functions: A Survey of some Theoretical and Practical Aspects of Genetic Algorithms. BioSystems, **39**; 263-278.

Balanced Random and Adaptive Interval Arithmetic for Systems of Linear Interval Equations

J. Žilinskas[1], I. D. L. Bogle[2]

[1]Faculty of Informatics,
Kaunas University of Technology,
Studentu 50-214b, LT-3031, Kaunas, Lithuania
Julius.Zilinskas@ktu.lt

[2]Centre for Process Systems Engineering, Department of Chemical Engineering,
University College London,
Torrington Place, London WC1E 7JE, UK
d.bogle@ucl.ac.uk

Abstract

The paper concerns interval methods – valuable tools for solving engineering problems – for finding outer approximations for the solution set of systems of linear interval equations. The paper shows how interval methods are used to analyse steady-state concentrations of systems of coupled reactors. The results of the experiments evaluating the outer approximations for the solution set of systems of linear interval equations using Gaussian elimination with standard and balanced random interval arithmetic are given. Adaptive interval arithmetic is proposed to overcome disadvantages of balanced random interval arithmetic.

1 Introduction

Interval arithmetic proposed in [1] is a valuable tool in engineering. Global optimisation using interval analysis [2] is a popular approach in process engineering [3], [4]. Uncertain values may be modelled using intervals [5] exploiting interval arithmetic as a tool for safety verification. In this paper we show how interval arithmetic may be used to solve a system of linear equations in which coefficients and right hand side values are unknown numbers ranging in certain intervals, for example when they are uncertain or are not exact and their ranges are known.

It is often necessary to find the solution of a system of linear equations in engineering applications. The examples of such applications [6] are in the analysis of steady-state concentrations of a system of reactors in chemical/biochemical engineering; determining stoichiometry of complex reaction systems; scheduling of blending and pooling operations at oil refineries; analysis of a statically determinate trusses in civil/environmental engineering; finding currents and voltages in resistor circuits in electrical engineering; solving spring-mass systems in mechanical/aerospace engineering.

However measured values never have infinite precision and the enclosures of the values of unknowns should be found instead of the real values with unknown accuracy. We demonstrate this using steady-state analysis of a system of reactors when measured flow rates and concentrations of system inputs are not exact because of measurement inaccuracy. Conservation of mass is employed to determine the steady-state concentrations of the system of reactors resulting in a system of linear interval equations with an interval coefficient matrix and an interval right hand side vector. The problem is solved finding interval vector enclosures for the solution set of a system of linear interval equations.

The guaranteed enclosure of the solution set of systems of linear interval equations may be found using the interval version of the classical real valued Gaussian elimination algorithm. A disadvantage of standard interval arithmetic is the dependency problem which makes it difficult to obtain tight intervals.

Balanced random interval arithmetic proposed in [7] tries to estimate tighter intervals at a cost of losing the guarantee of enclosure but with high probability of success. The disadvantage of balanced random interval arithmetic is that computations with this arithmetic uses more calculations than with standard interval arithmetic and that it is assumed that distributions of centres and radii of the evaluated balanced random intervals are normal which is not always true. To overcome the disadvantages of balanced random interval arithmetic, adaptive interval arithmetic is proposed.

We start with a background in interval arithmetic in section 2 and present standard interval and balanced random interval arithmetic. In section 3 we present systems of linear interval equations and show how different are the interval vector hull of actual solution set and the estimates of enclosure of the solution set found using Gaussian elimination with standard and balanced random interval arithmetic. The advantage of balanced random interval arithmetic is shown. In section 4 we show an example in chemical engineering where steady-state concentrations of a system of reactors should be found when measured flow rates and concentrations of system inputs are not exact. It is shown why interval methods should be used when coefficients of equations are not exact. In section 5 we propose a new adaptive interval arithmetic which we believe can help to overcome disadvantages of balanced random interval arithmetic. We conclude the paper with conclusions and acknowledgements in sections 6 and 7.

2 Balanced Random Interval Arithmetic

Interval arithmetic was proposed by Moore in [1]. Interval arithmetic operates with real intervals $[x]=[x_1,x_2]=\{x\in\Re|x_1\le x\le x_2\}$, defined by two real numbers $x_1\in\Re$ and $x_2\in\Re$, $x_1\le x_2$. For any real arithmetic operation $x\bullet y$ the corresponding interval arithmetic operation $[x]\bullet[y]$ is defined:

$$[x]\bullet[y]=\{x\bullet y|x\in[x], y\in[y]\}.$$

For example, the standard interval addition and subtraction operations are defined as:

$$[x]+[y]=[x_1,x_2]+[y_1,y_2]=[x_1+y_1,x_2+y_2],$$
$$[x]-[y]=[x_1,x_2]-[y_1,y_2]=[x_1-y_2,x_2-y_1].$$

The guaranteed lower and upper bounds for the function values in the region defined by intervals of variables can be estimated applying standard interval operations with intervals instead of real operations in the algorithm to calculate the function values. The evaluated bounds always include the full range of the function values in the defined region:

$$\{f(\mathbf{x})|\mathbf{x}\in[\mathbf{x}]\}\subseteq[f]([\mathbf{x}]),$$

where $f:\Re^n\to\Re$ is a function of real variables, $[f]:[\Re,\Re]^n\to[\Re,\Re]$ is a function of interval variables.

A disadvantage of standard interval arithmetic is the dependency problem: when a given variable occurs more than once in interval computation, it is treated as a different variable in each occurrence. This causes widening of computed intervals making it more difficult to obtain tight intervals. For example:

$$[x]-[x]=[x_1,x_2]-[x_1,x_2]=[x_1-x_2,x_2-x_1]\neq[0,0].$$

Inner interval arithmetic assumes that all operands are dependent and exchange the ends of one operand in every operation

$$[x]\bullet_{in}[y]=[x_1,x_2]\bullet[y_2,y_1],$$

for example

$$[x]+_{in}[y]=[x_1,x_2]+[y_2,y_1]=[x_1+y_2,x_2+y_1],$$
$$[x]-_{in}[y]=[x_1,x_2]-[y_2,y_1]=[x_1-y_1,x_2-y_2].$$

Using inner interval arithmetic

$$[x]-_{in}[x]=[x_1,x_2]-[x_2,x_1]=[x_1-x_1,x_2-x_2]=[0,0].$$

However in the general case the evaluated inner interval is tighter than the full range of the function values.

If it is known that operands in subtraction or division operation are dependent or are monotonic and have the same monotonicity inner interval operations may be used instead of standard interval operations. If it is known that operands in summation or multiplication are monotone and do not have the same monotonicity inner interval operations may be used instead of standard interval operations. The resulting interval would be tighter without losing the guarantee that the result includes the full range of the function values in the defined region. However it is difficult to find whether operands are dependent and to find their monotonicity.

Balanced random interval arithmetic proposed in [7] extending the ideas of [8], is obtained by choosing standard and inner interval operations at each step of the computation randomly with the predefined probabilities of standard and inner operations. A number of sample intervals are evaluated using balanced random interval arithmetic. It is assumed that the distributions of centres and radii of the evaluated balanced random intervals are normal. An approximate range of the function values in the defined region is evaluated using the mean values and the standard deviations of centres and radii of the evaluated balanced random intervals:

$$[\mu_{centres}\pm(3.0\sigma_{centres}+\mu_{radii}+3.0\sigma_{radii})].$$

If the evaluated range exceeds the standard interval range of the function values in the same region, the evaluated interval is intersected with the standard interval while not reducing the probability that the evaluated range contains all the function values in the region.

The result of balanced random interval arithmetic is equal to the result of standard interval arithmetic when the predefined probability of standard interval operations is equal to 1 and the probability of inner interval operations is equal to 0. The result of balanced random interval arithmetic is equal to the result of inner interval arithmetic when the predefined probability of standard interval operations is equal to 0 and the probability of inner interval operations is equal to 1. Balanced random interval arithmetic provides wider or narrower bounds depending on the predefined probabilities. The values used for the predefined probabilities depend on the balance required between tightness of resulting intervals and the probability that resulting intervals contain the full range of the function values.

For example, the ranges of a multidimensional scaling function (used in global optimization) with data from soft drinks testing in random sub-regions of the feasible region have been estimated using balanced random interval arithmetic and results are reported in [9]. It has been shown that when the predefined probability of standard interval operations is equal to 0.6, the estimated ranges of 99.5% of the sub-regions enclose the function values at uniformly distributed random points and the estimated ranges are 39.4% tighter than the ranges estimated using standard interval arithmetic.

3 Systems of Linear Interval Equations

A system of linear interval equations with interval coefficient matrix [**A**] and interval right hand side vector [**b**] is defined as the family of systems of linear equations in which the coefficients are unknown numbers ranging in certain intervals:

$$\{\mathbf{Ax}=\mathbf{b} \mid \mathbf{A}\in[\mathbf{A}],\mathbf{b}\in[\mathbf{b}]\}.$$

The solution set of system of linear interval equations, given by

$$\{\mathbf{x}\in\Re^n \mid \mathbf{Ax}=\mathbf{b},\mathbf{A}\in[\mathbf{A}],\mathbf{b}\in[\mathbf{b}]\}.$$

usually is not an interval vector and has a very complicated structure [10]. In this paper we are interested in finding interval vector enclosures (outer approximations)

$$[\mathbf{x}]\supseteq\{\mathbf{x}\in\Re^n \mid \mathbf{Ax}=\mathbf{b},\mathbf{A}\in[\mathbf{A}],\mathbf{b}\in[\mathbf{b}]\}.$$

for the solution set of system of linear interval equations.

To find the outer approximation of the solution set of systems of linear interval equations exactly, 2^k, where k is number of interval coefficients, systems of linear equations with real coefficients should be solved. If all the coefficients of the system of equations are intervals, $2^{n(n+1)}$ systems of linear equations with real coefficients should be solved, where n is number of equations.

Some contractors used to reduce the outer approximation of the solution set of system of linear interval equations are presented in [11]. We will use the Gaussian elimination contractor

$$[f_{GE}]([\mathbf{A}],[\mathbf{x}],[\mathbf{b}])=[\mathbf{x}]\cap[\phi]([\mathbf{A}],[\mathbf{b}]),$$

which is the interval version of the classical real valued Gaussian elimination algorithm, to find the outer approximations of the solution set of systems of linear interval equations. As an example system of linear interval equations we will use the example from [11]:

$$[\mathbf{A}]=\begin{bmatrix} [4,5] & [-1,1] & [1.5,2.5] \\ [-0.5,0.5] & [-7,-5] & [1,2] \\ [-1.5,-0.5] & [-0.7,-0.5] & [2,3] \end{bmatrix}, [\mathbf{b}]=\begin{bmatrix} [3,4] \\ [0,2] \\ [3,4] \end{bmatrix}. \quad (1)$$

The outer approximation of the solution set found by Gaussian elimination with standard interval arithmetic is

$$[f_{GE}]([\mathbf{A}],[\mathbf{x}],[\mathbf{b}])=\begin{bmatrix} [-1.819278874340440, & 1.168720384766975] \\ [-0.4140698379116950, & 1.725229626734419] \\ [\ 0.7002320584443479, & 3.420754348250934] \end{bmatrix}.$$

The repetition of the Gaussian elimination contractor does not reduce the approximation any more, therefore the contractor is said to be idempotent.

Gaussian elimination with balanced random interval arithmetic was used to evaluate the outer approximation of the solution set of example systems of linear interval equations (1). The balanced random interval arithmetic assumption that the distributions of centres and radii of the evaluated balanced random intervals are normal has been verified experimentally. Therefore, the outer approximation of the solution set of system of linear interval equations could be evaluated using Gaussian elimination with balanced random interval arithmetic.

The numerical results are summarised in Table 1 where the interval vector hulls of the actual solution set of systems of linear interval equations found solving 2^{12} real systems of linear equations and of the solutions of 10000 real systems of linear equations with real random coefficients uniformly distributed in interval coefficients, and outer approximations of the solution set found using Gaussian elimination with standard and balanced random interval arithmetic with the predefined probability of standard interval operations equal to 0.6 and 30 samples are given. The volume ratios are also given.

Table 1: Interval vector hulls and outer approximations of the solution set of example system of linear interval equations (1)

	Interval vector hull or outer approximation	vol. ratio
Actual solution set	$\begin{bmatrix} [-0.8095537397862980, & 0.6854663774403473] \\ [-0.2566259991586033, & 1.066982809721399] \\ [\ 0.9624398073836273, & 2.593953764078246] \end{bmatrix}$	0.186
Solutions with random coefficients	$\begin{bmatrix} [-0.3912270391600173, & 0.5406455552601220] \\ [-0.1498627227594909, & 0.7584771653940147] \\ [\ 1.041742733439600, & 2.189924178539334] \end{bmatrix}$	0.056
Solution with standard interval arithmetic	$\begin{bmatrix} [-1.819278874340440, & 1.168720384766975] \\ [-0.4140698379116950, & 1.725229626734419] \\ [\ 0.7002320584443479, & 3.420754348250934] \end{bmatrix}$	1
Solution with balanced random interval arithmetic	$\begin{bmatrix} [-1.194597036856818, & 1.168720384766975] \\ [-0.4140698379116950, & 1.329151515285030] \\ [\ 0.7002320584443479, & 3.238219680821866] \end{bmatrix}$	0.601

The volume of the outer approximation of the solution set of systems of linear interval equations found using Gaussian elimination with standard interval arithmetic is more than 5 times larger than the volume of the interval vector hull of the actual solutions set, more than 17 times larger than the volume of the interval vector hull of the solutions of systems with random coefficients and 1.66 times larger than the volume of the outer approximation of the solution set found using Gaussian elimination with balanced random interval arithmetic.

Gaussian elimination with balanced random interval arithmetic with different predefined probabilities of standard interval operations was used to evaluate the outer approximations of the solution set of example system of linear interval equations (1). The outer approximations were evaluated using 30 samples of interval outer approximations for $[x_1]$, $[x_2]$ and $[x_3]$. For each value of predefined probability 1000 outer approximations were evaluated. The mean volume reductions for each value of predefined probability were evaluated dividing the mean volume of the outer approximations evaluated using Gaussian elimination with balanced random interval arithmetic by the volume of the outer approximation evaluated using Gauss elimination with standard interval arithmetic. The enclosure ratios of the actual solution set and of the solutions with random coefficients were evaluated for each value of predefined probability showing how many outer approximations evaluated using Gaussian elimination with balanced random interval arithmetic enclose the actual solution set and how many of them enclose the solutions of 10000 systems with random coefficients. The volume reduction and enclosure ratios depending on predefined probability of standard interval arithmetic operations are shown in Figure 1.

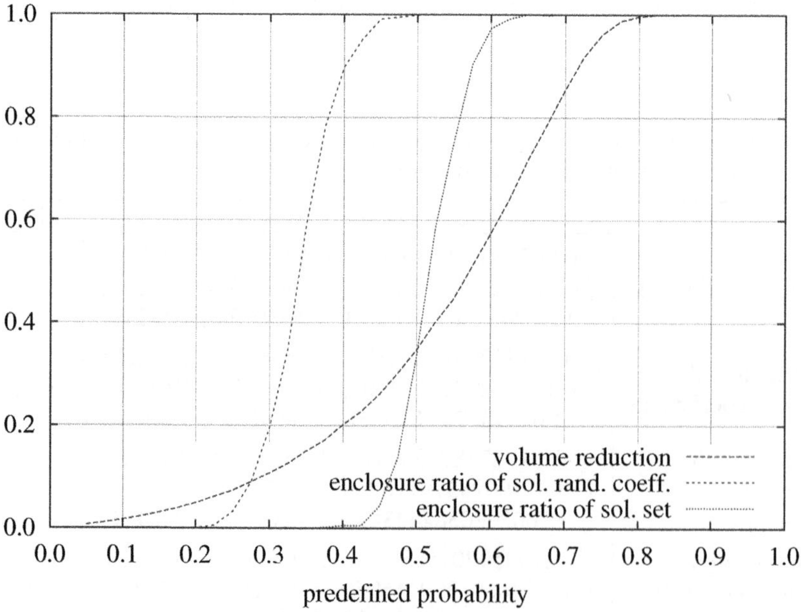

Figure 1: The volume reduction and enclosure ratios depending on predefined probability of standard interval arithmetic operations

The enclosure ratio of the actual solution set of the example system of linear interval equations is more than 0.999 when the predefined probability of standard interval operations is equal to or larger than 0.65. When the predefined probability of standard interval operations is equal to 0.65 the mean volume reduction is 0.72. The outer approximations of the solution set of system of linear interval equations evaluated using Gaussian elimination with balanced random interval arithmetic enclose the actual solution set with probability larger than 0.999 when the mean volume is up to 28% smaller than the volume of the outer approximation evaluated using Gaussian elimination with standard interval arithmetic.

The enclosure ratio of the actual solution set of the example system of linear interval equations is more than 0.97 when the predefined probability of standard interval operations is equal to or larger than 0.6. When the predefined probability of standard interval operations is equal to 0.6 the mean volume reduction is 0.57. The outer approximations of the solution set of system of linear interval equations evaluated using Gaussian elimination with balanced random interval arithmetic enclose the actual solution set with probability larger than 0.97 when the mean volume is up to 43% smaller than the volume of the outer approximation evaluated using Gaussian elimination with standard interval arithmetic.

The enclosure ratio of the solutions with random coefficients is more than 0.999 when the predefined probability of standard interval operations is larger than 0.50. When the predefined probability of standard interval operations is equal to 0.50 the mean volume reduction is 0.35. The outer approximations of the solution set of system of linear interval equations evaluated using Gaussian elimination with balanced random interval arithmetic enclose with probability larger than 0.999 all the solutions of 10000 systems of linear equations with random coefficients when the mean volume is up to 65% smaller than the volume of the outer approximation evaluated using Gaussian elimination with standard interval arithmetic.

4 An Example in Chemical Engineering

Systems of linear interval equations may be applied when coefficients or/and right hand side values are unknown numbers ranging in certain intervals, for example when they are uncertain or are not exact but their ranges are known.

We will demonstrate the application using a chemical engineering application – steady-state analysis of a system of reactors. Five reactors linked by pipes [6] are shown in Figure 2. Flow rates Q_{ij} (in cubic meters per minute) and concentrations of system inputs c_{0i} (in milligrams per cubic meter) are shown. Conservation of mass is employed to determine the steady-state concentrations c_i (in milligrams per cubic meter) of the system of coupled reactors. Five simultaneous mass-balance equations are used to characterise the system. They may be expressed in matrix form as system of linear equations **Ac**=**b**, where

$$\mathbf{A}=\begin{bmatrix} Q_{12}+Q_{15} & 0 & -Q_{31} & 0 & 0 \\ -Q_{12} & Q_{23}+Q_{24}+Q_{25} & 0 & 0 & 0 \\ 0 & -Q_{23} & Q_{31}+Q_{34} & 0 & 0 \\ 0 & -Q_{24} & -Q_{34} & Q_{44} & -Q_{54} \\ -Q_{15} & -Q_{25} & 0 & 0 & Q_{54}+Q_{55} \end{bmatrix}, \mathbf{b}=\begin{bmatrix} Q_{01}c_{01} \\ 0 \\ Q_{03}c_{03} \\ 0 \\ 0 \end{bmatrix}. \qquad (2)$$

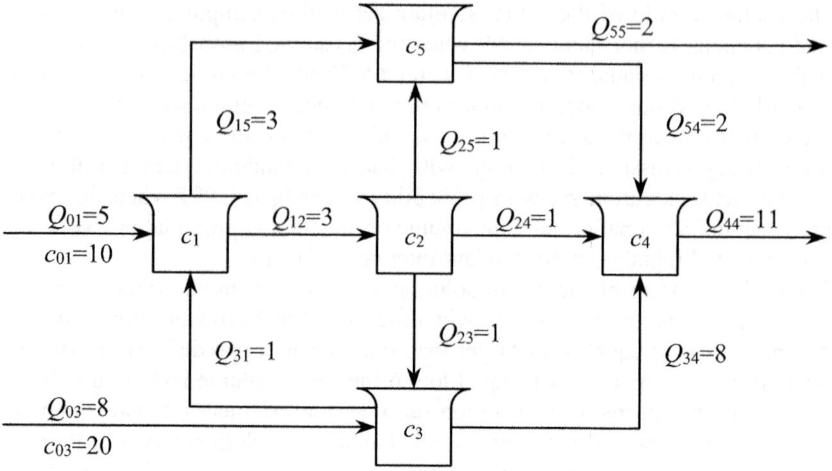

Figure 2: System of interconnected reactors

Real numerical methods can be used to solve this system of linear equations to determine the real valued steady-state concentrations (in milligrams per cubic meter) of the system of coupled reactors:

$$\mathbf{c}=\begin{bmatrix}11.50943396226414\\11.50943396226413\\19.05660377358488\\16.99828473413374\\11.50943396226412\end{bmatrix}.$$

However measured values never have infinite precision. Possible measurement accuracy of flow rates is ±0.5m³/min (between ±5% and ±50%). Possible measurement accuracy of concentrations is ±5mg/m³ (between ±25% and ±50%). In this case, measured values of flow rates and concentrations of the example may correspond to actual values Q_{01}=4.7, c_{01}=5, Q_{03}=8.3, c_{03}=24, Q_{15}=2.8, Q_{12}=3.2, Q_{31}=1.3, Q_{25}=1.2, Q_{24}=1.3, Q_{23}=0.7, Q_{55}=2.3, Q_{54}=1.7, Q_{44}=10.7, Q_{34}=7.7. The real valued steady-state concentrations (in milligrams per cubic meter) of the system of coupled reactors determined with these values

$$\mathbf{c}=\begin{bmatrix}8.861555848559036\\8.861555848559032\\22.82256545488790\\18.90826369610409\\8.861555848559027\end{bmatrix}.$$

are quite different than those found with integer data.

The values of flow rates and input concentrations may be specified as intervals and used in (2) formulating the system of linear interval equations with coefficients

$$[\mathbf{A}]=\begin{bmatrix}[5,7] & [0,0] & [-1.5,-0.5] & [0,0] & [0,0]\\ [-3.5,-2.5] & [1.5,4.5] & [0,0] & [0,0] & [0,0]\\ [0,0] & [-1.5,-0.5] & [8,10] & [0,0] & [0,0]\\ [0,0] & [-1.5,-0.5] & [-8.5,-7.5] & [10.5,11.5] & [-2.5,-1.5]\\ [-3.5,-2.5] & [-1.5,-0.5] & [0,0] & [0,0] & [3,5]\end{bmatrix},\ [\mathbf{b}]=\begin{bmatrix}[22.5,82.5]\\ [0,0]\\ [112.5,212.5]\\ [0,0]\\ [0,0]\end{bmatrix}.$$

Gaussian elimination with standard and balanced random interval arithmetic was used to evaluate the outer approximations of the solution set of system of linear interval equations to determine the intervals of steady-state concentrations of system of coupled reactors. The value of predefined probability of standard interval operations used for balanced random interval arithmetic is 0.65, and 30 samples were used. The results are shown in Table 2 where the interval vector hull of the solutions of 10000 systems of linear equations with real random coefficients uniformly distributed in interval coefficients and outer approximations of the solution set found using Gaussian elimination with standard and balanced random interval arithmetic are given. The volume ratios are also given. The volume of the outer approximation of the solution set of system of linear interval equations found using Gaussian elimination with standard interval arithmetic is more than 10 times larger than the volume of the interval vector hull of the solutions of systems with random coefficients and approximately twice larger than the volume of the outer approximation of the solution set found using Gaussian elimination with balanced random interval arithmetic.

Table 2: Interval vector hull and outer approximations of the solution set of system of linear interval equations for system of coupled reactors

	Interval vector hull or outer approximation	vol. ratio
Solutions with random coefficients	[4.360931013379671, 22.82629347452528] [3.041963813107369, 47.48411039388054] [11.77562473469834, 31.15880083780083] [9.273636159941116, 31.70379377435785] [3.103584133781611, 38.47848141555886]	0.097
Solution with standard interval arithmetic	[4.025844930417491, 28.16546762589931] [2.236580516898604, 65.71942446043180] [11.36182902584492, 38.88489208633099] [7.798859019794264, 56.51421719767059] [2.236580516898603, 65.71942446043182]	1
Solution with balanced random interval arithmetic	[4.025844930417491, 28.16546762589931] [2.236580516898604, 65.71942446043180] [11.36182902584492, 37.83444663330012] [7.798859019794264, 38.86590657347641] [2.236580516898603, 56.66961202717653]	0.526

The results of the experiments also show that if measurement accuracy of flow rates is $\pm 0.5 m^3/min$ and measurement accuracy of concentrations of inputs is $\pm 5 mg/m^3$, the real valued steady-state concentrations of system of coupled reactors found using real numerical methods are useless because steady-state concentrations may be in much larger intervals than the intervals of coefficients. Accuracy of the real valued solutions are worse than $\pm 10 mg/m^3$ and some of them are worse by $\pm 50 mg/m^3$. Interval methods for systems of linear equations should be used when coefficients of equations are not exact which is always the case in practical applications.

5 Adaptive Interval Arithmetic

The disadvantage of balanced random interval arithmetic is that computations with this arithmetic uses more calculations than with standard interval arithmetic, because the number of sample balanced random intervals should be evaluated. Another disadvantage is the assumption that distributions of centres and radii of the evaluated balanced random intervals are normal. The assumption is not true when computations involve a small number of arithmetic operations, as was shown in [9] for simple mathematical global optimization test functions.

To overcome the disadvantages only one sample interval should be evaluated. In this case computations would use similar number of calculations as standard interval arithmetic and the assumption would be not necessary because only one sample interval would be used to evaluate the result. To get results tighter than standard intervals, inner interval arithmetic operations should be used instead of some standard interval arithmetic operations. The decision which interval arithmetic operation should be used may be not random with predefined probabilities as when using balanced random interval arithmetic, but adaptive to the function evaluated. In this case the adaptation procedure may be run before the computations to decide which operations in the algorithm to calculate the function values should be implemented as inner interval operations instead of standard interval operations. After the adaptation procedure computations are performed similarly as for standard interval arithmetic, but with some operations changed to inner interval operations. If standard interval arithmetic produces results very similar to the exact range of function values as is the case for some simple mathematical global optimization test functions, the adaptation procedure may decide not to change any operations and use only standard interval operations.

The adaptation procedure was formulated as an integer programming problem. The number of variables in the problem is equal to the number of interval operations in the interval algorithm necessary to evaluate the ranges of function or to find the outer approximation of the solution set of system of linear interval equations. The values of variables show where standard and where inner interval operations should be used in the interval algorithm. The width of ranges of the function values or the volume of the outer approximation is minimised subject to the enclosure constraint (ranges of function should enclose the exact range of function values and outer approximation should enclose the solution set of system of linear interval equations) to find the optimal values of variables.

We use the first example system of the linear interval equations (1) to show that the use of adaptive interval arithmetic is efficient. 28 interval operations are performed in the algorithm to find the outer approximation of the solution set of system of 3 linear interval equations using Gaussian elimination.

The adaptation procedure which checks some random configurations suggested to perform 6th, 7th, 12th, 17th, 18th and 25th operations as inner interval operations and all others as standard interval operations. Using this configuration, the outer approximation of the solution set found is

$$\begin{bmatrix} [-0.8980341738905392, & 0.6882924658093841] \\ [-0.2986196435700629, & 1.072196268621887] \\ [\ 0.8793509369229011, & 2.607976170776108] \end{bmatrix}$$

which encloses the actual solution set and whose volume is only 16% larger than the volume of the interval vector hull of the actual solution set. The volume of this outer approximation is almost 5 times smaller than the volume of the outer approximation found using Gaussian elimination with standard interval arithmetic – when all operations are standard interval operations.

6 Conclusions

The use of balanced random interval arithmetic to estimate the enclosure of the solution set of systems of linear interval equations was demonstrated using an example system of linear interval equations.

The estimates of the enclosure of the solution set of the example system of linear interval equations evaluated using balanced random interval arithmetic with probability of enclosure larger than 0.999 are 28% more tight than the enclosure evaluated using standard interval arithmetic. The estimates obtained with probability of enclosure larger that 0.97 are 43% more tight than the enclosure evaluated using standard interval arithmetic.

It has been shown using a practical system of interconnected reactors that when coefficients of system of linear equations are not exact, interval methods should be used, because the real valued solutions obtained using real coefficients may be very poor when their accuracy is not known.

The estimates of the enclosure of the solution set of systems of linear interval equations for the practical system of interconnected reactors evaluated using Gaussian elimination with balanced random interval arithmetic are almost twice as higher as the enclosure evaluated with standard interval arithmetic.

The estimate of the enclosure of the solution set of the example system of linear interval equations evaluated using Gaussian elimination with proposed adaptive interval arithmetic encloses the actual solution set and its volume is only 16% larger than the volume of the interval vector hull of the actual solution set. This estimate of the enclosure is almost 5 times tighter than the enclosure evaluated using Gaussian elimination with standard interval arithmetic.

7 Acknowledgements

We wish to acknowledge the financial support through the Royal Society/NATO Postdoctoral Fellowship for the research "Stochastic Interval Methods for Process Optimisation".

References

1 Moore RE, (1966) Interval Analysis. Prentice Hall.

2 Hansen E, (1992) Global Optimization using Interval Analysis. Marcel Dekker.

3 Gau CY, Stadtherr MA, (2002) New Interval Methodologies for Reliable Chemical Process Modeling. Computers and Chemical Engineering, **26** (6); 827-840.

4 Byrne RP, Bogle IDL, (1999) Global Optimization of Constrained Non-convex Programs using Reformulation and Interval Analysis. Computers and Chemical Engineering, **23** (9); 1341-1350.

5 Adjiman CS, (1999) Safety Verification in Chemical Plants: A New Quantitative Approach. Computers and Chemical Engineering, **23** Supplement; S581-S584.

6 Chapra SC, Canale RP, (2002) Numerical Methods for Engineers: with Software and Programming Applications – 4th ed. McGraw-Hill.

7 Zilinskas J, Bogle IDL, (2003) On the Generalization of a Random Interval Method. In: European Symposium on Computer Aided Process Engineering – 13, Lappeenranta, Finland; 989-994.

8 Alt R, Lamotte JL, (2001) Experiments on the Evaluation of Functional Ranges using Random Interval Arithmetic. Mathematics and Computers in Simulation, **56** (1); 17-34.

9 Zilinskas J, Bogle IDL, (2003) Evaluation Ranges of Functions using Balanced Random Interval Arithmetic. Informatica, **14** (3); 403-416.

10 Neumaier A, (1990) Interval Methods for Systems of Equations. Cambridge University Press.

11 Jaulin L, Kieffer M, Didrit O, Walter E, (2001) Applied Interval Analysis. Springer-Verlag.

An Effective Real-Parameter Genetic Algorithm for Multimodal Optimisation

P. J. Ballester, J. N. Carter

Imperial College London, Department of Earth Science and Engineering,
RSM Building, Exhibition Road, London SW7 2AZ, UK
p.ballester, j.n.carter @imperial.ac.uk

Abstract

Evolutionary Algorithms (EAs) are a useful tool to tackle real-world optimisation problems. Two important features that make these problems hard are multimodality and high dimensionality of the search landscape.

In this paper, we present a real-parameter Genetic Algorithm (GA) which is effective in optimising high dimensional, multimodal functions. We compare our algorithm with a previously published GA which the authors claim gives good results for high dimensional, multimodal functions. For problems with only few local optima, our algorithm does not perform as well as the other algorithm. However, for a problem with very many local optima, our algorithm performed significantly better.

1 Introduction

Many real-world optimisation problems, particularly in engineering design, have a number of key features in common: the parameters are real numbers; there are many of these parameters; and they interact in highly non-linear ways, which leads to many local optima in the objective function. Clearly it is useful to have optimisers that have been shown to be effective at solving problems with these characteristics. It has been shown elsewhere [1] that Genetic Algorithms (GAs) are good at solving multimodal functions. In this paper, we describe and demonstrate a GA that appears to be good at solving problems where the objective function is characterised as being: high dimensional; real variable; continuous and smooth; many local optima.

Deb et *al.* [2] have recently produced a comprehensive review of optimisation methods. They included in their study: real parameter GAs, Self-Adaptative ESs (Evolution Strategies), DE (Differential Evolution) and GMs (Gradient Methods). All of these methods were tested on a set of high dimensional analytical test functions. They concluded that a real-parameter GA known as G3-PCX had the best overall performance. It is worth noting that G3-PCX was shown to have better convergence than GMs on some unimodal functions. G3-PCX also obtained the best results, in the published literature, for a 20-variable Rastrigin function.

An important observation made in this review is that: 'in most real-world problems, the knowledge of the exact optimum is usually not available, and the performance of an Evolutionary Algorithm (EA) on a symmetric initialisation may not represent the EA's true performance in solving the same problem with a different initialisation or other problems'. Consequently it is important that algorithms are tested with skewed initialisations, so as to give a better indication of their performance on real-world problems.

Ballester and Carter [1] investigated the performance of a variety of GA formulations over a set of 2-variable multimodal functions. It was found that GAs which used random selection with crowding replacement strategies were robust optimisers. In this paper, we will demonstrate one of these GAs, which is also effective for high dimensional functions.

We arrange the rest of the paper as follows. Section 2 describes the structure of the GAs tested in this paper. In Section 3, we explain the experimental setup. Results are presented and a comparison with G3-PCX made in Section 4. Lastly, we present our conclusions and discuss future work in Section 5.

2 GA Description

Our real parameter GA uses a steady state population model. In each generation, two parents are selected from the current population to produce λ children through crossover. Offspring and current populations are then combined so that the population remains at a constant size.

This GA combines the following features: parental selection is not fitness biased, a self-adaptative crossover operator, implicit elitism and locally scaled probabilistic replacement. We will refer to this GA as SPC-vSBX (Scaled Probabilistic Crowding with vSBX crossover). Below we describe the details of SPC-vSBX's selection, replacement and crossover schemes:

2.1 Selection

We use uniform random selection, without replacement, to select two parents from the current population. Unusually for a GA, fitness is not taken into account during the selection process.

2.2 Replacement

We use a scaled probabilistic crowding scheme for our replacement policy. First, NREP individuals from the current population are selected at random. These individuals then compete with the offspring for a place in the population.

In the probabilistic crowding scheme [5], the closest preselected individual ($\boldsymbol{x}^{cst}$) enters a probabilistic tournament with the offspring ($\boldsymbol{x}^{ofp}$), with culling likelihoods (survival, if we were in a maximisation problem) given by

$$p(\boldsymbol{x}^{ofp}) = \frac{f(\boldsymbol{x}^{ofp})}{f(\boldsymbol{x}^{ofp}) + f(\boldsymbol{x}^{cst})} \quad p(\boldsymbol{x}^{cst}) = \frac{f(\boldsymbol{x}^{cst})}{f(\boldsymbol{x}^{ofp}) + f(\boldsymbol{x}^{cst})} \tag{1}$$

where $f(\boldsymbol{x})$ is the objective function value for an individual $\boldsymbol{x}$.

If the differences in function values across the population are small with respect to their absolute values, these likelihoods would be very similar in all cases. The scaled probabilistic crowding replacement is introduced to avoid this situation. It operates with culling likelihoods

$$p(\boldsymbol{x}^{ofp}) = \frac{f(\boldsymbol{x}^{ofp}) \quad f_{best}}{f(\boldsymbol{x}^{ofp}) + f(\boldsymbol{x}^{cst}) \quad 2f_{best}} \qquad p(\boldsymbol{x}^{cst}) = \frac{f(\boldsymbol{x}^{cst}) \quad f_{best}}{f(\boldsymbol{x}^{ofp}) + f(\boldsymbol{x}^{cst}) \quad 2f_{best}} \tag{2}$$

where f_{best} is the function value of the best individual in the offspring and selected group of NREP individuals.

This replacement scheme has several beneficial features. The fittest individual does not always win, which helps to prevent premature convergence. Crowding schemes such as this promote the creation of subpopulations that explore different regions of the search space. This has been shown [1] to be beneficial for creating multiple optimal solutions and to increase the effectiveness in finding the global minimum. It implements elitism in an implicit way. If the best individual in either offspring or current parent population enters this replacement competition will have probability zero of being culled.

2.3 Crossover

In this work, we use a version of the Simulated Binary Crossover (SBX) [3] [4] called vSBX [1]. These crossovers are self-adaptative in the sense that the spread of the possible offspring solutions depends on the distance between the parents, which decreases as the population converge.

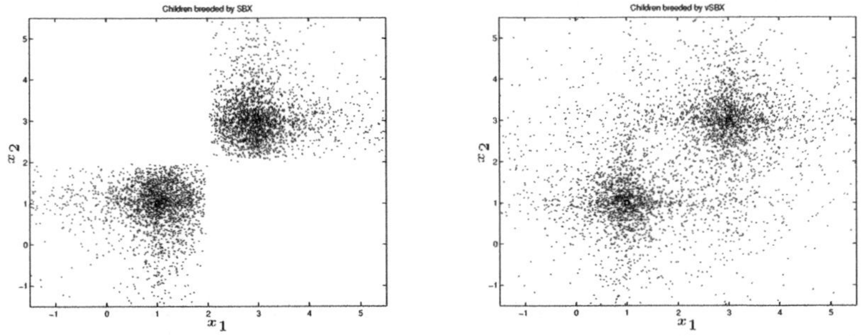

Figure 1. Children bred from parents $\boldsymbol{x}^{(1)} = (1\ 1)$ and $\boldsymbol{x}^{(2)} = (3\ 3)$ for a) SBX ($= 1$) and b) vSBX ($= 1$)

In SBX, children have zero probability of appearing in some regions of the parameter space, as shown in Figure 1. vSBX does not exclude any regions, while preserving the good SBX properties. This may allow a better exploration of the search space. It should be noted that SBX and vSBX preferentially search along the coordinate directions. This may give an advantage on test functions where minima are aligned along coordinate directions.

In vSBX, the first decision is which set of equations to use to generate the offspring. Using a single random number w $[0, 1]$, we use the form $y_j^{(1)}$ if $w < 0.5$

and $y_j^{(2)}$ if $w \quad 0.5$. We also need a set of random numbers $u_j \quad [0,1]$, one for each coordinate direction. The expression to generate the coordinates of $y_j^{(1)}$ are

$$y_j^{(1)} = \begin{cases} \frac{1}{2}((1+\beta_j)x_i^{(j)} + (1 \quad \beta_j)x_j^{(2)}) \; 0 < u_j \bullet 0.5 \\ \frac{1}{2}((3 \quad \beta_j)x_j^{(1)} \quad (1 \quad \beta_j)x_j^{(2)}) \; 0.5 < u_j \bullet 1 \end{cases} \tag{3}$$

The expression to generate the coordinates of $y_j^{(2)}$ are

$$y_j^{(2)} = \begin{cases} \frac{1}{2}((1 \quad \beta_j)x_j^{(1)} + (1+\beta_j)x_j^{(2)}) \quad 0 < u_j \bullet 0.5 \\ \frac{1}{2}(\; (1 \quad \beta_j)x_j^{(1)} + (3 \quad \beta_j)x_j^{(2)}) \; 0.5 < u_j \bullet 1 \end{cases} \tag{4}$$

for each expression β_j is given by

$$\beta_j = \begin{cases} \left(\frac{1}{2u_j}\right)^{\frac{1}{\;+1}} & 0 < u_j \bullet 0.5 \\ \left(\frac{1}{2(1 \quad u_j)}\right)^{\frac{1}{\;+1}} & 0.5 < u_j \bullet 1 \end{cases} \tag{5}$$

The larger the value of η the more concentrated is the search around the parents.

3 Experimental Setup

We use the same experimental setup as in Deb *et al.* [2], allowing a direct comparison with their results. The stopping criteria are: either a maximum of 10^6 function evaluations or an objective value of $10 \quad ^{20}$ is obtained.

The benchmark consists in five analytical 20-variable functions given in Table 1: Ellipsoidal (f_{elp}), Schwefel (f_{sch}), Generalized Rosenbrock (f_{ros}), Rastrigin (f_{rtg}) and a rotated Rastrigin function (f_{rrtg}).

Table 1. Test functions. M=20 in all cases.

Function	Type	Global Minimum
$f_{elp}(\boldsymbol{x}) = \sum_{j=1}^{M} jx_j^2$	Unimodal	$f_{elp}(\boldsymbol{x}=\boldsymbol{0}) = 0$
$f_{sch}(\boldsymbol{x}) = \sum_{j=1}^{M} \left(\sum_{k=1}^{j} x_k\right)^2$	Unimodal	$f_{sch}(\boldsymbol{x}=\boldsymbol{0}) = 0$
$f_{ros}(\boldsymbol{x}) = \sum_{j=1}^{M-1} (100(x_j^2 - x_{j+1})^2 + (x_j - 1)^2)$	Multimodal	$f_{ros}(\boldsymbol{x}=\boldsymbol{1}) = 0$
$f_{rtg}(\boldsymbol{x}) = 10M + \sum_{j=1}^{M} (x_j^2 - 10\cos(2 \quad x_j))$	Multimodal	$f_{rtg}(\boldsymbol{x}=\boldsymbol{0}) = 0$
$f_{rrtg}(\boldsymbol{y}) = 10M + \sum_{j=1}^{M} (y_j^2 - 10\cos(2 \quad y_j))$ $\boldsymbol{y} = \boldsymbol{A}\boldsymbol{x}$ with $A_{j,j} = 4 \; 5$, $A_{j,j+1} = 3 \; 5$ (j odd), $A_{j,j-1} = -3 \; 5$ (j even), $A_{j,k} = 0$ (the rest)	Multimodal	$f_{rrtg}(\boldsymbol{x}=\boldsymbol{0}) = 0$

The Ellipsoidal and Schwefel functions are unimodal. The Generalized Rosenbrock has been regarded as a unimodal function, but there is evidence [2] suggesting that it contains several minima in high dimensional instances. The Rastrigin function is highly multimodal. It has a minimum in any point with each component x_j taking an integer value. Lastly, the rotated Rastrigin is a non-separable, highly multimodal function without local minima arranged along the axis.

As an experiment has some dependence on the initial conditions, we repeat each experiment, each time with a different initial population. We do not initialise the population symmetrically around the global minimum, all variables are initialised at random within [10, 5].

4 Discussion of the Results

Our algorithm, SPC-vSBX, contains four tunable parameters: N, λ, NREP and η. In this study, we fix $\eta = 0.01$ and NREP=2.

Table 2. Performance comparison between SPC-vSBX and G3-PCX over the test functions. The best, median and worst columns refer to the number of function evaluations required to obtain a value of 10^{-20}. If the target is not reached then the best found function value within 10^6 evaluations is given. 'Success' refers to how many runs reach the target accuracy (unimodal) or end up within the global basin (multimodal). '?' accounts for information not specified in the original study [2].

Model	Crossover	(N,)	Function	Best	Median	Worst	Best Found	Success
SPC	vSBX-0.01	(6,1)	Elp	49,084	50,952	57,479	10^{-20}	10/10
G3	PCX-(0.1,0.1)	(100,2)	Elp	5,826	6,800	7,728	10^{-20}	10/10
SPC	vSBX-0.01	(6,1)	Sch	260,442	294,231	334,743	10^{-20}	10/10
G3	PCX-(0.1,0.1)	(150,2)	Sch	13,988	15,602	17,188	10^{-20}	10/10
SPC	vSBX-0.01	(12,1)	Ros	10^6	-	-	10^{-4}	48/50
G3	PCX-(0.1,0.1)	(150,4)	Ros	16,508	21,452	25,520	10^{-20}	36/50
SPC	vSBX-0.01	(20,3)	Rtg	260,658	306,819	418,482	10^{-20}	6/10
SPC	vSBX-0.01	(40,3)	Rtg	639,102	721,401	800,754	10^{-20}	10/10
G3	PCX-(?,?)	(?,?)	Rtg	10^6	-	-	15.936	0
SPC	vSBX-0.01	(75,3)	Rot. Rtg	10^6	-	-	8 955	0
G3	PCX-(0.1,0.1)	(300,3)	Rot. Rtg	10^6	-	-	309 429	0

In table 2, we compare SPC-vSBX and G3-PCX. G3-PCX reports the best results in the literature for the first four functions, except the Ellipsoidal (the BFGS quasi-Newton GM with a mixed quadratic-cubic polynomial line search approach achieved a solution in the order of 10 ^{24} in 6,000 function evaluations [2]). We observe that over the unimodal functions SPC-vSBX has no advantage.

In the 20-variable Rosenbrock function, there are two known local minima [2] with function values of 3.986624 and 65.025362. G3-PCX found solutions better than 10 ^{20} in 36 out of 50 runs, but in the other 14 got stuck in the best local minimum. SPC-vSBX only found a best solution of 10 ^{4} in 50 runs. It found solutions below the best local minimum (ie. within the global basin) in 48 out of 50 runs.

The more surprising results were obtained on the highly multimodal Rastrigin function. SPC-vSBX could find a solution better than 10 ^{20} in 6 out of 10 runs, whereas G3-PCX was reporting an overall best solution of 15.936 within the prescribed limits. In the other 4 runs, SPC-vSBX always found one of the best local minima with value 0.9949591. It is also observed that by incrementing N (N=40), all the runs reached the required accuracy at a cost of taking longer to converge. Moreover, within the limit of 10^6 function evaluations, SPC-vSBX-(20, 3) was able to find the global minimum of a 50-variable Rastrigin function. The best, median and worst function values in 10 runs were $5.280 \quad 10^{\ 6}$, $1.258 \quad 10^{\ 2}$ and 1.019, respectively.

Lastly, we tested the algorithms on the rotated Rastrigin function. This time we use the G3-PCX code downloaded from the KanGAL website, which can be found at *http : //www.iitk.ac.in/kangal/soft.htm*. We ran G3-PCX with many different N and λ and we found a best value of 309.429. The best solution found with SPC-vSBX was 8.955.

5 Conclusions and Future Work

We have presented a GA (SPC-vSBX) which has been shown to be effective in optimising high dimensional real-variable functions. Of particular importance are the results obtained with the Rastrigin function, the hardest of all tested in terms of number of local minima, where SPC-vSBX has the best performance reported in the literature.

Since vSBX preferentially searchs along the coordinate directions, it may give an advantage on test functions with minima aligned with the axis. Therefore, we also tested the algorithms on a rotated Rastrigin function. SPC-vSBX significantly outperformed G3-PCX.

In future work, we will test SPC-vSBX performance in other hard multimodal test functions as well as in real-world optimisation problems.

References

1. Ballester, P.J. and Carter, J.N. (2003) Real-parameter Genetic Algorithms for Finding Multiple Optimal Solutions in Multi-modal Optimization. Proceedings of the Genetic and Evolutionary Computation Conference, Ed. Erick Cantú-Paz et al. (Lecture Notes in Computer Science, Springer), 706-717.
2. Deb, K., Anand, A. and Joshi, D. (2002) A Computationally Efficient Evolutionary Algorithm for Real-parameter Optimization. Evolutionary Computation **10**(4), 345-369.
3. Deb, K. and Agrawal, S. (1995) Simulated Binary Crossover for Continuous Search Space. Complex Systems **9**(2), 115-148.
4. Deb, K. and Kumar, A. (1995) Real-coded Genetic Algorithms with Simulated Binary Crossover: Studies on Multi-modal and Multi-objective Problems. Complex Systems **9**(6), 431-454.
5. Mengshoel, O.J. and Goldberg, D.E. (1999) Probabilistic Crowding: Deterministic Crowding with Probabilistic Replacement. Proceedings of the Genetic and Evolutionary Computation Conference, Ed. W. Banzhaf et al. (Morgan Kauffmann), 409-416.

Dynamic Graph Drawing with a Hybridized Genetic Algorithm

Bruno Pinaud, Pascale Kuntz, RØmi Lehn

Ecole polytechnique de l'universitØde Nantes
Rue Christian Pauc - BP 50609
44306 Nantes Cedex 3
France
{bruno.pinaud, pascale.kuntz, remi.lehn}@polytech.univ-nantes.fr

Abstract

Automatic graph drawing algorithms, especially those for hierarchical digraphs, have an important place in computer-aided design software or more generally in software programs where an ef cient visualization tool for complex structure is required. In these cases, aesthetics plays a major role for generating readable and understandable layouts. Besides, in an interactive approach, the program must preserve the mental map of the user between time t 1 and t. In this paper we introduce a dynamic drawing procedure for hierarchical digraph drawing. It tends to minimize arc-crossing thanks to a hybridized genetic algorithm. The hybridization consists of a local optimization step based on averaging heuristics and two problem-based crossover operators. A stability constraint based on a similarity measure is used to preserve the likeness between the layouts at time t 1 and t. Computational experiments have been done with an adapted random graph generator to simulate the construction process of 90 graphs. They con- rm that, because of the actual algorithm, the arc crossing number of the selected layout is close to the best layout found. We show that computation of the similarity measure tends to preserve the likeness between the two layouts.

1 Introduction

Graph drawing algorithms are now generally included in Computer-Aided Design (CAD) software. A major advantage of graphs is that they can be used both as theoretical models for inherent relations between handled components, and as ef cient visualization tools. They make access to complex structures easier without getting bogged down in abstract concepts. Among the numerous graph representations [1], hierarchical digraphs, also called layered digraphs, have a specially important place in CAD. In those graphs, vertices are arranged on vertical (Figure 1) or horizontal layers, and arcs

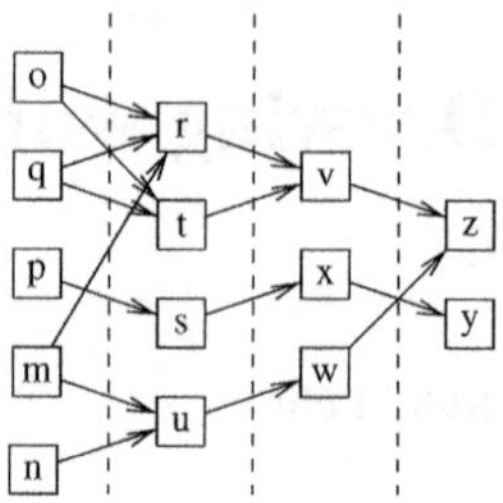

Figure 1: Example of a layered digraph with vertical layers

linking vertex pairs are represented by oriented line segments going in the same direction. These diagrams are particularly well-adapted to represent task decompositions and ow models in engineering planning.

When producing automatic layouts, the key words which guide the process are clarity and intelligibility. Although these notions call for subjective factors which closely depend on the audience, different criteria, modeled by combinatorial constraints, are commonly advocated by the information visualization community. Beside the physical constraints inherent to the medium (*e.g.* standard size sheets, computer screens), aesthetics plays a fundamental role. Their optimization aims at facilitating both readability and memorization of the information contained in the graphs. Frequently adopted aesthetic criteria include: minimization of the drawing area, minimization of the arc-crossing number, minimization of the sum of the arc lengths, *etc.* However, recent experiments have con rmed that the minimization of arc crossing is by far the most important criterion [2]. Hence, we here retain this one.

Minimizing arc crossing in a hierarchical layout could seem intuitively easier than the general problem of minimizing edge crossing on a plane since the choice of geometric coordinates for vertices is replaced by a choice of vertex ordering on each layer. Yet, it remains *NP*-complete [3]. Moreover, the explosion of the interactivity degree of CAD software has generated an additional constraint. Whatever the application eld, numerous components and their associated links may be dynamically added or deleted during the rst design stages. Consequently, the graph being visualized often changes over time. However, few algorithms take the interactivity with the user into account. If a modi cation is performed, the algorithm runs again and produces a new drawing which may be thoroughly different from the previous one. As noted by Papakostas *et al.* [4], this is a waste of human resources to continually re-analyse the entire drawing and also of computation resources to re-compute the entire layout after each modi - cation . Hence, the layout must not only remain readable over time, but the user's 'mental map' must also be preserved as much as possible [5].

In this paper, we tackle this multiobjective problem with a genetic algorithm (GA). GAs have already been applied to different static graph drawing problems with promising results (e.g. [6], [7], [8]). And, a recent statistical analysis of tness landscapes associated with different local operators for the hierarchical layout problem has con- rmed the presence of numerous local optima [9]. This advocates a highly stochastic approach, like GAs, able to move away from neighbours of a local optimum 'late' in the search process. Moreover, due to their intrinsic parallelism, GAs are particularly well-adapted to easily take the additional stability constraint into account with a low cost. At each time t, a set of potentially good drawings according to the aesthetics is generated with a GA. Then a drawing is selected among them according to its resemblance to the previous one presented to the user at time t 1.

The rest of the paper is organized as follows. Section 2 brie y presents related works. Section 3 introduces a formal de nition of the problem. Section 4 describes the optimization process based on a GA; we develop new speci c problem operators and a local hybridization with the so-called averaging heuristics. Section 5 presents the experimental results and a comparison with Tabu Search.

2 Related works

Numerous deterministic heuristics for the static hierarchical layout problem follow the layer-by-layer sweep scheme: the vertices of each layer are reordered to reduce crossings while holding the vertex orderings on the other layers. Various strategies have been proposed for reordering (see [10] for a chronological review). The most commonly used are the sorting methods and the averaging heuristics which include the popular barycenter heuristics from Sugiyama [11], the median heuristics [3], and their variants. Sorting methods exchange vertices using crossing numbers in a way similar to classical sorts. Averaging heuristics are based on the idea that arc crossings tend to be minimized when connected vertices are placed facing each other. Consequently, vertices are arranged according to their neighbour average positions *e.g.* the arithmetic mean or the median.

These approaches are still certainly the most employed in software. However, it is well-known that their probability of getting down in local optima is far from being negligible. In an attempt to improve the quality of the results, metaheuristics have known a particular interest in graph drawing since the mid 90's. Tabu Search [10] and GRASP (Greedy Randomized Adaptive Search Procedure) [12] have been recently developed for the hierarchical drawing. Numerical comparisons on graphs with various densities (including those observed in real-life applications) show that both these search processes signi cantly outperform the deterministic layer-by-layer sweep approaches.

In comparison with static drawings, the literature on dynamic drawings is still in its infancy. In the hierarchical case, the rst proposed heuristics are closely linked to the Sugiyama heuristics [13]. Also the multiobjective optimization has been tackled by different approaches: constraint propagation followed by constraint relaxation in order to cope with inconsistencies [14], or penalty assignment to vertex changes from t 1 to t and linear programming to compute the vertex coordinates on each layer [15].

3 Multiobjective problem description

In the following, we consider a hierarchical acyclic digraph $G = (V, A)$ with a vertex set V and an arc set A. Let $L =$ $l_1, l_2, ..., l_K$ be a set of K layers and a given partition

$V_1, V_2, ..., V_K$ of V on L in classes with respectively $n_1, n_2, ..., n_K$ vertices. The vertex assignment inside the layers is here supposed to be done. The vertex ordering on each layer l_k is de ned by $\pi_k : l_k \quad 1, 2, ..., l_k$, where $\pi_k(u) = i$ means that the vertex $u \quad l_k$ is on the ith position on l_k. The drawing of G is completely de ned by the set of the orderings $\Pi_G = \quad \pi_1, \pi_2, ..., \pi_K$ on each layer l_i of L. The crossing number associated with Π_G is denoted by $c(\Pi_G)$. Note that, without loss of generality, we can restrict ourselves to proper graphs with arcs span equal to 1 (the span of an arc $(u, v) \quad A$ with $u \quad L_i$ and $v \quad L_j$ is $j \quad i$). In practice, every arc with a span λ greater than 1 has to be replaced by a path of $\lambda \quad 1$ dummy vertices on every consecutive layers.

When the drawing problem is included in a dynamical process, two constraints should be simultaneously satis ed: the drawing Π_{G_t} proposed at each step t must minimize $c(\Pi_{G_t})$ and must remain as close as possible to the previous $\Pi_{G_{t-1}}$ to preserve the user's mental map. To evaluate the likeness between different drawings, we resort to a measure $\delta\left(\Pi_{G_t}, \Pi_{G_{t-1}}\right)$ based on the number of inverted vertex pairs between $t \quad 1$ and t:

$$\delta\left(\Pi_{G_t}, \Pi_{G_{t-1}}\right) = 1 \quad \frac{1}{K} \sum_{k=1}^{K} \frac{C_k(t \quad 1, t)}{P_k(t \quad 1, t)} \tag{1}$$

where $P_k(t \quad 1, t)$ is the number of common vertex pairs for both drawings, and $C_k(t \quad 1, t)$ is the number of vertex pairs which have been inverted between the drawing at $t \quad 1$ and the one at t. If $\delta\left(\Pi_{G_t}, \Pi_{G_{t-1}}\right)$ is close to 1, then the number of inverted pairs is small and the drawing at $t \quad 1$ is consequently preserved in the new drawing at t. More complex measures could be investigated in the future [16].

4 Graph optimization

The optimization process of the drawings is based on a hybridized genetic algorithm (HGA). And, the stability constraint (maximizing $\delta\left(\Pi_{G_t}, \Pi_{G_{t-1}}\right)$) is only taken into account in the selection of the solution in the last computed population. More precisely, at a given step t, HGA is applied to graph G_t, and among the best solutions computed by HGA, we select the one $\widehat{\Pi_{G_T}}$ which is the closest to the previous drawing $\widehat{\Pi_{G_{t-1}}}$. We rst present the problem-based operators developed for the static drawing problem. Then we detail the integration of the dynamics.

4.1 The hybridized genetic algorithm

The HGA developed for minimizing $c(\Pi_G)$ follows the basic scheme with two speci c characteristics. We have introduced problem-based crossovers and a local hybridization which signi cantly improves the computation time. Crossover is inspired from the following remark: the combination of two well-adapted drawings of sub-graphs can produce a better drawing (Figure 2). The combination may be applied between layers (inter-layer crossover) or inside each layer (intra-layer crossover).

The inter-layer crossover is a unique point crossover between layers. The intra-layer crossover aims at combining vertices of a same layer. However, combining blocks inside a layer meets a well-known dif culty for ordinal codings which is to de ne

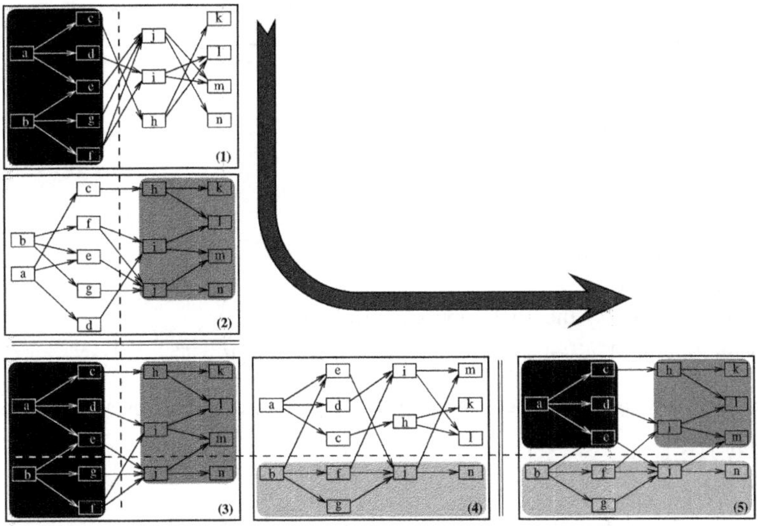

Figure 2: Improvement of an initial layout with successive sub-graph combinations. The inter-layer combination of (1) and (2) produces (3) and the intra-layer combination of (3) and (4) produces (5) which is a global optimum.

a crossover which guarantees a feasible solution [17]. The intra-layer crossover is a generalization of the Order Crossover 1 [17] for multi-permutations. A pivot whose random position is normalized by the layer cardinality is de ned for each layer of the parents. For the rst child, vertex positions of the rst parent are retained above the pivot, and below it, positions of missing vertices are completed by those of the second parent according to the vertex ordering. And vice-versa for the second child.

The hybridization step is based on local search with the barycenter and median operator described in [9]. The selection is determined by a classical roulette wheel based on the arc crossing number as the tness function. The mutation switches two randomly chosen vertices inside a layer. Each of these operators is applied with a given probability.

4.2 Integration of the stability constraint

While the graph is small (V < 20), we do not consider the dynamic constraint and the best solution computed by HGA at each step t is retained. Indeed, in this case, changes in the drawing aspect due to vertex and arc additions are very important and we cannot consider that a 'stable structure' exists to be preserved.

Algorithm 1 describes the integration of the stability constraint in the dynamic drawing process. At each step t, we select the drawing which maximizes the similarity among the 25% of the population composed of the best drawings produced by HGA. If the same similarity value is reached for different drawings, we keep the one which minimizes c.

Algorithm 1 Pseudo-code of the interactive drawing procedure

$t \quad 0$;
while $V_{G_t} \; < 20$ {
  Apply HGA on Π_{G_t};
  Display the best drawing $\widehat{\Pi_{G_t}}$;
}
$t \quad t+1$;
while termination criterion {
  Apply HGA on Π_{G_t};
  Put in *solution_array* the nal population of HGA;
  For each drawing $\Pi^i_{G_t}$ of *solution_array* compute $\delta\left(\Pi^i_{G_t}, \widehat{\Pi_{G_{t-1}}}\right)$;
  Select the drawing $\widehat{\Pi_{G_t}}$ which minimizes $c\left(\Pi^i_{G_t}\right)$ and maximizes $\delta\left(\Pi^i_{G_t}, \widehat{\Pi_{G_{t-1}}}\right)$.
  Display $\widehat{\Pi_{G_t}}$;
  $t \quad t+1$;
}

5 Computational experiments

As far as we know, unlike the classical static drawing problem, there are no test case bases available for large scale numerical experimentations. Experiments are often done with human beings and focus on a panel of known responses. Although this approach is obviously necessary for the validation, it is faced with different intrinsic limitations: availability of the subjects, variations of their own behaviours and restrictions of the evaluated drawing number.

To obtain quantitative measures of validation on large graph sets, we have developed a graph generator adapted to the dynamic problem. We rst brie y present it. Then, we describe the results obtained for the dynamic evolution of the drawings of 90 graphs. A comparison is given with Tabu Search (TS) which is to this day one of the best-known metaheuristics for the static problem [10].

5.1 Graph generator

The graph generator is composed of two major steps:

1. the initialization phase generates random connected graphs of small size. This phase re ects the beginning of numerous design processes where the graphs considered may be quite small and completed along the process.

2. at each step t, once a solution has been computed by the interactive drawing procedure for G_t, a new graph G_{t+1} is generated by adding randomly placed vertices, arcs and layers to G_t. For each generation, we ensure the preservation of the graph connectivity by adding at least one arc between the new vertices and the existing ones.

The generation process stops when the maximal vertex number (here set to 80) is reached or when $c\left(\Pi_{G_t}\right)$ is greater than 500. The maximal number of layers is set to 15, the maximal number of vertices to add after each step has been set to 2. These sizes are typical of real-life applications where vertices are often represented by labelled boxes and can rarely exceed 70 or 80 on a standard size sheet for an exhaustive visualization.

This generator has been used to simulate the dynamic evolution of a set of 90 graphs which corresponds to a total of 1400 layouts. The mean density value de ned by $d(G_t) = m/\sum_{k=2}^{h} l_{k\ 1} \quad l_k$ is 0.65 (std. dev. 0.05).

5.2 Results

In computational experiments, we have chosen the following genetic operator probabilities: 0.05 for mutation and 0.2 for each crossover. The population size has been xed to 100. Preliminary experiments have shown that larger populations give sightly better results, but this choice is a compromise between the quality and the computational cost. The HGA stops at each step t when no improvement has occurred after 100 generations. HGA is coded in the C language.

Figure 3-a shows the average crossing number evolution for the whole set of graphs built by the generator. We here retain the crossing number of the selected solution and compare it with a drawing optimized by TS. As expected, the crossing number linearly increases with the vertex number. Except for one case, HGA solutions are always better than TS solutions. And the improvement is all the greater as the vertex number increases.

The algorithm naturally implies that the selected drawing at each step t (which takes into account the stability constraint) is close to the best solution of HGA. This is con rmed by Figure 3-b which shows the difference between the crossing number of the selected solution and the crossing number of the best HGA solution. The mean difference is equal to 3.14% (std. dev. 2.42%) for HGA. However, it is signi cantly higher for TS: 9.64% (std. dev. 4.28%).

To better understand the integration of the stability constraint in the dynamic drawing procedure, we have studied the evolutions of the similarity index $\delta\left(\Pi_{G_t}, \Pi_{G_{t-1}}\right)$. Computing the mean value of the similarity index on the whole set of graphs is a nonsense. We prefer to show a gure of this evolution with a representative graph (Figure 4). This gure shows the evolution of δ for the best HGA solutions and the selected solutions. The similarity is signi cantly greater for the selected solutions. The important variations which appear from time to time are due to the dynamics: depending on the positions of the added vertices and edges, graphs $G_{t\ 1}$ and G_t may be quite different and consequently require a new drawing. In this case, the subset of the best HGA generated solutions for the crossing constraint contains drawings of G_t different from those of $G_{t\ 1}$. However, overall, our strategy tends to preserve the likeness.

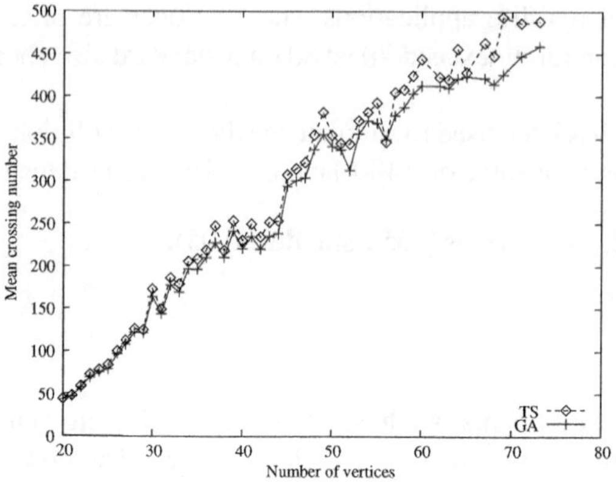

(a) Average crossing number for the selected solution. Comparison with TS

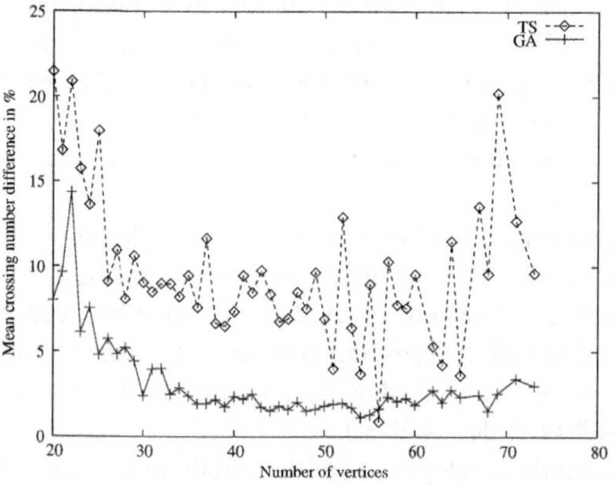

(b) Mean difference of the crossing number of the HGA selected solution and the best TS solution with the best static HGA solution

Figure 3: Analysis of the drawing quality for the dynamic evolution of a set of 90 pseudo-random graphs (total of 1400 layouts).

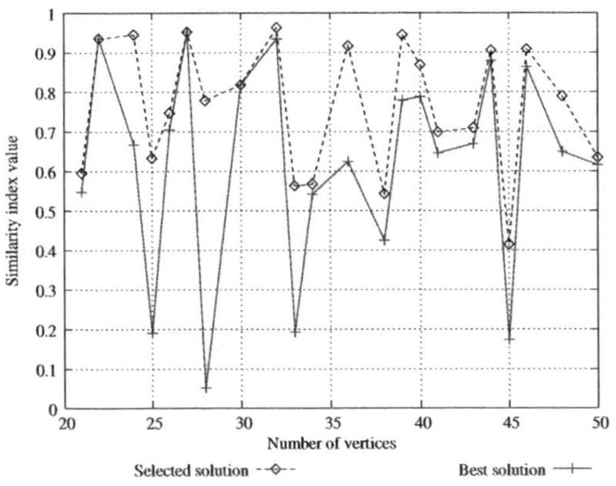

Figure 4: Evolution of the stability constraint for the selected solutions and the best HGA solutions.

6 Conclusion and perspectives

We developed a procedure for dynamic hierarchical digraph drawing based on a hybridized genetic algorithm. Computational experiments were done with a random graph generator adapted to this problem. They showed that quality-wise the solution displayed to the user (*i.e.* computed to be visually similar to the previous drawing) was not much debased from the best layout found. Moreover, our method also outperforms one of the best known metaheuristics for the static problem.

After these promising results, we still have to compare our algorithm with some widely used methods for multiobjective optimizations such as a real multiobjective genetic algorithm or more generally with Pareto-based approaches [18]. There is also another interesting point to study further, as noted in Section 3, the similarity measure. We need to use some more complex measures by, for instance, taking into account the degree of each vertex. A vertex with a high degree might produce numerous arc-crossings contrary to a vertex with a lower degree. In the construction phase of a new drawing, moving this vertex far from its original position will result in important changes in the structure of the graph. This consequently generates a new drawing quite dissimilar to the previous one which is not what we aim at.

For carrying out such experiments, we need a graph generator which can reproduce the behaviour of a human being. In the present version the addition of new vertices is done completely at random. When an individual uses a real application, he does not have a random behaviour. He may intensify a part of the graph, leaving another one not very closely connected ; he may also add vertices only in the rst layer and not in the others, and so on. We need to enhance the generator with some human behaviours such as a set of scenarios of real graph constructions done by a person.

References

[1] G. Di Battista, P. Eades, R. Tamassia, and I.G. Tollis. *Graph drawing - Algorithms for the visualization of graphs.* Prentice Hall, 1999.

[2] H. Purchase. Effective information visualization: a study of graph drawing aesthetics and algorithms. *Interacting with computers*, 13(2), 2000.

[3] P. Eades and N. Wormald. Edge crossings in drawings of bipartite graphs. *Algorithmica*, 11:379 403, 1994.

[4] A. Papakostas, J.M. Six, and I.G. Tollis. Experimental and theoretical results in interactive orthogonal graph drawing. In *Proc. of Graph Drawing'96*, volume 1190 of *Lecture Notes in Computer Sciences*, pages 371 386. Springer Verlag, 1997.

[5] P. Eades, W. Lai, K. Misue, and K. Sugiyama. Preserving the mental map of a diagram. In *Proc. of Compugraphics*, pages 24 33, 1991.

[6] L.J. Groves, Z. Michalewicz, P.V. Elia, and C.Z. Janikow. Genetic algorithms for drawing directed graphs. In *Proc. of the 5th Int. Symp. on Methodologies for Intelligent Systems*, pages 268 276. Elsevier, 1990.

[7] A. Ochoa-Rodr guez and A. Rosete-Su rez. Automatic graph drawing by genetic search. In *Proc. of the 11th Int. Conf. on CAD, CAM, Robotics and Manufactories of the Future*, pages 982 987, 1995.

[8] J. Utech, J. Branke, H. Schmeck, and P. Eades. An evolutionary algorithm for drawing directed graphs. In *Proc. of the Int. Conf. on Imaging Science, Systems and Technology*, pages 154 160. CSREA Press, 1998.

[9] P. Kuntz, B. Pinaud, and R. Lehn. Elements for the description of tness landscapes associated with local operators for layered drawings of directed graphs. In M.G.C. Resende and J.P. de Sousa, editors, *Metaheuristics: Computer Decision-Making*, volume 86 of *Applied optimization*, pages 405 420. Kluwer Academic Publishers, 2004. ISBN 1-4020-7653-2.

[10] M. Laguna, R. Marti, and V. Valls. Arc crossing minimization in hierarchical digraphs with tabu search. *Computers and Operation Research*, 24(12):1175 1186, 1997.

[11] K. Sugiyama, S. Tagawa, and M. Toda. Methods for visual understanding of hierarchical systems. *IEEE Trans. Syst., Man, Cybern.*, 11(2):109 125, 1981.

[12] R. Marti. Arc crossing minimization in graphs with GRASP. *IIE Trans.*, 33(10):913 919, 2001.

[13] J. Branke. Dynamic graph drawing. In D. Wagner M. Kaufmann, editor, *Drawing Graphs: methods and models*, pages 228 246. Springer, 2001.

[14] K.F. B hringer and F.N. Paulisch. Using constraints to achieve stability in automatic graph layout algorithms. In *Proc. of CHI'90*, pages 43 51. ACM, 1990.

[15] S.C. North. Incremental layout in DynaDAG. In *Proc. of Graph Drawing'95*, volume 1027 of *Lecture Notes in Computer Science*, pages 409 418. Springer-Verlag, 1996.

[16] S. Bridgeman and R. Tamassia. A user study in similarity measures for graph drawing. *J. of Graph Algorithms and Applications*, 6(3):225 254, 2002.

[17] D. Whitley and N. Yoo. Modeling simple genetic algorithms for permutation problems. In L. D. Whitley and M. D. Vose, editors, *Fundations of Genetic Algorithms III*, pages 163 184. Morgan Kaufmann, 1995.

[18] Carlos A. Coello Coello. A comprehensive survey of evolutionary-based multiobjective optimization techniques. *Knowledge and information systems*, 1(3):269 308, 1999.

Quantum-Inspired Evolution Algorithm: Experimental Analysis

F. Alfares[1], M. Alfares[2], I.I. Esat[1]

[1]Mechanical Engineering Department,
Brunel University
Uxbridge, Middlesex, UB10 3PH, UK
Fouzan.Alfares@brunel.ac.uk
Ibrahim.Esat@brunel.ac.uk

[2]Department of Electronic Engineering Technology,
College of Technological Studies
P.O. Box: 42325 Shuwaikh, 7065, State of Kuwait
mfares@paaetms.paaet.edu.kw

Abstract

Quantum computing mimics behaviour of atoms in processing information. Unfortunately due to restrictive rules of processing imposed by quantum behaviour only few successful algorithms have been developed in quantum computing. Quantum inspired algorithm is a concept, which employs certain elements of quantum computing to use in a wider class of search and optimisation problems. The main parts of a quantum-inspired algorithm are the qubits (quantum equivalent of bits) and the gates. Qubits hold the information in a superposition of all the states, while the quantum gates evolve the qubit to achieve the desired objective, which is, in optimization the maximum or the minimum. The paper addresses the ability of the Quantum-Inspired Evolution Algorithm (QIEA) to solve practical engineering problems. QIEA, which is developed by authors, is based on their previous work and it is improved to test a series of unitary gates. A set of experiments were carried out to investigate the performance of QIEA as for speed, accuracy, robustness, simplicity, generality, and innovation. To assess effectiveness of a new algorithm, there are a set of guidelines proposed by [1]. Based on these guidelines, the paper selected three test functions to carry out a benchmark study. The paper also presents a comparative study between QIEA and classical Genetic Algorithms (GA) and Particle Swarm Optimization (PSO) techniques in order to assess the proposed QIEA.

1 Introduction

According to Moore's law, the number of transistors per square inch on integrated circuits will be doubled approximately every two years. This law has withstood for about 30 years, so by about year 2015 to 2020 the size of transistors will reach atomic size where quantum mechanics laws become dominant. Feynman, [2] recognised that sub atomic level behaviour of transistor can not be simulated or modelled by using classical computers. He suggested building a quantum computer in order to simulate a quantum system; he also proposed the structural framework to achieve this.

Quantum computing is the new era in computer science. The new developments in this field attracted many researches from different disciplines. Research in quantum computing incorporates quantum mechanics theory, computer science and information theory. The research in quantum computing can be divided into two main streams, building a quantum computer, and building quantum algorithms.

From engineering point view, the question is how can engineering discipline benefit from this fast growing research field? There are two possible approaches to solve engineering problems [3]. The first approach is to use available quantum algorithms such as Shor's period finding or Grover's amplitude amplification algorithms in solving problems with similar characteristics. The problem with quantum computing is that, they need to follow certain rules. For example they cannot retrieve information from qubits, as this cause 'collapse' of superposition states. Therefore algorithms must be designed in such a way to ensure that qubit contains only the final result when it collapses.

The second approach is to build algorithms, which are inspired from quantum behaviour, in particular information-evolution, and employing some elements of quantum computing such as qubits and quantum gates but not restricted by quantum computing rules.

In order to use the available quantum algorithms, such as of Shor and Grover, one should investigate the way that these two algorithms behave. If cyclic behaviour is inherent in a given engineering problem, it may be possible to associate this with the quantum gate's characteristics. In other word for engineering problems with cyclic or recurrent behaviour, quantum computing may be relevant. It may be possible to associate quantum gate behaviour (such as quantum fourier transform) with a class of engineering problems. For example this may be useful in extracting information from time series. Gates perform the main operation in a quantum algorithm. Gates processes qubits and if one to observe behaviour of gates (say in case of quantum fourier transform), it would be possible to see that information transfer between different phase plane takes place. The amount of information transfer depends on the importance (or magnitude) of the data being processed. This observation explains the success of quantum search, especially when it applies to databases.

Hogg proposed a framework for structured quantum search [4] by associating the cost (objective function) with the gate behaviour. This work is extended by Hogg & Portnov [5] who produced a quantum algorithm for combinatorial optimization. Cyclic behaviour of gates and the transfer of information between phase planes points to the possible that engineering problems, such as signal processing, diagnostics, forecasting may benefit from such algorithms.

The second approach is to develop algorithms inspired from quantum computing. By borrowing one or more parts of a quantum computing such as qubits and quantum gates, one can develop a quantum-inspired evolution algorithm. What influences the development of this was quantum gate behaviour, which results in evolutionary updating of qubits. This is the main mechanism for optimisation.

Moore and Narayanan [6] proposed a framework for general quantum-inspired algorithms. Han and Kim [7] developed an evolution algorithm which stores information in qubits and used rotation quantum gate to search for a solution to a knapsack problem and named it genetic quantum algorithm. This, in fact, is not genetic algorithm since it does not contain the main parts of GA like selection or crossover or even mutation processes. Later Han and Kim [8] changed the name to quantum-inspired evolution algorithm studying the performance of the algorithm with respect to the angles of the rotation gates. An improved approach to the 'genetic' quantum algorithm proposed by Han and Kim [7] was proposed by Zhang et al. [9]. It appeared that Zhang was not aware of the fact that Han and Kim already re-named their approach. Zhang's implementation involves maintaining the population giving the best solutions while replacing the others after certain number of evolution steps. It was argued that this was done to avoid the solution being trapped in a local optimum. This is reported as claimed, as one would expect to be contrary to common sense expectation. Several algorithms combining Genetic Algorithms and Quantum effect were also reported; like Narayanan and Moore [10] developed a genetic algorithm that uses the theory of many universes of quantum mechanics to modify the crossover scheme. Also Rylander et al. [11] presented a quantum genetic algorithm (QGA) that exploits the quantum effects of superposition and entanglement. Moreover, Li and Zhuang [12] presented a genetic algorithm based on the quantum probability representation. Narayanan and Manneer [13] proposed a quantum-inspired neural network algorithm. All these quantum-inspired algorithms were applied to NP-hard workbench problems like travelling salesman and knapsack. This represent a rather narrow section of engineering problems and if one to evaluate the effectiveness of quantum algorithms, one would need wider problem definition, guidelines (such as those proposed by [1]) and benchmarks.

The main contribution of the paper is its study of the effectiveness of quantum-inspired algorithms in solving engineering problems. This investigation is an extension to their previous work reported by [3]. In particular, the algorithm is extended to use a number of gates. This enables the effectiveness to be studied with respect to gate selection. The paper also compares the algorithm against other methods such as GA and PSO.

The paper structure begins with an introduction to quantum computing and this is followed by brief explanations of the quantum-inspired algorithm structure. A short description of QIEA with GA and PSO is presented. Then, the test functions used to evaluate the QIEA are presented with discussion of the results. Finally, conclusions of the experiments and the future work are addressed.

2 Quantum-Inspired Algorithms

Quantum-inspired algorithm consists mainly of qubits and quantum gates. Qubits preserve the information in a superposition of all the states, while the quantum gates evolve the information to reach the desired objective, which is in optimization, the maximum or the minimum. The core work of Grover's search algorithm [14] is to separate the required solution from the other data that are simultaneously presented in the quantum register. This separation is achieved by reinforcement of the amplitude of the desired state by changing the difference in the phase angles of the selected and the other states. This process evolves for ($\sqrt{N}$) times after which a measurement is taken. Reading qubit collapses the superposition of all states to the solution state that has the highest amplitude. Hogg [15] has extended the work to build a quantum optimization algorithm by associating the cost function with the amplitude amplification.

The major components of any quantum algorithm consist of quantum bits, quantum registers and quantum gates. Quantum bits (qubits) encode the variables within it. Quantum registers are a collection of qubits that acts like the memory in classical computer sense. And quantum gates perform unitary transformations and measurements; transformations are time reversible. The measurement selects one variable to be evaluated and destroys the others (collapse).

In the following, brief description for the quabits and quantum gates are given.

2.1 Qubits

Qubits represent the information in bits, which are a unit vectors in complex Hilbert space. In a quantum computer, the logic circuitry and time steps are classical, only the memory bits that hold the variables are in quantum superpositions [16]. A classical computer has bits that exist in a state of one or zero, but quantum bits "qubits", exist in a state of one and zero simultaneously. This means that a large, even amount of data could be encoded in amplitudes of a single qubit by choosing α and β as the probabilities (amplitudes) of choosing the value of 0 or 1. The mathematical representation of a qubit is given as:

$$|\psi\rangle = \alpha|1\rangle + \beta|0\rangle$$

$|\alpha|^2$ gives the probability that the qubit will be found in '0' state and $|\beta|^2$ gives the probability that the qubit will be found in the '1' state. Normalization of the state to unity guarantees that an infinite amount of classical information can be encoded in a single qubit, however only one data could be retrieved from the qubit.

$$|\alpha|^2 + |\beta|^2 = 1$$

Reading the qubit's state a result of either 0 or 1 will be obtained. Only one bit of information can be extracted. Upon measurement it erases all the information in the qubit except for the single bit that the measurement reveals.

The time evolution of a quantum system is usually approximated by the Schrödinger equation. The Schrödinger equation is described by a linear partial differential equation with the property that any linear superposition of its eigen-functions is also a solution. The general state may be written as a superposition of eigen states according to the superposition property of quantum mechanics.

2.2 Quantum Gates

Quantum gates operate on normalized vectors in a finite-dimensional complex Euclidean space as divergent to classical gates that operate on bits taking values in the set [0, 1]. In principle, any quantum computer can be viewed as an assembly of elementary quantum gates.

The Schrödinger's equation governs the dynamics of a quantum system. When jumping from a state to another state, the orthogonality is preserved by linear unitary transformation and in the same time it is considered rotations of a complex vector. The unitary transformations are reversible. These transformations are called quantum gates, since they allow only one action to pass through them. And also they are universal (gates are called universal gates if they can be used to create any logic circuit, such as the NAND gate in classical circuits) in which Deutsch et al. [17] have shown that almost every gate operates on two or more qubits is a universal gate.

The definition of quantum gate is analogues to classical reversible gates by means of unitary operators acting on the qubit basis states. The overall phase of a quantum system makes no difference to the probability of measuring particular basis state although any relative phase difference does affect measurements, which depend on the interference between the two basis states. Hadamard is a transformation, which takes a basis state and transforms it into a linear combination of the two basis states. This transformation is very useful in describing quantum behaviour. By applying a linear number of operations to the quantum register it will generate a superposition of (2^n) number of terms. The ability to create such superpositions is one of the key properties that give quantum parallel processing its power. Han and Kim [7] used the rotation gate in their work. However, this paper applies different types of gates, to study the general behaviour of the algorithm and to verify the universality issue in quantum gates.

There are many types of quantum gates used in building quantum algorithms. A matrix representation of quantum gates is given below.

$$\begin{bmatrix} \frac{1}{\sqrt{2}} & \frac{-1}{\sqrt{2}} \\ \frac{1}{\sqrt{2}} & \frac{1}{\sqrt{2}} \end{bmatrix} \begin{bmatrix} \frac{1}{\sqrt{2}} & \frac{1}{\sqrt{2}} \\ \frac{1}{\sqrt{2}} & \frac{-1}{\sqrt{2}} \end{bmatrix} \begin{bmatrix} 0 & 1 \\ 1 & 0 \end{bmatrix} \begin{bmatrix} 1 & 0 \\ 1 & -1 \end{bmatrix} \begin{bmatrix} \cos\theta & \sin\theta \\ -\sin\theta & \cos\theta \end{bmatrix}$$

$$(a) \qquad\qquad (b) \qquad\qquad (c) \qquad (d) \qquad\qquad (e)$$

Figure 1: Matrix form of gates
Square Root of NOT (a), HADAMARD (b), Quantum NOT (c), Z (d), and Rotation (e).

3 Quantum-Inspired Evolution Algorithm, Genetic Algorithm, and Particle Swarm Optimization

This section will introduce a short description to the applied techniques in this study. The techniques are quantum-inspired evolution algorithm; genetic algorithms and particle swarm optimization technique. The algorithm used for this study is based on the work of Han and Kim [7], however it is modified to include different type of gates.

3.1 Quantum-Inspired Evolution Algorithm

The flow chart of QIEA is shown in Figure 2. The main parts of the algorithm are the Make, Evaluate, and Update processes. The evolution stage of the algorithm concentrate mainly on the 'Update Q(t)' procedure which can be described as a set of qubit chromosomes Q(t) updated by multiplying it with some appropriate quantum gates U(t), (Quantum gates are reversible gates and can be represented as unitary operators acting on the qubit basis states: $U''U = UU''$, where U'' is the hermitian adjoint of U. There are several quantum gates, such as NOT gate, Controlled NOT gate, Rotation gate, Hadamard gate, etc. [18] which is formed by using the binary solutions P(t) and the stored best solution. The QIEA maintains a population of qubit individuals $Q(t) = \{q_1^t, q_2^t, \ldots, q_n^t\}$, at generation t, where n is the size of population, and q_j^t is a qubit individual defined as

$$q_j^t = \begin{bmatrix} \alpha_{j1}^t & \alpha_{j2}^t & \ldots & \alpha_{jm}^t \\ \beta_{j1}^t & \beta_{j2}^t & \ldots & \beta_{jm}^t \end{bmatrix}$$

where m is the number of qubits, i.e., the string length of the qubit individual, and $j = 1, 2, \ldots, n$, and $P(t) = \{x_1^t, x_2^t, \ldots, x_n^t\}$ at generation t. One binary solution, x_j^t, $j = 1, 2, \ldots, n$, is a binary string of the length m, and is formed by selecting each bit using the probability of qubit, either $|\alpha_i^t|^2$ or $|\beta_i^t|^2$, $i = 1, 2, \ldots, m$, of q_j^t. Each solution x_j^t is evaluated to give some measure of its fitness. Afterwards, the best solution is stored and the algorithm starts the loop to update the solutions to move towards the best, during the process if a better solution than the best is found then it replaces it with the new one.

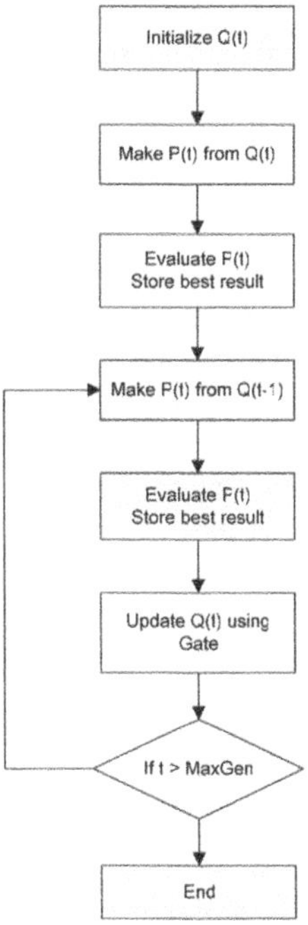

Figure 2: Flow chart of the QIEA

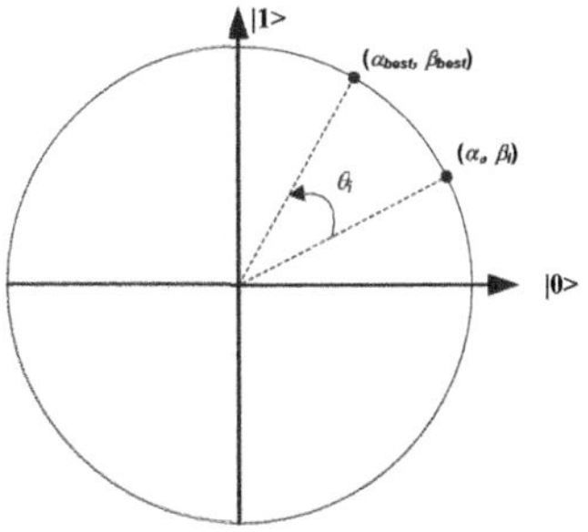

Figure 3: Movement of one solution to the best

The algorithm uses the objective function information to enhance the search. The process evolves the weakest solution to move in direction of the best solution. This information modifies the angle of the rotation by a small amount, as in rotation gate, in order not to pass a good solution. This can be seen in Figure 3.

Only two quadrants from a circle are required to move the solutions to the best solution among the search population, one in the first quadrant and the other in the third quadrant. The values used here for rotation gate are the same as suggested by Han and Kim [8] and take the values of (0.01π) and (-0.01π) for the first and third quadrant. The movement of qubit vectors towards the optimum solution is controlled by certain rule. The rule is stated as: if the fitness value of the current vector is less than the fitness value of the current best and if the bit value is opposite of best vector bit then update the qubit using the quantum gate.

3.2 Genetic Algorithms

Genetic Algorithms (GA's) are typically heuristic methods. They are based in one hand on a heuristic gradient ascension method (selection & crossover) and, in another hand, on a semi-random exploration method (mutations). Advantages of GA's are that they are simple to understand and to implement, and give fast solution (at least sub-optimal). Disadvantages are that they tend to fail with the more difficult problems and need good problem knowledge to be tuned. GA's are inspired from biological processes (i.e. cells' division, DNA, crossover, mutation). The underlining idea is to generate successive sets of solutions (generations), making each new generation inheriting properties from the best solutions of the precedent. In order to perform such a step, GA selects the best solutions and mixes them together (crossover). For further details, refer to [19].

3.3 Particle Swarm Optimization

Kennedy and Eberhart [20] proposed Particle Swarm Optimization (PSO) as an algorithm simulates the social behavior of organisms such as bird flocking and fish schooling. The core effect of the algorithm depends on collaboration of species onto achieving their goals as a group. In a PSO system, multiple candidate solutions coexist and collaborate simultaneously. Each solution candidate, called a 'particle', flies in the problem search space (similar to the search process for food of a bird swarm) looking for the optimal position to land. A particle, as time passes through his quest, adjusts its position according to its own 'experience', as well as according to the experience of neighboring particles.

Tracking and memorizing the best position encountered enhances particle's experience. For that reason, the PSO algorithm possesses a memory (i.e. every particle remembers the best position it reached during the past). PSO system combines local search methods (through self experience) with global search methods (through experience of neighboring particles), attempting to balance exploration and exploitation. Two factors characterize a particle status on the search space: its position and velocity. Kennedy et al. [21] explored several models to manipulate these factors to accurately resemble the dynamic of the social behavior of birds, before reaching to the equations of velocity and new position, which achieve good performance on optimization problems. The algorithm is simple and uses vectors to describe the movement from one point to another. Motion is described in terms of the relative position of current location to the global and local best positions. Figure 4 shows a simple PSO algorithm adapted from [22].

```
Let P be the size of the PSO population.
Let PSO[i] be the position of the ith particle of the PSO population; this represents a
candidate solution for the problem
Let fitness[i] be the cost function of the ith particle.
Let V[i] be the travelled distance (or velocity) of the ith particle.
Let Gbest be an index too global-best position.
Let Pbest[i] be the position of the local-best position of the ith particle.
Let Pbest_fitness[i] be the local-best fitness for the best position visited by the ith
particle.
Step 1 (Initialization): For each particle i in the population:
          Step 1.1: Initialize PSO[i] randomly.
          Step 1.2: Initialize V[i] randomly.
          Step 1.3: Evaluate fitness[i].
          Step 1.4: Initialize Gbest with the index of the particle with the best fitness
                    among population.
          Step 1.5: Initialize Pbest[i] with a copy of PSO[i]  ∀i ≤ P.
Step 2: Repeat until a stopping criterion is satisfied:
          Step 2.1: Find Gbest such that fitness[Gbest] fitness[i] ∀i ≤ P.
          Step 2.2: For each particle i: Pbest[i] = PSO[i] iff  fitness[i] > Pbest_fitness[i]
                    ∀i ≤ P.
          Step 2.3: For each particle i: update V[i] and PSO[i] according to the following
                    equations
                    V[i] = w*V[i] + c1*rand()*(Pbest - PSO[i]) + c2*rand()*(Gbest - PSO[i])
                    PSO[i] = PSO[i] + V[i]
          Step 2.4: Evaluate fitness[i] ∀i ≤ P
```

Figure 4: Simple PSO algorithm

4 Results and Discussion

The paper addresses the applicability of the Quantum-Inspired Evolution Algorithm (QIEA) in solving engineering problems. The main study issues of QIEA are speed, accuracy, robustness, simplicity, and generality. Also a comparison with GA and PSO is performed to bear out the QIEA performance. All algorithms are encoded in MATLAB running on Pentium 4 (1.7 MHz) PC processor. All algorithms including GA and PSO maintain a population of 20 and number of generations = 200. GA used in the experiment consists of stochastic universal sampling selection procedure with single point crossover. The probability of crossover (P_c) = 0.7 and probability of mutation (P_m) = (P_c / Length of chromosome). For PSO, 'arbitrary constants' c_1 and c_2 are = 2, 'inertia weight' w is between 0.1 and 0.9. Three test functions were chosen to test the performance of the QIEA:

1. Levy test function 5 [23].

$$f(x,y)=\sum_{i=1}^{5} i.\cos((i-1).x+i).\sum_{j=1}^{5} j.\cos((j+1).y+j)+(x+1.42513)^2+(y+0.80032)^2$$

$$for\ -10 \le x \le 10\ ,\ -10 \le y \le 10.$$

2. Test function used as an example in [19].

$$f(x_1,x_2) = 21.5 + x_1 \sin(4\pi x_1) + x_2 \sin(20\pi x_2)$$

$$for\ \ -3.0 \le x_1 \le 12.1\ \ \&\ \ 4.1 \le x_2 \le 5.8$$

3. Branin-RCOS Test Function from [19].

$$f(x_1,x_2) = a\,(x_2 - b\,x_1^2 + c\,x_1 - d)^2 + e\,(1-f)\cos(x_1) + e;\ \text{where}\ a = 1,\ b=5.1/(4\pi^2),$$

$$c = 5/\pi,\ d = 6,\ e =10,\ f=1/(8\pi)\ \ \text{for}\ -5 \le x_1 \le 10\ ,\ 0 \le x_2 \le 15.$$

The comparison with PSO is for the three test functions, while GA was used for comparing performance on test function 1 and 3.

Table 1: Results of the Algorithms

	Test Function 1*		Test Function 2		Test Function 3**		
	Elapsed Time (Sec)	Optimum Solution	Elapsed Time (Sec)	Optimum Solution	Elapsed Time (Sec)	Optimum Solution	Generation
GA	**6.400**	-128.8285	--	**38.8275**•	11.827	**0.39790**	149
PSO	11.897	-171.6040	9.674	38.7323	15.693	0.39957	124
QIEA1	9.854	-32.7560	10.966	38.4815	14.16	0.40079	42
QIEA2	11.937	-134.6120	10.575	38.5690	14.09	**0.39795**	145
QIEA3	21.381	**-174.537**	19.458	38.3367	**7.671**	0.40154	**41**
QIEA4	8.873	-71.5650	**8.7820**	38.8049	7.782	0.40193	122
QIEA5	22.913	**-172.9000**	16.194	**38.8498**	23.664	0.40209	176

1 SRN Gate 2 HADAMARD Gate 3 QNOT Gate 4 Z Gate 5 Rotation Gate

*Global minimum at (-1.3068,-1.4248) with value = -176.1375

**Global minimum at (-π, 12.275), (π, 2.275) and (9.42478, 2.475) with value = 0.397887

• P_c=0.25, P_m=0.01, Generations = 1000.

4.1 Speed

In comparing elapsed time (time to obtain optimum), The QIEA algorithms showed varied performance in comparison with each other. This is important in illustrating that different gates perform differently. However compared to GA and PSO, QIEA algorithms perform well. As far as speed of convergence concerned, most QIEA algorithms exhibits very high convergence rate at early stage of search. This can be observed from the Figures 5, 6 and 7. This behaviour of convergence indicates potential of these methods. Therefore issue to be resolved in exploiting the power of these methods is association between gate and problem type. The main search heuristics of QIEA is based on preserving the best solution in updating population this plays a significant role in speeding the execution. Comparing the five gates used in QIEA has shown that Z gate is, on average, is the fastest among the other gates, while the rotation gate is the slowest.

4.2 Accuracy

In searching for the answer QIEA generally outperforms GA and PSO. For test functions 1 and 2 QIEA outperforms the others, while for the test function 3 QIEA results are the same as GA. It is worth indicating that the accuracy is gate dependent. Our results identify that QIEA algorithms are powerful and offer very good potential, what is missing, as in the case of speed, association of gate type and problem type. The best results among QIEA algorithms are given by QNOT gate. Due to the movement of fixed steps (Figure. 3), the state vector traverses the solution space and locking to the any minima on its path. Although this results in fast convergence, as explained above, makes it prone to getting trapped in local minimum. This behaviour can be observed in Table 1, for example, for test function 1, where objective function values in two cases (-32.756 and -71.565) are far from the solution and represents local minima trapping. There is no part in the algorithm that enables the search engine to overcome this local trapping. An addition to the evolution engine of QIEA is required to let the algorithm escapes from being trapped. This addition (or improved algorithm), may lead to a costly penalty in speed of the algorithm. Also from the results, GA and PSO behave in a stable manner and reach the required solution but with different accuracy.

4.3 Robustness

Robustness is a measure of the performance of an algorithm is affected with various factors such as stability of solution, convergence, the influence of initial population, numerical accuracy. Most of QIEA algorithms are sensitive to the starting initial probabilities given to all qubits. The algorithm with rotation gate is sensitive to size of the angle used; as it becomes bigger it may pass a good solution or even the optimum. That is why the value of changing angle is small ($0.01\pi = 0.0314159$ rad $= 5.654862$ degrees). For other gates action is either flipping the sign or replacing the value of the probability for the qubit, this may take the solution away from near optimum solution as it jumps away, and this is clear for the case of Z-gate. To generalise this it can be said that QIEA is sensitive to the type of gate, which contradict the original quantum algorithm principals, which states that all of quantum gates are universal. So it can be stated that the QIEA is not a robust algorithm for the mentioned reasons. Although this seems to be a disadvantage as we judge robustness by classical measures, which does not account for uncertainty inherent in quantum computing.

4.4 Simplicity

The algorithm is simple to implement and does not require a complex detailed steps. It consists from representation as a qubits, evolving the qubits based on their cost functions for different types of gates that only affects the required solutions to move towards the best solution by only small amount of movement each time.

4.5 Generality

The algorithm shows good results for some cases but behave poorly for others, so it cannot be generalized as a tool to solve different types of problems. Even the results obtained by [7] does not reflect the true generalization of the algorithm to solve combinatorial problems.

In general for the speed, it is demonstrated that the algorithm is very fast in execution and faster than GA and PSO. However for reaching accurate answer, the algorithm, for the test cases, does not always find the optimum and it is prone to getting trapped in local minima. In terms of assessing the overall performance of gates, Z gate was the worst while QNOT with Rotation gates were the best.

However, results are promising, not because it provides generalisation, which is a concept of classical computing and behavioural uncertainty of quantum algorithm, should not be judged by, but because quantum computing, even at this early stage development, competes well with some matured methods.

5 Conclusions

A series of experiments have been conducted to verify the ability of quantum-inspired evolution algorithm to solve different types of problems. This is a step towards applying the algorithm to real life and more complex engineering problems. The results are encouraging, which over perform some well-known and well-established optimization techniques, such as GA and PSO. The paper as well as presenting some interesting findings, it concludes that further research should be conducted to overcome some problems inherent with the algorithm, such as premature convergence to local optimum and a number of other difficulties associated with the uncertainty principle which effect the behaviour of the algorithm.

References

1. Barr, R., Golden, B.L., Kelly, J.P., Resende, M.G. and Stewart, W.R., (1995) Designing and Reporting on Computational Experiments with Heuristic Methods. Journal of Heuristics. **1**. 9-32.

2. Feynman, R., (1982) Simulating physics with computers. International Journal of Theoretical Physics. **21**. 467-488.
3. Alfares, F. and Esat, I.I. (2003).Quantum Algorithms; How Useful for Engineering Problems. in Proc. of the Seventh World Conference on Integrated Design & Process Technology. Austin, Texas, USA. 669-673.
4. Hogg, T., (1998) Highly Structured Searches with Quantum Computers. Physical Review Letters. **80**. 2473-2476.
5. Hogg, T. and Portnov, D.A., (2000) Quantum Optimization. Information Sciences. **128** (3). 181-197.
6. Moore, M.P. and Narayanan, A., (1995) Quantum-Inspired Computing. Department of Computer Science, University of Exeter, Technical Report No. 344; http://www.dcs.exeter.ac.uk.
7. Han, K.-H. and Kim, J.-H. (2000).Genetic quantum algorithm and its application to combinatorial optimization problem. in Proc. of the 2000 Conference on Evolutionary Computation. Piscataway, NJ: IEEE Press, 2; 1354 -1360.
8. Han, K.-H. and Kim, J.-H., (2002) Quantum-inspired evolutionary algorithm for a class of combinatorial optimization. IEEE Transactions on Evolutionary Computation. **6** (6). 580 -593.
9. Zhang, G., Jin, W. and Li, N., (2003) An Improved Quantum Genetic Algorithm and Its Application. LNAI. **2639**. 449-452.
10. Narayanan, A. and Moore, M. (1996).Quantum-inspired genetic algorithms. in Proc. of the 1996 IEEE Conference on Evolutionary Computation (ICEC '96). Nayoya University, Japan: IEEE, 61 -66.
11. Rylander, B., Soule, T., Foster, J. and Alves-Fos, J. (2000).Quantum Evolutionary Programming. in Proc. of the Genetic and Evolutionary Computation Conference (GECCO-2000), 373-379.
12. Li, B. and Zhuang, Z.-Q., (2002) Genetic algorithm based-on the quantum probability representation. LNCS. **2412**. 500-505.
13. Narayanan, A. and Manneer, T., (2000) Quantum artificial neural network architectures and components. Information Sciences. **128** (3). 231 -255.
14. Grover, L.K. (1998).Framework for fast quantum mechanical algorithms. in Conference Proceedings of the Annual ACM Symposium on Theory of Computing, 53-62.
15. Hogg, T., (2000) Quantum Search Heuristics. Physical Review A (Atomic, Molecular, and Optical Physics). **61** (5). 052311/1-7.
16. Grover, L.K. (1999).Quantum Mechanical Searching. in Proceedings of the Congress on Evolutionary Computation. Piscataway, NJ: IEEE Press, 2255-2261.
17. Deutsch, D., Barenco, A. and Ekert, A., (1995) Universality in Quantum computation. Proc. of Royal Society London A. **449** (1937). 669-677.
18. Hey, T., (1999) Quantum computing: an introduction. Computing & Control Engineering Journal. **10** (3). 105-112.
19. Michalewicz, Z., (1999) Genetic Algorithms + Data Structures = Evolution Programs. 3rd, revised and extended ed, Berlin: Springer-Verlag.

20. Kennedy, J. and Eberhart, R.C. (1995).Particle swarm optimization. in Proceedings of the IEEE International Conference on Neural Networks, 1942-1948.
21. Kennedy, J., Eberhart, R.C. and Shi, Y., (2001) Swarm Intelligence, San Francisco: Morgan Kaufmann Publishers.
22. Salman, A., I., A. and Al-Madani, S., (2002) Particle swarm optimization for task assignment problem. Microprocessors and Microsystems. **26** (8). 363-371.
23. Levy, A., Montalvo, A., Gomez, S. and Galderon, A., (1981) Topics in Global Optimization, New York: Springer-Verlag.

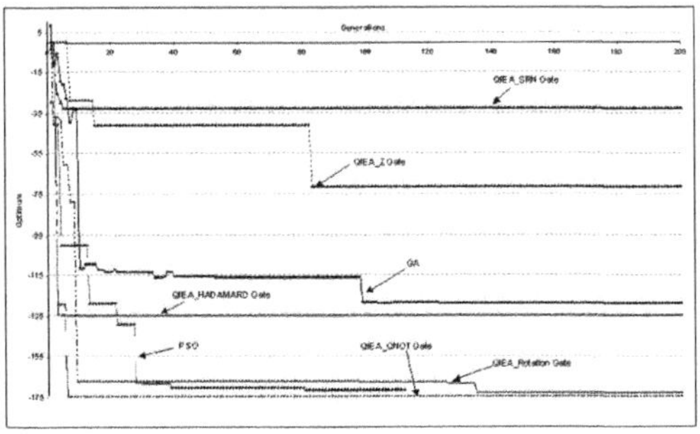

Figure 5: Comparison between the Algorithms: GA, PSO and QIEA (5 Gates) for Test Function 1

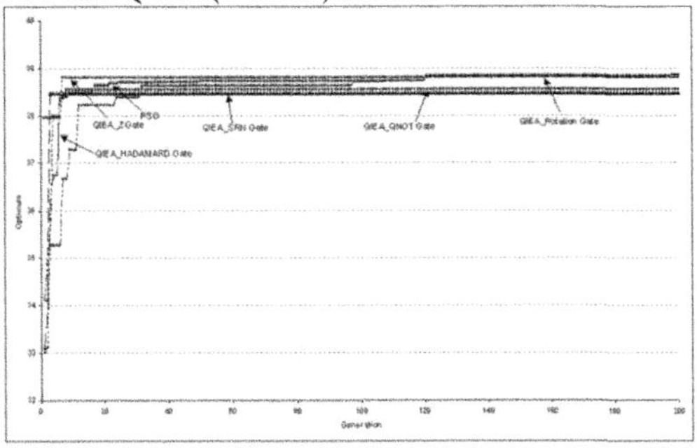

Figure 6: Comparison between the Algorithms: GA, PSO and QIEA (5 Gates) for Test Function 2

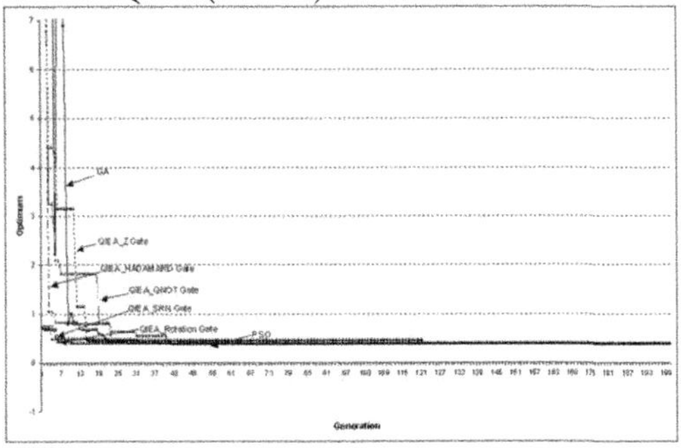

Figure 7: Comparison between the Algorithms: GA, PSO and QIEA (5 Gates) for Test Function 3

Author Index

GPSR Compliance

The European Union's (EU) General Product Safety Regulation (GPSR) is a set of rules that requires consumer products to be safe and our obligations to ensure this.

If you have any concerns about our products, you can contact us on ProductSafety@springernature.com

In case Publisher is established outside the EU, the EU authorized representative is:

Springer Nature Customer Service Center GmbH
Europaplatz 3
69115 Heidelberg, Germany

Batch number: 10755053

Printed by Printforce, the Netherlands